T0310271

Indoor Positioning

Indoor Positioning

Technologies and performance

Nel Samama
Electronics and Physics Department
Institut Mines-Telecom, France

Published by John Wiley & Sons, Inc., Hoboken, New Jersey.
Published simultaneously in Canada.

For general information on our other products and services or for technical support, please contact our Customer Care Department within the United States at (800) 762-2974, outside the United States at (317) 572-3993 or fax (317) 572-4002.

Wiley also publishes its books in a variety of electronic formats. Some content that appears in print may not be available in electronic formats. For more information about Wiley products, visit our web site at www.wiley.com.

Library of Congress Cataloging-in-Publication Data:

Names: Samama, Nel, 1963- author.
Title: Indoor positioning : technologies and performance / Nel Samama, Electronics and Physics Department, Institut Mines-Telecom, France.
Description: Hoboken, New Jersey : John Wiley & Sons, Inc., [2019] | Includes bibliographical references and index. |
Identifiers: LCCN 2019010513 (print) | LCCN 2019017747 (ebook) | ISBN 9781119421856 (Adobe PDF) | ISBN 9781119421863 (ePub) | ISBN 9781119421849 (hardback)
Subjects: LCSH: Indoor positioning systems (Wireless localization)
Classification: LCC TK5103.48323 (ebook) | LCC TK5103.48323 .S27 2019 (print) | DDC 006.2—dc23
LC record available at https://lccn.loc.gov/2019010513

Cover design by Wiley
Cover image: Photo by Tobias Fischer on Unsplash.

Set in 10/12pt WarnockPro by SPi Global, Chennai, India

Printed in the United States of America

V10010736_060319

Contents

Preface

This preface gives some ideas about the way this book has been written: the main philosophy and how it has been designed. It is intended to provide an overview of indoor positioning technologies and systems. Note that as it deals with indoor, it is mainly oriented toward pedestrians or objects.

The two main reasons for the book are to take stock of the real performances, i.e. in fact the limitations, of the various indoor positioning technologies and its corollary, to show that it is already possible to produce many systems, meeting real needs, on the simple condition of very slightly changing the angle of our vision of positioning. This vision is indeed the result of a long process that makes us understand positioning only in the form proposed by the GPS. Thus, all solutions should follow the same mode, forgetting, for example, that continuous positioning is only necessary in very specific cases.

I also felt that there was a need for clarification: how can we understand that this area, which has been promised a high turnover for years, is only implemented in limited cases. The book was chosen to comment on the various indoor positioning technologies, according to a classification that mixes the physical techniques used. As a result, this may seem a little strange to purists, but the objective is to better understand the reasons for the limited level of penetration of these solutions in everyone's daily life. This break with more traditional presentations is also an attempt to "move" the points of view and angles of approach.

The way to treat technologies is to consider them in an "elementary" way, i.e. individually. This allows us to extract the performance and see what complementarities each would need in order to extend its performance. Chapter 12 will provide some insights into the various ways in which these individual technologies can be combined.

A fundamental aspect when it comes to indoor location, and in particular when talking about the citizen and his mobile phone, is of course respect for everyone's privacy. This point is not addressed in this book, which is intended to be technically oriented, but must be present in everyone's mind.

The last chapter of the book could have been the introduction if the only targeted audience were "specialists." Thus, after a few hours of "navigation" through the book, the reader will easily be able to move on to this last chapter, which summarizes, giving simple examples, all the current difficulties of defining an acceptable system.

The book was written as a broad discussion on the field and technologies rather than a technical reference book for academic use. It is intended for those who wish to understand the reasons for the relative stagnation of deployments or who wish, in the short term, to carry out such a deployment. Practical aspects therefore play a significant role.

Discussion also means exchange. In this way, readers' comments are encouraged.

Paris, February 2019

Nel Samama

Acknowledgments

In the acknowledgments to my previous book, *Global Positioning*, I thanked all those who had to endure my many absences: my family members of course, and also my colleagues at work. It seems to me that I told them I would never do it again.

I reiterate here my sincere gratitude for so much self-sacrifice on their part, often leaving me in my corner quietly when they understood that it was necessary. I would have no credibility if I said I would never write a third book, but I hope they know how much I have understood and appreciated the efforts they have made for me. Without mentioning them by name because they will easily recognize themselves, I thank them for their constant help and support.

A big thank you to the two Alexandres, Alexandre Vervisch-Picois and Alexandre Patarot, whose comments have greatly improved several chapters.

Last but not least, as for Global Positioning, a very special thanks to Dick who made many corrections to the English language of the book: he is once again certainly the only person who will ever read the complete book twice!

Introduction

Main Objectives of this Book

Abstract

This introduction explains the main reasons that led me to write this book. The writing of a book is exciting but very time-consuming: I learned this from a previous experience. Thus, for me, such an adventure is the result of an observation, and the format I have chosen follows quite logically (in my opinion). The observation is that indoor positioning is a long-standing quest of many actors: industrialists, small and medium-sized enterprises, institutions, academics, researchers, etc. The economic outlook, although probably often overestimated, has been described for many years as exceptional. What is surprising to me is that it always remains at a very high level. However, no "viable" solution (we will come back to this term later) is really available today on a scale commensurate with what the stakes seem to be (economic but also applicative). From a technical point of view, I did not find the problem complex to state, but it was probably because of my daily "immersion" in it. So what is the problem? Why do I have such difficulty getting my interlocutors to understand the field, whatever it may be, in order to solve a very real problem? Why is it that after so many years and so much effort we do not have "indoor GPS"? This introductory chapter gives some information on the path that led me to the writing of this book as well as on the format chosen for the latter, which is more a discussion than a purely technical work.

Keywords *Introduction; indoor positioning; indoor GPS; indoor problem*

I have been involved in the field of indoor positioning for about 20 years now. In the early years, it was the daily excitement. There was not a day without a new need emerging: pedestrian guidance of course, but management of production and animal welfare in hangars; piloting drones for structural analysis in arsenals; continuity of the car navigation function in covered areas, tunnels, or car parks; monitoring of firefighters in their interventions, etc. Many calls for projects were available and the fact that we were working in the field was in itself rewarding. This is quite classical in the world of research and development: it is the normal life of an applied research subject. Where things seem different to me is that despite all this, despite the very many solutions proposed in extremely varied technical directions, despite the scale of potential markets that has never been denied, there is still not a catalog of acceptable technical and technological solutions. All deployments are unitary while the need seems global.

A first reason, which is not sufficient but "enlightening," is undoubtedly our over-reliance on "technology," which will solve any problem if the need and especially the markets are present. The global positioning system (GPS) was capable of an incredible achievement. The second element follows from this: as the markets are described as gigantic, some have been tempted to sell the chickens before they are hatched, without any real technical solution. This has made it possible both to recover R&D funding, which is sometimes very substantial, and to make real technological progress. However, these have not been enough. One thing leading to another, investors have become more reluctant and that is how we have witnessed (and still witness) the succession of financing and slowdown phases of the latter. Some major projects, such as the European Galileo program, are sometimes one-off accelerators.

However, the main parameters, both technical and applicative, remained simple enough from my point of view. The past tense used in the previous sentence is fundamental in my decision to write this book: I was obliged to note my inability to convey my message of "simplicity" of the technical problem to my interlocutors. In such a case, it is probably necessary to question yourself a little in order to move forward. As far as technology is concerned, my contribution is based on this book, which has enabled me to understand that what I thought was simple is not really so simple, even if it is based on only a few basic broad lines but which are expressed in a multitude of details. On the application level, however, my initial feeling was reinforced during the writing process: it is in fact the needs that are very poorly formalized, preventing "technicians" from making useful progress. Some will object that the latter are not supposed to solve a specific practical problem but should be used, in their reflections, by all without segmentation. Okay, but then let us find "intermediate structures" combining research and development because these two aspects are essential to progress in this field. That, I believe, is the root cause of our current problem. Leaving those who have a technological solution (or think they do), the power to guide responses to (generally poorly expressed) needs has led us to the current impasse.

I would thus advocate, stronger than ever, the need for exchanges built between the various actors, institutional, financial, technical, and application in order to bring out the main classes of what we are looking for. This would make it possible both to know what we want, but above all to give strong (and potentially profitable) objectives to technical actors. The efforts made so far are gradually being dispersed, replaced by new techniques that do not know better where to go. Leaving the choice of the parameters to optimize to the technicians is to allow them to conclude favorably in the contexts that suit them best, and experience shows us that this has not been, in any case so far, the right one.

Attempt to Clarify the Problem

First of all, throughout the book, a semantic difference is made between technique and technology. The first term applies to the "mathematical" (or geometric) approach that leads to the position of a terminal (any object, person, or entity). Examples include triangulation, trilateration, or the determination of the slope of a Doppler curve when the transmitter moves relative to the terminal. Technology is then a specific way of carrying out measurements. Thus, for example, the trilateration technique is implemented by the GPS technology, but also by the UWB (ultra-wideband radio) one.

The aim of the book is to list the many technologies available today. The scope is large, without of course being totally exhaustive. These technologies are described in their "elementary" functioning, i.e. without coupling them together and without implementing overly elaborate associated treatments. The objective is thus to remain at a relatively "physical" level in order to characterize their intrinsic potential. However, Chapter 12 presents some current approaches to coupling, merging, or hybridizing technologies and provides an opportunity to discuss some fundamental elements of these approaches, and especially the links between these approaches and the "physics" of technologies.

About 40 technologies are discussed. In the presence of such a list, the question arises of the organization and classification, or grouping, of them: how to support the reader in his journey of understanding? My choice was to classify them by "range," not by technique. Indeed, in the relative complexity of the problem, I tried to put myself in the shoes of a reader of this book in order to answer a practical question about deploying a system under its own conditions. It seems to me that the first question to ask, long before knowing what we are going to implement, is the geographical scope of our problem.

This being said, the selected organization does not solve all the problems. First of all, classification is not always very simple to decide because some technologies have the particularity of being able to be implemented according to different modalities, which would position them in different categories. Then, because in other cases, classification may depend on the way the technology is implemented. All this will be discussed in the corresponding chapters.

After the first chapter introducing a history of indoor positioning, we will discuss the problem more fully in Chapter 2. Chapter 3 then presents the description of the techniques (as defined above) and we will come to Chapter 4, which is a lengthy discussion on a significant set of parameters and criteria for indoor positioning. We will take time to identify all the potential problems in this area and conclude with a set of summary tables of all technologies, arranged in alphabetical order according to all the criteria selected.

We will then become fully aware, by using these tables, of the complexity of the problem. By complexity, we will see that it is not really a question of technical complexity, but more of an application and practical implementation complexity, of extra technical constraints if we can say so. In fact, it is the accumulation of the latter that makes the problem almost insoluble, except in limited cases. Thus, it is then suggested, in the following discussion, to look for a solution not only on a technical or technological point of view but also of the potential revision of the said constraints.

All this only makes sense if, and doubt is allowed, the markets for indoor positioning are real and not fantasized. However, we will not enter into this debate.

Nevertheless, Chapter 4 is certainly the most important one in the book for those who wish to understand the field without a priori and who wish to get their own idea on the issues of indoor positioning and continuity of the positioning service.

Comments for a Deployment in Real Conditions

The following Chapters, from 5 to 11, organized according to the range of the technologies will allow some technical descriptions, but more specifically discussions on the strengths and weaknesses of these technologies. The comments are then proposed as part of a real deployment of the technology, and their main objective is to highlight the important points to be considered. The overall idea of the book is not to discourage but, on the contrary, to make it clear that among the many approaches available, it is likely that the solution to a given deployment exists, but that it is necessary to be able to understand the limitations in order to adapt the system requirements objectively. Disillusionment with the true capabilities and performance of a system is just counterproductive, i.e. it generates disappointment and frustration, often leading to a diversion from the whole field of positioning. It is this attitude that the book seeks to avoid by announcing, as objectively as possible (I hope), what can be expected from the technologies discussed, and what cannot.

It is useful to understand that the classification used is not free from potential criticisms. The boundary between two chapters is not necessarily very clear and certainly questionable. However, it allows you to make choices and quickly understand some of the basic issues of an indoor positioning solution.

The modern world is looking at how to analyze and process the massive amount of data that are becoming increasingly available. The world of positioning is no exception to this strong trend. Thus, based on the observation that one technology alone is not able to meet the various needs, the current direction is based on the coupling, more or less subtle, of several technologies. The basic principle is that if two complementary technologies are combined, it is potentially possible to obtain the best performance by combining the two. This is undeniable in theory but still poses some practical problems. As a result, many approaches exist and will be briefly described in Chapter 12. The book's approach consists in giving the main current leads and references on the subject, and also to propose a new discussion on the limitations (and perhaps fantasies?) of these approaches. The fact, for example, that they effectively provide performance improvement, sometimes significant, in many cases is very real and cannot be questioned. However, it would be dangerous not to understand that in complex cases (i.e. when "elementary" technologies are at the limit of their fields of use), these approaches no longer present the previous performance gain. Chapter 12 will provide an opportunity to return to this point in more detail.

The same is true for Chapter 13 on mapping. This is an absolutely fundamental area when it comes to positioning and localization. However, it is often necessary to make this map much more than just an image, as has been done in the field of road mapping. Indeed, it is essential to associate with each element of the map attributes that will allow it to be able to provide the required services. For example, it will be necessary to be able to say whether an element is a displacement zone or not, whether it is possible to pass through a partition because there is a door or an opening in this partition. Similarly, when calculating a route, it is important to characterize a corridor by a high speed of movement compared to that possible in a room, even if the latter has several doors (the route must go around the room and especially not cross the room if the latter is a meeting room, for example). The same applies to areas where one changes from one floor to another, whether it is by elevators or stairs. An attribute "persons with reduced mobility" is again essential for the service to be acceptable. All this requires a specific approach, of a type similar to what was done a few decades ago for the outside world. In the indoor case, the size of a building's mapping is reduced (compared to road mapping), but involves new functionalities, such as floor management or bidirectionality of all spaces.

The last chapter is a description of what some daily situations could be if a real continuity of service (position) existed. It is a matter of imagining, in a realistic way, how the current organization could be modifiable in order to offer more flexibility in everyone's schedules. The basic idea is that our lives are mainly governed by the rhythm of the passing of time (a notion well assimilated today and especially shared by all on the same basis, the watch, which is individual, portable, and available everywhere and all the time). Imagining that the position of every person is equally shared and available everywhere and all the time, how this would simplify everyone's daily life, while reducing unnecessary expenses.

Conclusion

The book seems to me to be useful for anyone wishing to approach the subject of indoor positioning or the continuity of the localization function. In particular, application and service developers are often confronted with the implementation of software blocks whose finesse of use they do not always perceive as required. We then end up with inefficient systems. This is often due to a lack of knowledge of the mechanisms and limitations of the technologies deployed, which have a significant impact on the way data are processed. I hope that this book will help everyone to better understand the real technical and application challenges.

1

A Little Piece of History ...

Abstract

In this chapter, we briefly look back at the evolution of geographical positioning. Our intention is to show that indoor positioning is indeed a very recent need that has come about due to the spread of modern mobile-connected terminals and owners wanting to receive numerous so-called services, many of which are greatly enhanced when associated with the user location. The benefit of many of them is 10-fold when associated with the user location. Thanks to the Global Positioning System, the famous GPS, this association was made possible in the early 1990s. Unfortunately, this fantastic system has been unable to meet the performance required indoors, where a "typical" urban citizen spends the majority of his or her time (*The term "typical" will appear sometimes in the book. Although experience has shown such "typical" persons, objects, or environments do not exist, we will use this term to appoint a classical situation*).

Keywords   *History; longitude problem; navigation; clocks; Harrison*

As soon as human beings decided to explore new territories, or even just to move within new territories, they needed a way to locate themselves in their environment.

1.1   The First Age of Navigation

The origins of navigation are as old as man himself. The oldest traces have been found in Neolithic deposits and in Sumerian tombs, dating back to around 4000 years CE The story of navigation is strongly related to the history of instruments, although they did not have a rapid development until the invention of the maritime clock, thanks to John and James Harrison, in the eighteenth century. The first reason that pushed people to "take to the sea" is probably related to both the quest for discovery and the necessity of developing commercial

Indoor Positioning: Technologies and Performance, First Edition. Nel Samama.

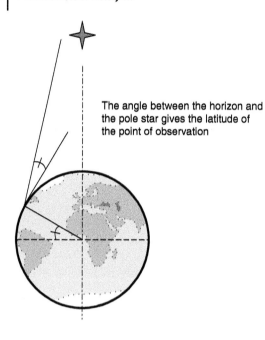

Figure 1.1 Determining latitude with the pole star.

The angle between the horizon and the pole star gives the latitude of the point of observation

activities. In the beginning, navigation was carried out without instruments and was limited to "keeping the coast in view." It is likely that numerous adventurers lost their lives by trying to approach what was "over the horizon."

The astronomical process used for positioning was quite inaccurate, and hence, frequent readjustments were required. The localization was even more complex because of the lack of maps. Nowadays, the situation of indoor positioning is in the same state: accuracy is not at the desired level, and frequent readjustments are needed. Moreover, one of the most important problems is the lack of indoor maps allowing navigation (i.e. not just an image). This very hot topic is dealt with in Chapter 13.

Unfortunately, astronomical positioning was only able to give the latitude of the point, as can be understood from Figure 1.1. The longitude problem would remain unresolved for centuries: will it be the same for indoor positioning?[1]

A first remark can be made at this stage: positioning at the epoch was not continuous in time and space, contrary to what we are looking for today. However, is it really essential indoors?

1.2 Longitude Problem and Importance of Time

The so-called longitude problem was much more difficult to solve and took almost three centuries. During this period, significant progress occurred

---

1 In fact, technological solutions already exist, but this is the combination of the perceived constraints that the solution should address, which is too stringent for current technologies.

concerning instruments and maps, but nothing for determining the longitude. As early as 1598, Philipp II of Spain offered a prize to whoever might find the solution. In 1666, in France, Colbert founded the "Académie des Sciences" and built the Observatory of Paris: one of his first goals was to find a method to determine longitude. King Charles II also founded the British Royal Observatory in 1675 in Greenwich to solve this problem of finding the longitude at sea. Giovanni Domenica Cassini, a professor of astronomy in Bologna, Italy, was the first director of the French academy and in 1668 proposed a method of finding the longitude based on the observations of the moons of Jupiter: this work followed the observations made by Galileo[2] concerning these moons using an astronomical telescope. It had been known from the beginning of the sixteenth century that the time of the observation of a physical phenomenon could be linked to the location of the observation; thus, knowing the local time where the observations were made compared to the time of the original observation (carried out at a reference location) could give the longitude. Cassini established this fact with the moons of Jupiter after having calculated very accurate ephemeris. Unfortunately, this approach needs the use of a telescope and is not practically applicable at sea.

On 11 June 1714, Sir Isaac Newton confirmed that Cassini's solution was not applicable at sea and that the availability of a transportable timekeeper would be of great interest. It has to be noticed that Gemma Frisius also mentioned this around 1550, but it was probably too early. On 8 July 1714, Queen Anne offered, by Act of Parliament, a £20 000 prize[3] to whoever could provide longitude to within half a degree. The solution had to be tested in real conditions during a return trip to India (or equivalent), and the accuracy, practicability, and usefulness had to be evaluated. Depending on the success of the corresponding results, a smaller part of the prize would be awarded.

The development of such a maritime timekeeper took decades to be achieved but finally had an impact on far more than navigation. The history of Harrison's clocks is quite interesting, and time is really the fundamental of modern satellite navigation capabilities. We have seen that Isaac Newton himself confirmed that the availability of a transportable maritime clock would be the solution to the longitude problem: the realization of such a clock, however, was not so easy. The main reason is that the clock industry was fundamentally based on physical principles dependent on gravitation (the pendulum). This was acceptable for terrestrial needs, but of no help in keeping time when sailing. Thus, a new system had to be found.

The reason that time is of such importance is because of the Earth's motion around its axis. As the Earth makes a complete rotation in 24 hours, it means that every hour corresponds to an eastward rotation of 15°. Thus, let us suppose that one knows a reference configuration of stars (or the position of the sun or the moon) at a given time and for a given well-known location (e.g. Greenwich). If you stay at the same latitude, then you will be able to observe the same

---

2 Hence, the name of the European GNSS.
3 This amount is equivalent to more than $15 million today.

configuration but at another time (later if you are eastward and earlier if westward): the difference in times directly gives the longitude, as long as the time of the reference location (Greenwich in the present example) has been kept. The longitude is simply obtained by multiplying this difference by 15° per hour, eastward or westward. The method is very simple and the major difficulty is to "keep" the time of the reference place with a good enough accuracy, i.e. with a drift less than a few seconds per day. Pendulums, although of good accuracy on land, were unable to provide this accuracy at sea, mainly because of the motions of the ship and changes in humidity and temperature.

John Harrison built four different clocks, leading to numerous innovative concepts. After almost 50 years of remarkable achievements (August 1765), a panel of six experts gathered at Harrison's house in London and examined the final "H4" watch. John and William (his son) finally received the first half of the longitude prize. The other half was finally awarded to them by the Act of Parliament in June 1773. Certainly more important is the fact that John Harrison was finally recognized as being the man who solved the longitude problem.

One of the most famous demonstrations of Harrison's clocks' efficiency was given by James Cook during the second of his three famous voyages in the Pacific Ocean. This second trip was dedicated to the exploration of Antarctica. In April 1772, he sailed south with two ships: the *Resolution* and the *Adventure*. He spent 171 days sailing through the ice of the Antarctic and decided to sail back to the Pacific islands. He returned to London harbor in June 1775, after more than 40 000 nautical miles. During this voyage, he was carrying K1, Kendall's copy of Harrison's H4. The daily rate of loss of K1 never exceeded eight seconds (corresponding to a distance of two nautical miles at the equator) during the entire voyage: this was the proof that longitude could be measured from a watch.

Indoor positioning is almost in the same situation as that of the longitude determination in the early eighteenth century: it seems to be quite close, but there is indeed no satisfactory solution. Hopefully, it will take less than 50 years to find an acceptable approach.

1.3 Link Between Time and Space

The perception of time has changed quite a lot over the centuries until the current omnipresent availability of a precise time that can thus be shared by everybody. By briefly analyzing the evolution of the effects of this availability of time on people's life, some parallels are drawn concerning possible changes induced by the availability of positioning.

1.3.1 A Brief History of the Evolution of the Perception of Time

At the very beginning, time and space were notions that people felt: the number of days of walk needed to reach a given place and drawing simple maps of places. This was achieved long before writing was available.

With the augmentation of the diversity of his activities, human being has increased both his living space and the need to measure time in order to better organize commercial activities, for example. The lunar calendar appeared to help in this task: the observation of the phases of the moon was enough to give a date. Unfortunately, this was limited to activities such as agriculture, which relies more on annual cycles. Then, solar calendars appeared that allowed the collective organization of the activities of the society. The notion of year and months was already present. Furthermore, it was quite precise for seasonal activities. Further improvements were rapidly required in order to divide the day into time units to organize the activities within a given day. The initial approaches were based on the sundial, but the obvious problem is that the duration of a unit of time is not the same in every season: thus, a daytime unit lasted longer in summer than in winter. Ingenious water clocks (clepsydras) were imagined to solve this problem: in addition, this made the time available at night. Time became available: the next steps were to make it both transportable and synchronized from place to place.

The monks were the first to develop "clocks" in order to synchronize their religious practices. The first achievements were based on rings and gongs. Here, the interesting point is the fact that it allowed for synchronization for a whole group of people (those that heard the bells): knowing the precise time is absolutely not required.[4] Universal time was nevertheless not yet a worry as life revolved around local affairs. Furthermore, the night remained "another world," but it was acceptable to use the Sun for time. The evolution was, however, to develop clocks that were able to "ring" at various times of the day, even without a dial and hands. The most advanced such clocks were also able to ring at night in order to organize the whole life of the village.

The next step in the management of time measurement and restitution was the advent of mechanical dials that allowed people to "locate" themselves within the day. Representations are used (often based on religious or astronomical symbols) in such a way that even those that did not read were able to understand the time. All the mechanisms used at that time were based on gravitational effects meaning that it was not possible to use them at sea (this leads us back to the beginning of the chapter).

Meanwhile, Western countries started to expand around the world where difficulties appeared for commercial activities and synchronization. The first trains are in operations, but clocks are still synchronized on the Sun midday and time drifts are "visible." Trains raised the need for a coordinated universal time, and this was the starting point for time zones.

The industrial revolutions brought about a change in attitudes toward time: the work was no longer related to the task, but to a given amount of time, new relationships were created between employers and employees, and new claims arose concerning the rights of workers who sometimes organized strikes in

---

4 Note that this notion could be interesting in the case of positioning: there is no permanent need for knowing precise positioning, as long as it is possible to know the path to be followed and maybe the time it will take to reach the next stopping place.

support of their claims. Industry realized that "time is money" and life itself became defined in relation to time. In addition, time became a global notion, shared worldwide. This globalization raised the (paradoxical?) need for an individual timekeeper[5]: everybody needed to be synchronized with the rest of the world, or at least with his professional and personal neighborhood.

Over the past few centuries, time has clearly increased its ascendancy on human activities. Financial transactions are nowadays fundamentally based on time, and the Internet and all telecommunication networks must be synchronized. Almost every action is quantified in time (and hence in money): at work, this is clear, and also for travel, either professional or personal, leisure, entertainment, etc. In the development of time measurement, one has also faced the disappearance of the mechanisms that were the visual part of the time passing. Some displays, if not all, no longer have hands but give digital values.

1.3.2 Comparison with the Possible Change in Our Perception of Space

The representation of the Earth has also changed quite a lot over the centuries. As time was being synchronized around the world, there was also a need for more accurate representation of the world in terms of maps, routes, etc. Note that although many different needs are at the origin of this requirement, time is certainly one of the most important. As the world's activity is largely based on time, it is very important to be able to evaluate the time needed for any given trip, either of people or of goods.

If we try to make a comparison between the evolution of time measurement and the evolution of positioning systems, it is certainly possible to say that positioning is today in the situation time faced more than 150 years ago with the advent of portable clocks. This was this technical feat that allowed the appropriation of time by everybody. The equivalent in positioning is now available with satellite-based positioning systems (thanks to the pioneer global positioning system [GPS]). A few features are similar between the first portable watches and basic GPS receivers of today: the similar approach of needing an identical referential worldwide, the availability of a personal local measurement, and the possibility to "synchronize"[6] with anybody else using a similar device. In addition, time and position are closely linked in GNSS (global navigation satellite systems) and this feature will help bring together the two aspects.

In conjunction, there is another technical achievement that is fundamental for the dissemination of portable positioning devices and their incorporation into everybody's life: telecommunications. When someone uses the time read

---

5 The same phenomenon is visible today with the Internet and the "permanent connectivity" feature: as globalization is not achievable for people, there is the need for individual devices allowing globalization.

6 "Synchronize" either relates to time or to position.

on a wristwatch, it is automatically shared with others because the uniqueness of the common referential is enough. This is absolutely not the case for positioning: even considering a shared geographical referential, the position is a specificity of one person. In order to share these data with others, there is the need to communicate this information. This is why the advent of both positioning and telecommunication are bound to provide a wide development of positioning (maybe on a similar scale to what happened for time).

In the scope of this evolution, it is possible to consider that positioning could be profitable in domains such as ubiquity, or in other words, the automatic discovery of anyone's environment, or also in group management. For ubiquity, it is clear that if the positions of all people and objects were easily available, in all possible environments and at almost no expense, the environment of everybody could be discovered. The telecommunications required are available today, but not positioning (and this book deals with the most difficult aspect: indoors). An extension of this could be that people would need to define themselves with some criteria that would lead to belonging to a group of like-minded people. The above-mentioned discovery of environments could then be to find, from a geographical point of view, people or objects that belong to your group (or any other group). This is currently being implemented in the Social Networks communities with applications such as "find a friend" or "find a point of interest." The idea is to extend these applications to everything in the scope of the so-called Internet of Things (IoT). The indoor positioning of objects and people is therefore a fundamental feature.

When compared to the evolution of time and its impact on society, it is even possible to imagine that positioning could be used in many other ways (considering positioning as the combination of positioning and telecommunications). Knowing how people are moving around in the city,[7] it is possible to organize the "waves" of movement and then to define the policies to be followed by the town council in terms of roads, infrastructure, and public transport, for example. This aspect relating to flow management is also a strong concern for public buildings such as airports or museums, for example. This leads us to transportation. The health and safety authorities could also use positioning-related devices: emergency calls are already in use, but one can imagine that the above-mentioned group management approach could be part of the management of any emergency call. For example, if somebody fell ill in the street or in a building, an alert could be transmitted to people who are geographically close and who have been identified as competent in this medical field. This raises the problem of the definition and the access to the corresponding information files, and also to privacy issues, but could be one direction of future developments.

---

7 The proposed concept can easily be applied to a country or even to the world, as well as to smaller structures like a district or inside a company.

The current problem of "Data," either geographical or not, and personal privacy is a fundamental one which must be dealt with urgently if one wants to provide valuable, but acceptable, services to users based on their location.

1.4   The Radio Age

The wish to communicate over long distances was described long before the radio conduction phenomenon was discovered. The first related facts are dated fourth and fifth century CE (by optical means) using fires on top of mountains, serving as "communication relays." This approach was still used by the first optical telegraphs in the seventeenth century. Of course, the main disadvantage of such a system lies in the fact that transmission is limited to the optical line of sight and requires good "air conditions," i.e. no fog. This problem led to the development of the electrical telegraph.

On the 24 November 1890, Edouard Branly discovered the phenomenon of "radio conduction": an electrical discharge (generated by a Hertz oscillator) had the effect of decreasing the resistance of his "tube." It appeared that electrical propagation was possible without cables. Further works showed that "adding" a metallic rod to the generator improved the range of the transmission (i.e. the detection was also possible further away from the generator): Alexander Popov was just about to invent antennas. The transmission path grew from a few tens of meters to 80 m. In 1896, Popov succeeded in transmitting a message over 250 m (the message was composed of two words: "Heinrich Hertz").[8]

At the same time, Guglielmo Marconi, who was deeply influenced by the publications of Faraday and the life of Benjamin Franklin, felt that it should be possible to establish a transmission over a few kilometers. After a lot of works, he transmitted the letter "S" coded in Morse ("· · ·") over 2400 m at the end of 1895. In September 1896, by using a kite as an antenna, Marconi achieved a 6 and then a 13 km radio path. In May 1897, a transmission of 15 km was demonstrated between two English islands (Steep Holm and Flat Holm), followed by similar performances in Italy in the La Spezia harbor. Marconi founded, on 20 July 1897, the Wireless Telegraph and Signal Company. In March 1899, the first trans-channel message was sent between South Fireland (Great Britain) and Wimereux (France): the addressee was Edouard Branly. With antenna heights of 54 m, this 51 km transmission was achieved with a global performance of 15 words per minute. In July, a 140 km path was achieved between a sea position and the coast. After this new success, Marconi was almost certain that trans-horizon radio paths were possible.

---

8 For more details, see the exciting "Comment BRANLY a découvert la radio," Jean-Claude Boudenot, EDP Sciences (in French!).

In October 1900, Marconi started drawing up the plans of the Poldhu station (in Cornwall, United Kingdom), which was planned to be the transmission station for the first trans-Atlantic transmission. The chosen site in North America was Signal Hill, in Newfoundland, still a British colony at this time. This station was ready for experimentation on the ninth of December. From this date, it was decided that Poldhu would send the letter "S" ("·· ·") each day between 11:30 and 14:30, Signal Hill time (the need for synchronization is definitely a fundamental aspect). On 12th July, the signal was received at 12:30, through a path of 1800 miles (3500 km), including the Earth's curvature!

Coming back to navigation, it was only a few years later (1907) that radio electric signals were used, by transmitting time signals. As already described, knowing the time at a specific location is fundamental in calculating the longitude. Until then, this was achieved through the use of Harrison's clocks. The radio transmission was a fantastic improvement, especially in terms of accuracy, as the signal is transmitted at the speed of light, thus greatly increasing the accuracy of the "time transfer." The corresponding improvement of positioning is around 10 times better. The second application of radio electric waves was to use the signal as a new landmark that no longer needed to be in visible line of sight. The first such system was implemented on board a ship in 1908, together with a movable antenna that could give an indication of the bearing of the transmitter. This was the first dedicated radio navigation system. Note that many elements of positioning systems (angle measurements, time synchronization, need for ephemeris, etc.) were already present at this time.

In addition, the new radio beacons allowed positioning using measurements based on electrical properties such as the amplitude of currents or voltages, for instance. This was going to simplify the automation of the navigation systems as electrical engineering was rapidly progressing. Some approaches are still used today for positioning, especially indoors.

1.5 First Terrestrial Positioning Systems

Thus, the first systems were based on radio goniometry[9]: by having a rotating antenna and by carrying out the detection of the maximum power, it was possible to determine the direction of a landmark. The radio compass was one of the most advanced forms of radio goniometrical systems. Another approach was that used for radio lighthouses. As determining both the identification and the orientation of the transmitters had to be easy to obtain, the technique consisted of having a couple of antennas radiating complementary signals (for instance, the equivalent of A "·—" and N "—·" in Morse). When a receiver is in both main

---

9 Goniometry is the way of measuring the angle of rotation of the aerial of a wireless system in order to obtain the direction of arrival of the radio wave.

radiating lobes, the signal received is continuous. In 1994, more than 2000 radio lighthouses were available all around the world.

As local time generators (oscillators or atomic clocks) were developing rapidly, new uses of radio signals were imagined. This was the case of so-called hyperbolic systems. The basic principle states that all locations having the same difference of signal travel time to two fixed points, for instance, two radio transmitters, lie on a geometrical figure, which is a hyperbola when dealing in two dimensions (more generally, the mathematical locations are defined by a quadric). The focal points of this hyperbola are the transmitters. As signal processing capabilities increased, such time difference estimations and measurements became possible. Note that the synchronization at the mobile receiver's end is thus avoided as long as time differences are carried out. The basic idea was then to obtain two such differences in order to allow the calculation of the intersection point of the resulting two hyperbolae (see Figure 1.2): this approach leads to a theoretical single point in a two-dimensional space.

The first system that used this technique was the Decca,[10] which came into operation at the end of the Second World War. It worked within the frequency band of 70–128 kHz, allowing approximately 450 km of operational range. The resulting accuracy was in the range of a few hundred meters, depending on the

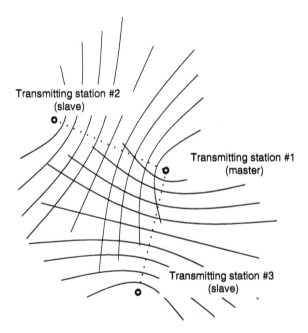

Figure 1.2 Representation of the hyperbolic approach.

---

10 Proposed by the Decca Navigator Company.

propagation conditions. The new era of radio electric signals allowed a rigorous evaluation of accuracy, a very important parameter.

The enhanced-LOng RAnge Navigation system (e-Loran) is also a hyperbolic system that added new features concerning the modulation scheme, based on pulse trains forwarded by each master and slave station.[11] These first terrestrial systems provided "local" area coverage, even though this coverage can be quite large (this is the case for LORAN). However, some people imagined an even more ambitious project that would be the ultimate version of a terrestrial system with a global coverage: the Omega system. It was made up of eight stations using very low frequency (VLF) band in order to have a complete coverage of the Earth. It was still a hyperbolic approach: each station transmitted sequentially, always in the same order for about one second (the duration of emission is specific to each station). The emission consisted of pure continuous waves (no modulation scheme) at 10.2, 11.33, and 13.6 kHz, respectively. The global accuracy was generally better than 8 km.

The major reason for the poor accuracy of the above-mentioned systems is included in the propagation modeling (this point has constantly driven the evolution of modern systems). The reader should notice that this aspect is also the main source of difficulties for indoor positioning when dealing with radio systems, but not only. Indeed, a large majority of approaches are limited by propagation aspects.

1.6 The Era of Artificial Satellites

In the late 1920s, physicians and mathematicians showed that it was theoretically feasible to imagine artificial satellites launched from the Earth's surface and orbiting the Earth. Of course, a lot of research was still required, but it was thought possible. On 4 October 1957, the Soviet Union launched Sputnik-1 (see Figure 1.3), called the "basketball," weighing 183 pounds, on an elliptic orbit with a 98 minute revolution period.

To prove that a satellite was actually orbiting the Earth, it was planned that it should transmit a signal. Sputnik used a 400 MHz carrier frequency with sound modulation data. In such a way, once demodulated, it was possible to "hear" Sputnik.[12] Nothing was really known about this flight: the orbit, the speed of the satellite, the duration of the transmission, etc. Therefore, it was a fantastic opportunity to carry out some tests. Among others, George C. Weiffenbach and William H. Guier, members of the Applied Physics Laboratory of

---

11 The master station is the one that masters the time. The slave stations have to be synchronized with the master station.
12 What was then "hearable" can be listened to at: https://www.youtube.com/watch?v=r-bQEiklsK8.

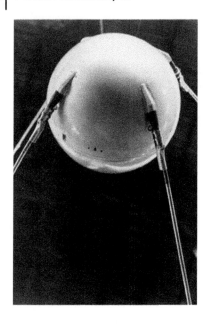

Figure 1.3 Sputnik, called the "basketball." Source: Courtesy of NASA.

the Johns Hopkins University, carried out such investigations. They succeeded in determining the Sputnik's orbit by analyzing the Doppler shift[13] of the signal while the satellite was in radio visibility, i.e. for about 40 minutes of the 108 minutes of a complete revolution of Sputnik.

The method they used to achieve such a goal was of fundamental importance as it is the starting point of all modern satellite navigation systems. The measurement was the Doppler shift, the unknown variable was the orbit of the satellite, and another piece of data was the actual location of the place of observation (i.e. the laboratory). After about three weeks of observation and a few calculations, they finally showed that it was possible to calculate the orbit, knowing both the Doppler shift and the exact location where the measurements were carried out. It has to be remembered that, in 1957, this was at the height of the Cold War between the Soviet Union and the United States. The US Army, and more specifically, the US Navy, had a difficulty concerning the positioning of its fleets cruising in northern oceans. These ships were equipped with missiles to which precise missions could be allocated. The problem was that, although the guidance of such a missile was controlled by an inertial system of high quality, the starting location of the flight was still obtained through the use

---

13 The Doppler shift is the physical phenomenon that shifts the frequency of any wave transmitted, depending on the relative speed between the transmitter and the receiver. Let us define D as the distance between the transmitter and the receiver: the frequency received is increased when D decreases and decreased when D increases. Note that this phenomenon is a physical time compression of the signal and applies to all waves (sound, radio, light, etc.).

of terrestrial systems, i.e. not very accurate. A more accurate system would be of great help for this specific purpose.

Frank McClure, head of the department, made a suggestion: would it be possible to invert this problem? That is, would it be possible to be able to calculate the location of the observation point, knowing the orbit of the satellite, by carrying out the same measurements as those achieved to define the Sputnik orbit, i.e. the Doppler shift of the received signal. Thus, the problem of satellite positioning was solved, thanks to Sputnik and led to the Navy Navigation Satellite System (NNSS), or "TRANSIT" program, which was launched in 1958, directed by Richard Kershner.

The first satellite was launched in September 1959, and before the end of 1964 (an amazingly short time for anyone working on modern projects), 15 launches had been carried out with 8 more for research purposes. These eight were related to the program and concerned the following:

- the establishment of a network of terrestrial surveillance stations,
- the determination of the terrestrial gravity, which is of primary importance in order to predict the orbits of a satellite over a long period (12 hours in the case of TRANSIT),
- the definition of terrestrial and maritime receivers.

The TRANSIT system became operational for the Navy in 1964. The mean accuracy obtained was typically in the range of 200–500 m.

The limitations of TRANSIT were the starting point for the specifications for the second generation of US satellite-based positioning system, which were as follows:

- *Availability*: 24 hours a day, 365 days a year, for all the covered locations and whatever the meteorological conditions (we mentioned this point concerning the terrestrial systems, for which the propagation conditions are of great concern). This last point has fundamental implications and modern systems are still spending a great deal of effort on improvements to propagation-related matters.
- *Accuracy*: three-dimensional positioning with delivery of speed (real speed vector in three dimensions too) and precise time (one has to remember that time delivery was the first application of radio signals to maritime and navigation domains in the early years of the twentieth century).
- *Coverage*: the whole planet should be covered with an extension to space (low and medium Earth orbit satellites usually position themselves by using GPS signals).

1.6.1 GPS System

As the TRANSIT system was made operational in 1964 for the US Navy, the early works on what would become, in 1973, the GPS program started with tests on both the CDMA scheme (code-division multiple access) and the PRN code approach (pseudorandom noise). These two techniques, widespread nowadays in radio systems, and more specifically in wireless telecommunications, were quite innovative concepts. In 1967, the US Navy started the TIMATION program to assess the effect of relativity, both special and generalized, on a satellite-based atomic clock.[14] In 1973, the programs related to satellite navigation from both the US Navy and the US Air Force merged into an official "Navigation Technology Program" called "NAVSTAR GPS," sometimes referred to as "Navigation Satellite with Time and Ranging Global Positioning System."

After the first stage of research programs, phase II started, in 1978, with the launching of the first four NAVSTAR satellites. From 1978 to 1985, 11 satellites were launched (called block I), and from 1989 to 1997, the 28 block II/IIR operational satellites followed. In 1985, seven satellites were available allowing about five hours a day of positioning. The 24 nominal satellites were in orbit in 1994 and the GPS system was declared operational in 1995.

The major difference with the TRANSIT system is that it is now based on a trilateration technique,[15] i.e. multiple distance measurements are carried out in order to allow the receiver to calculate a fix (TRANSIT was based on Doppler shift measurements).

1.7 New Problem: Availability and Accuracy of Positioning Systems

Our current topic, indoor positioning, has not been solved with GNSS. Moreover, the availability of these systems on a large scale led to questions concerning the continuity of the positioning service in all kinds of environments. In addition, the advent of high-performance portable telecommunication terminals has brought about the need for very versatile positioning systems. The very low cost, the ease of integration, and the lack of alternative systems have also led to a large dissemination of GNSS chipsets, leading in turn to a more and more frequent use of them. Thus, the way positioning is achieved with GNSS has indeed become a standard and it is quite difficult to suggest other visions (for example, a positioning that would not be continuous in time and space).

For about 20 years, various techniques and technologies have been developed, evaluated, and sometimes implemented in order to cope with this continuity. It does not seem to be the end of the story as no approaches have demonstrated that they answer the question adequately. The problem

---

14 GPS is the first widespread system that must implement both theories of relativity, in order to obtain accurate positioning. Neglecting these effects would lead to a 10 km error per day!
15 This technique is described in Chapter 3.

indeed seems linked with the expectations of the users, building managers, or ordinary citizens, who seek accurate technological solution at no cost. This is the starting point of a more complete discussion concerning the real problems of indoor positioning, dealt with in the next chapter.

Bibliography

1 Boorstin, D.J. (1983). *The Discoverers*. New York: Random House.

2 Gardner, A.C. (1958). *Navigation*. Bungay: Hodder and Stoughton Ltd.

3 Guier, W.H. and Weiffenbach, G.C. (1998). Genesis of satellite navigation. *Johns Hopkins APL Technical Digest* 19 (1): 178–181.

4 Ifland, P. (1998). *Taking the Stars, Celestial Navigation from Argonauts to Astronauts*. Newport News, VA/Malabar, FL: The Mariners' Museum/The Krieger Publishing Company.

5 Kaplan, E.D. and Hegarty, C. (2017). *Understanding GPS: Principles and Applications*, 3e. Artech House.

6 Kennedy, G.C. and Crawford, M.J. (1998). Innovations derived from the transit program. *Johns Hopkins APL Technical Digest* 19 (1): 27–35.

7 Parkinson, B. (1995). A history of satellite navigation. *Navigation – Journal of The Institute of Navigation* 42 (1): 109–164.

8 Parkinson, B.W. and Spilker, J.J. Jr. (1996). *Global Positioning System: Theory and Applications*. American Institute of Aeronautics and Astronautics.

9 Pisacane, V.L. (1998). The legacy of transit: guest editors introduction. *Johns Hopkins APL Technical Digest* 19 (1): 3–10.

10 Sobel, D. (1996). *Longitude*. London: Fourth Estate Limited.

11 Sobel, D. (1998). A brief history of early navigation. *Johns Hopkins APL Technical Digest* 19 (1): 11–13.

2

What Exactly Is the Indoor Positioning Problem?

Abstract

What are the main reasons that indoor positioning is not yet available, if not everywhere, at least wherever one would like to have it? This chapter gives a first "simple" answer and Chapter 4 will provide a much more detailed and well-argued one. As mentioned in this chapter, the main reasons are not really technical but rather due to the perceived needs that the different actors have in mind. When translated into technical constraints, no more solutions exist, not because it is not possible but because it is almost impossible to find an acceptable compromise between all the constraints in terms of cost, performance, ease of use, availability of technologies, etc. Thus, the main message at this stage is that there are a lot of possibilities for indoor positioning, but one has to take care not to accumulate too large a range of constraints, from economical ones to physical ones. We should concentrate on the real needs in order to provide industry and research with real specifications, unless the problem is in fact elsewhere: the reduced utility of indoor positioning – of course, we do not believe that, hence this book.

Keywords *Indoor positioning problem; perceived needs; compromises*

Indoor positioning is a very important topic, mainly in terms of continuity of services. This leads to many theoretical and experimental works in this field using a large range of technologies, from purely global navigation satellite system (GNSS) approaches to networks of physical sensors or wireless local area telecommunication networks (WLAN). Among all these technologies, the GNSS-based ones present the advantage of making better use of the satellite receiver, which is considered to be the "best" solution for outdoor applications (even with the current limitations in urban canyon environments). Thus, technologies such as high-sensitivity global positioning system (HS-GPS) or assisted-global positioning system (A-GPS) have been widely investigated within the satellite community: the results are interesting but do not give a

Indoor Positioning: Technologies and Performance, First Edition. Nel Samama.
© 2019 The Institute of Electrical and Electronics Engineers, Inc. Published 2019 by John Wiley & Sons, Inc.

definitive answer to indoor positioning. Pseudolites and repeaters are now solutions that could help in a final system with good accuracy and wide coverage: studies are being carried out and show encouraging results for both approaches but are still far from maturity. Because of the large deployment of WLANs for communication purposes, a great deal of work is being carried out on location finding with WLANs in order to find a way to "complement" the outdoor GNSS-based systems with indoor wireless local area network (WLAN including, among others WiFi and Bluetooth)-based positioning. Among other technologies that will be described later in this book, one has to highlight the ultrawide band (UWB) technology, based on radar concepts and now implemented for proximity high data rate communication. Because it uses a time-based approach, it is possible to see it as being a good accurate candidate for indoors.

The very important aspects one has to keep in mind are both the availability and accuracy required. GNSS allow a global coverage and almost permanent availability. This is very good, but in the case of indoors, where such systems do not yet provide equivalent performance, the questions are as follows: does one need the same (or higher) level of accuracy, and what kind of availability is required? Also, is the permanent location finding capability of GNSS compulsory? The next stages are going to show that the specifications are very important in indoor positioning systems, especially as almost all technical requirements can be achieved, by one or another technology. The difficulty arises when one wants to combine technical requirements: accuracy and simplicity, terminal cost, infrastructure cost, autonomous mode, etc. The user requirements are also of uppermost importance, even though one can consider that future applications will certainly be put forward by imaginative people, not yet expert in positioning technologies. Thus, most current indoor applications in the telecommunications domain, like location-based services (LBS), do not need a permanent positioning capability but require such a positioning on demand with a reduced delay. Accuracy depends clearly on the application: not very high for service finding and quite precise (to within a few meters) for navigation purposes. This requirement should certainly be even more stringent indoors because of the usually reduced size of the places concerned.

2.1 General Introduction to Indoor Positioning

With the advent of greater mobility, a heavy need for localization has emerged. This is true not only for automotive applications but also for personal needs, thus leading to the necessity of having a technical solution to indoor positioning. This latter point appears as of prime importance for telecommunication-related applications, as revealed by the efforts of USA concerning the Emergency Call E911[1]. The European Union provides such an Emergency Call, the E112, but has decided not to put any legal constraints on the call localization: operators are asked to make their best effort to provide a good location[2]. These regulations required developments, specifically in the areas not covered by GNSS. As a confirmation of this, the Galileo program[3] included a specific domain called the "local elements" that includes specifically the indoor domain. It is then quite clear that indoor positioning was a challenging technical aspect of global navigation. If it is clear that GNSS are the right candidates for global positioning in the places it works well, i.e. where the sky is free enough for the receiver to acquire enough satellites, a lot of possibilities exist concerning both urban canyons and indoor environments. There were typically two directions taken: the first one relied on the use of satellite navigation constellation signals in order to reduce the number of different electronic systems required to achieve the positioning function. The second one tried to implement a different technology indoors and the final system will be made up of the integration of GNSS for outdoors and this newly developed one indoors.

2.1.1 Basic Problem: Example of the Navigation Application

Let us take the navigation function as an example of the limitation of current systems. This allows a guidance application to be provided and has to be available in different environments, namely outdoors and indoors. Let us also compare two of the major positioning technologies in use: the telecommunication network cell identification (the so-called Cell-Id described in Chapter 10) and a GNSS trilateration method described in Chapter 11. Table 2.1 summarizes the simple situation regarding the continuity of the associated service.

The proposed navigation service is achievable with neither of these two technologies because of the lack of coverage in the case of GNSS and of the lack

Table 2.1 The "navigation" function and the continuity of service.

Localization technology	Cell-Id	GNSS
Indoors	Yes	No
Outdoors	Yes	Yes
Navigation function	No	Yes
Continuity of service	No	No

---

1 The FCC regulation states that succeeding communications to E911 should be located with an accuracy of 50 m 67% of the time and 150 m 95% of the time.
2 Directive 2002/22 on E112.
3 The European GNSS.

of positioning accuracy in the case of the Cell-Id approach. If one wants to propose such a service to a pedestrian, this is a real problem (although this is a very simple representation of reality, which is far more complex and is explained in Chapter 4). In comparison, the automotive domain is easier in terms of real constraints on the positioning engine: no power restrictions, location that can only be on predefined "tracks" (i.e. roads), constant "attitude" of the platform, etc., leading to a globally satisfactory GNSS-based system. This is very different when dealing with a pedestrian, who is the typical target of LBSs and applications. Note that some in-vehicle GNSS systems add inertial technologies and advanced map-matching to overcome the drawbacks of limited coverage GNSS. Definitely, the direct transposition from car to pedestrian navigation is not straightforward, even if some trials have been carried out in this area: inertial concepts, specific map matching to define indoor tracks, etc.

As a matter of fact, the localization that is a fundamental brick of all navigation-related applications, of all LBSs and of all applications requiring location data, should exhibit the following characteristics:

- Be available in various types of environments (countryside, urban area, indoors, etc.)
- Give an accuracy that clearly depends on the application. It is clear that accuracy of 1 m is not needed for numerous applications, whereas it is insufficient in other applications.
- Allow the continuity of service as a basic concept.

2.1.2 The "Perceived" Needs

The specifications of the localization are thus very different regarding the various possible applications. Furthermore, there are no current technologies that cover a large range of specifications. Considering just three requirements: accuracy, indoor, and outdoor needs, Table 2.2 shows the diversity of specifications, considering a classification by main domains. Of course, one knows that other requirements are of prime importance, such as the cost of infrastructure and terminals, etc. (see Chapter 4 for a deeper analysis).

This table is not very "accurate": the accuracy figures are very loose and the environmental requirements are also not clear. Unfortunately, this is the reality: the technical needs are tremendous and the situations cover a very large range of possibilities. Furthermore, this is only a very small part of the real complexity. For instance, let us take any line in Table 2.2: it is still possible to divide it into a lot of new lines with more precise specifications. An example of this could be given by the "Tourism" line: in this domain, many applications are already working in a location-based way. Navigation is then possible from one point of interest to the next. It is quite easy to imagine that if the localization engine is also working indoors, then the extension to museum visits or even to

Table 2.2 Specification by main domains.

Domains	Accuracy	Indoors	Outdoors
Assistance	≈100 m	Not compulsory	Essential
Comfort	<100 m	Not compulsory	Useful
Displacements	1–100 m	Useful	Essential
Games	1–100 m	Not compulsory	Useful
Health	1–100 m	Important	Important
Services	1–100 m	Useful	Essential
Tourism	1–100 m	Useful	Useful
Transport	1–10 m	Important	Essential
Emergency	1 m	Essential	Essential
...	...	...	...

mall navigation would be quite immediate. Of course, this new feature would be considered as an improvement (which would be true). For museum visits, the technical requirements could be an accuracy of around 1 m, but also the output of the absolute orientation of the terminal, so that it could be possible to determine whether the user is looking at a given sculpture or has his back turned to it. Current approaches just deal with a rough positioning giving the "room" the terminal is in. This is enough for today's applications, which indeed do not really use positioning.

The real problem is that it is possible to divide all the lines up in such a way and that there is no current technology comparable to GNSS for outdoors. For outdoor applications, this exhibits such good global performances that it can cover a very large range of needs. Indoor technologies are unfortunately not so versatile, nor are the indoor application specifications. This complexity is certainly the reason for the limited extension of current LBSs.

Other classifications are possible, as for instance one that relies on places where the services could be proposed. Table 2.3 shows such a summary. Once again, the main conclusions remain the same. Many studies have been carried out in order to define the right classification but have not led to any simplification of the initial problem and there are actually no real specifications that can be applied to a group of applications.

The tentative classifications proposed in Chapter 4 show that the real problem is indeed complex. It seems to be quite simple, but in reality is not. Thus, there is now an urgent need for advanced discussions between users, industry, and research in order to organize future developments. Without these exchanges, things will only progress at the instigation of some and not necessarily in a technically optimal way or for good reasons, i.e. to the benefit of the whole community of potential users.

Table 2.3 Specification by main places.

Places	Accuracy	Indoors	Outdoors
Airport/station	≈10 m	Essential	Essential
Country/mountain	<100 m	Not compulsory	Useful
Mall	< a few meters	Essential	Not compulsory
Conference center	< a few meters	Essential	Not compulsory
Warehouse	≈1 m	Essential	Useful
Sea/port	1–100 m	—	Essential
Museum	< a few m	Useful	—
Attraction park	≈10 m	Useful	Useful
Road	≈10 m	—	Essential
Lane	≈10 m	—	Essential
Storage zone	<10 m	Not compulsory	Essential
...	...	...	...

2.1.3 Wide Range of Possible Technologies

A first analysis of the possible technologies is carried out in this section, leading to the highlighting of major trade-offs that have to be made. The technologies to be considered are taken from the following categories:

- Networks of sensors (infrared, ultrasound, pressure sensors, etc.)
- Mobile telecommunication networks (4G, 5G, etc.)
- Additional sensors (odometer, accelerometer, gyroscope, magnetometers, etc.)
- Wireless local area networks (WLAN: Bluetooth, WiFi, UWB, etc.)
- Image based (pattern matching, image processing, etc.)
- GNSS based (pseudolites, repeaters, etc.)

The "networks of sensors" category includes not only infrared, ultrasound, sensors but also pressure sensors distributed throughout the building. The main disadvantage of these technologies is the need for a wide deployment of a heavy infrastructure, which is balanced by the greater accuracy sometimes achievable (down to a few centimeters). Nevertheless, this is no longer seen as a real candidate for mass market deployment, although actual implementations have been set up. Under certain conditions, it can be seen as an interesting solution for complementing another system, i.e. floor determination for a pressure sensor.

Concerning the "mobile networks of telecommunication," much has been said about the possibility of implementing technologies (see Chapter 10) such as TDOA (time difference of arrival) or E-TDOA (enhanced time difference of arrival) and even AOA (angle of arrival). It is known that none of these technologies has a real accurate positioning potential, except the Cell-Id[4]. The reason is that for telecommunication purposes, the Cell-Id is a built-in facility (thus nothing more has to be done for positioning). However, for all the other technologies, there is a specific strong need for the positioning approaches: a minimum of three base stations have to be seen from the mobile terminal. This is of course not the way a mobile network is set up unless in big cities where the need for high data rates for a large number of people involves a high redundancy.

"Additional sensors" are all that one can imagine in order to allow positioning at the terminal end using autonomous means. For example, a pedestrian navigation module (PNM) uses a combination of GPS and inertial sensors, together with pedestrian behavior models for walking, that finally allow pedestrian navigation in many environments. The principle is to use GPS when available, and to switch to inertial navigation when GPS is not available. The main difficulty is that it requires very accurate modeling of the pedestrian movements and is limited to a certain duration. With time, the accuracy degrades if not recalibrated.

Concerning the WLAN method, one can consider that the infrastructure required is free, as long as it has been deployed anyway for other purposes (mobile Internet access or wireless telecommunication). Indeed, this is not absolutely true when thinking in terms of usual methods. If the technique is to make time measurements, then you need to seriously upgrade the time reference, compared to current WLAN time capabilities. If you use received power levels as the main data, then once again you will need to increase the number of "access points" to a level higher than that needed for telecommunication purposes. The received signal strength (RSS) method consists in establishing data bases of received power from various base stations for a given place. As a matter of fact, it is required to set up a grid over the whole place in consideration with a 1-meter (for example) step in both X and Y directions. Then, measurements of the RSS from base stations are carried out. In this first approach, it is interesting to know about the accuracy achievable, versus the number of base stations. First results show that one needs at least three base stations to achieve a 3–4 m accuracy. In fact, you will certainly need more than that and at least 5 to achieve such a goal in a real environment. The principle of positioning is to search the databases (one for each base station) for the corresponding values, considering a 1–2 dB uncertainty, of the RSS measurements. This gives possible areas for each base station. Then, by merging these various areas, one finally obtains the position. This leads to some questions. What happens if the real environment changes: new walls or the moving of desks and cupboards. The other difficulty is the orientation and inclination of the terminal: a mobile terminal is, by definition, handheld in a

---

4 The case of the Matrix approach (see Chapter 10) is the only exception.

position that cannot be predicted. This position has a significant impact on the received power level. Then, the database value research can lead to a wrong area. This aspect is not so important because of the number of measurements that are made (typically 5), but is still there. Therefore, the WLAN approach still has to be upgraded to be a really valuable solution. Alternative approaches have also been investigated such as "symbolic WLAN," a sort of enhanced Cell-Id technique for WLAN (fully described in Chapter 8).

Cameras are today widely available in mobile terminals and can actually help for positioning purposes. Many different aspects have been dealt with, ranging from pattern matching techniques of an image with a database of images to SLAM (simultaneous localization and mapping), which uses successive images in order to determine the path followed by the camera. Note that this technique is quite efficient as it can also determine the path in unknown and no-calibrated environments. The main global problem with optical-based technologies is always associated with propagation aspects, light being easily stopped by any obstacle.

For the GNSS-based technologies, the most famous one is the assisted-GNSS (A-GNSS). As already stated, it does not work satisfactorily in "deep" indoor conditions. The other well-known GNSS-based technology is the one that makes use of pseudosatellites (pseudolites): the basic idea is to create a local terrestrial constellation of a few satellites (generators, for instance). This is a good idea and the only difficulty arises in the synchronization required between pseudolites (Chapter 11 will address the current achieved performances of this technology in indoor environments).

Another more versatile GNSS-based technology uses the so-called "GNSS transmitters," which can be seen as a cheap local element that could also have network functionality[5] (such as Galileo differential stations). A repeater is a simple component that includes an outdoor antenna to collect GNSS signals, a microwave amplifier, and an indoor transmitting antenna to transmit the signals. The implementation of a system that uses three such repeaters has been carried out and the results in various indoor configurations show an average accuracy in the 1–2 m range. Current results have been obtained with a single-frequency L1 standard GPS receiver. Galileo and modernized GPS exhibit very interesting features such as multiple civil frequencies, pilot tones, and more sophisticated codes and modulations than the existing systems. The indoor positioning could take advantage of this situation and further improve the accuracy, availability, and integrity. The main difficulty, also true for the new generations of GNSS transmitters, is mainly due to propagation perturbations because of the environment. Although they are less difficult than for optical technologies, they are nevertheless of concern.

---

5 Different versions exist, from the so-called "repeaters," to the so-called "Repealites" and the so-called "Grin-Locs."

Table 2.4 Specification by technologies.

Technologies	Indoors	Outdoors
Network of sensors	1–5 m	Not suitable
RF ID	<1 m	<1 m
WLAN	few meters	Not suitable
UWB	≈10 cm	Not suitable
Cell-Id	500 m to 10 km	100 m to 10 km
E-OTD (2G)/TDOA (3G)	≫200 m	<100 m
GNSS	Not available	≈5 m ↘
A-GNSS	10 m to not available	≈5 m ↘
Pseudolites	≈10 cm	≈5 m ↘
Repeaters	≈1–2 m	≈5 m ↘
Inertial	<1 m (time dependent)	<1 m (time dependent)
Image pattern recognition	≈1–2 m	A few meters
SLAM	<1 m	A few meters
…	…	…

It is now possible to draw up a table of technologies, comparable to those given in Tables 2.2 and 2.3. Table 2.4 thus shows all the achievable performances: it can then be seen that almost all accuracies are possible, but it is once again the other specifications that will allow the choice of the resulting final technical solution.

2.1.4 Comments on the "Best" Solution

The scientific and industrial communities consider it important to give an answer to the problem of positioning continuity from outdoors to indoors. This should be achieved for the large diversity of possible environments that are bound to be faced by users. Many technologies, as described above, have been developed in order to achieve such a goal, but it would be interesting to analyze the real application requirements in positioning, specifically indoors. If we think in terms of navigation within an office building, it is certainly enough to propose a simplified WLAN positioning system. If we want to implement a navigation system in an exhibition hall, a 2D repeater approach could be an accurate candidate. Last but not least, a guidance system in railway stations or airport terminals would probably take advantage of a combination of these two technologies, depending on the specific requirements of the users (staff or customers).

The various systems available for indoor positioning can be classified in many ways, depending on the criteria such as the localization type (symbolic, relative, or absolute), the coverage (indoors, outdoors, or ideally both), the fact that the infrastructure calculates the position of the mobile terminal, or that the calculation is carried out at the receiver, etc.

Systems as different as bar codes, magnetic detectors, imaging systems, or infrared, ultrasound, or radio systems, etc., have been imagined and implemented. The latter, based on radio wave propagation, is based on techniques such as direction of arrival, flight time measurements, and received power levels. This is the case of systems such as 4G/5G, "active badge," GNSS, and many others.

2.1.4.1 Local or Global Coverage

The idea of the coverage is comparable to that of the infrastructure discussed in the following pages. A global coverage means that there is no need for local components, thus involving no additional cost to the primary system. For GNSS constellations, it means that no local elements are required in order to provide the positioning in all conditions. This is clearly not the case as urban areas and indoors are not well covered: thus, there is a need for local elements. Nevertheless, the coverage of the GNSS is really global. This should also be the case, even with some limitations in the coverage, of 4G or 5G systems[6].

When considering indoor positioning, the large range of possible technologies makes this differentiation quite interesting. The main reason for that is the fact that one would appreciate getting this indoor feature for free, or almost free. Thus, all the technologies that will require a local infrastructure are bound to be "less interesting" than others that will not, even at the cost of reduced performance. GNSS-based technologies, such as high-sensitivity GNSS (HS-GNSS) or A-GNSS, are, according to this point of view, very attractive approaches[7]. The same would apply to inertial-based positioning, as the system is then operational in all environments without any calibration or deployment. On the other hand, technologies such as UWB, infrared, or ultrasound based are purely local systems, also requiring an important local deployment. Of course, this is a drawback with these approaches. However, this can sometimes easily be balanced by the low cost of the basic elements required and the simplicity of the deployment. This latter comment is particularly applicable to the case of radio frequency identification (RFID), for instance, where the tag is only a few cents, but the coverage is very limited. In between global and proximity coverage, there are those technologies that use global signals or receivers but which still require the addition of "augmentation" devices, such as pseudolites

---

6 it is not our present purpose to go over the respective advantages and disadvantages of terrestrial- and satellite-based systems
7 Even if in the case of A-GNSS, there is the need for specific equipment at the base station end indeed.

or repeaters. The coverage is clearly local for these additional components. Finally, there are some systems that, even if they require local deployment (and thus coverage), are subject to a large spread in the coming years for purposes other than positioning. The case of WLAN is one of them. The coverage of the indoor system is quite clearly local, but the deployments are so widely spread that the actual coverage is, if not global, much larger than local.

2.1.4.2 With or Without Local Infrastructure

From this first analysis, several directions of work have been imagined and developed in the scientific community, based on a local terrestrial infrastructure intended to reproduce indoor conditions equivalent to outdoors. The above-mentioned conditions are typically spatial diversity and power at a similar level. In fact, it is possible to follow two approaches: with or without infrastructure. The latter case is illustrated by the so-called high sensitivity GNSS (HS-GNSS): the idea is to decrease the detection level of the satellite signal down to -150 or $-160\,$dBm (as compared to the typical $-130\,$dBm of outdoor-received power level). Although this approach yielded real improvements, it does not appear to be the final answer to indoor positioning. 4/5G technologies, on the other hand, can also be considered as "no added infrastructure": unfortunately, they do not provide accuracy to the required level for many applications. WLAN could also be considered as a "no added infrastructure" solution, but once again, we found that the telecommunication deployment is not enough for positioning to a sufficient accuracy.

Thus, one should take into consideration the need for a local infrastructure. The main point is then to find the "lightest" possible infrastructure: either the least expensive or the easiest to deploy. The coverage and the complexity of the system are then of great importance.

2.2 Is Indoor Positioning the Next "Longitude Problem"?

There are some similarities between the longitude problem that took centuries to be solved and the indoor positioning one, mainly concerning the fact we would like to have a solution, but the wish is so great that it seems to prevent us from taking a step back in order to address the full complexity of the problem. Apart from this, there are also many differences.

The historical side of the longitude problem has been quickly described in Chapter 1. Let us come back to the technical aspects. Many theories were proposed, among which three directions were privileged (the modern techniques used for positioning purposes, whatever the technology deployed, have their mathematical and physical origins here):

1. Variation of terrestrial magnetic field;
2. Measurements of distances from specific stars to the moon; and
3. Maritime clock.

The first method is based on the observation that the magnetic terrestrial field varies from one point to another. The idea was then to imagine that it would be possible, by drawing the "map" of the magnetic field all over the globe, to have a direct correspondence between the local field measured and the location where this measurement is carried out. We now know that it is not true and that, for instance, the magnetic "north" is subject to variation over time. Nevertheless, even if John De Castro demonstrated that it was untrue as early as at the beginning of the sixteenth century, experimental works were carried out until the middle of the seventeenth century before this direction was abandoned. In 1638, the mathematician Henry Gellibrand showed that there were extensive local variations in the magnetic field over time, i.e. the variation at a given fixed location is huge compared with that needed for positioning. Furthermore, Edmund Halley undertook a measurement campaign in 1698 that finally came to the conclusion that longitude determination was not possible with this approach. It has to be noticed that this approach was in fact before the Queen Ann challenge and probably has a link to it, as a solution was neither available nor foreseeable at this time. Note also that a few indoor positioning solutions are using this approach by analyzing the variation of the indoor magnetic field, and more precisely its fluctuations. We shall discuss these technologies in Chapter 11.

The second method is the logical evolution of the nautical navigation art of the end of the seventeenth century. Measurement instruments were good enough to allow acceptable navigation and angle resolution improved a little bit. Galileo and Cassini had shown that angle measurements could lead to the location finding by observing the moons of Jupiter. Although this is not applicable at sea, developments were carried out to show that the transposition to the moon of Earth was possible (seventeenth and eighteenth centuries): the use of both measurements and tables are required. Unfortunately, the accuracy of angle measurement was not good enough (2–3°). Thus, new techniques appeared, together with the evolution of accuracy. Tables referred now to the angular distances from 15 well-known stars to the moon and the advent of the sextant made location finding possible. The accuracy remained nevertheless insufficient and the calculations were complex. The first use of this approach is quoted around 1767, concomitant with important improvements in sextant technology (from 1770 onward).

The third method took decades to achieve its development but finally had an impact on much more than just navigation: the maritime clock. The history of the Harrison clocks is quite interesting and was the starting point of modern satellite navigation. We know about the story (see Chapter 1), but let us discuss briefly the technical aspects.

John Harrison was born in Foulby in 1693. His first clock, built in 1713, had a mechanism that was made entirely from wood. He also worked for a while with his younger brother James and their first major project was a turret clock that required no lubrication. In 1726, John and James designed two precision clocks, to see how far they could push the capabilities of their design. By inventing a pendulum rod made of alternating wires of brass and steel, Harrison solved the problem of the pendulum length that varied according to the temperature, slowing or accelerating the clock. Thus, Harrison's clocks achieved an accuracy of one second in a month. A maritime timekeeper had to exhibit the same kind of accuracy in a very different environment. During this period, being unable to meet the Board of Longitude, he contacted Edmund Halley who facilitated a meeting with Georges Graham, a member of the Royal Society. The reason Graham accepted to meet with Harrison was probably linked to the fact that Graham's earlier works were carried out in the watch-making domain.

Four trials were required for John Harrison to achieve the final so-called H4 (Figure 2.1), which was quite different from the previous timekeepers (H1 presented in Figure 2.2 and H2 and H3) because it was of the size of a large pocket watch. William, John's son, sailed for the West Indies with H4, aboard the ship *Deptford* on 18 November 1761: the watch appeared to be only 5.1 seconds late on its arrival in Jamaica[8]. A second trial was carried out on a trip to Barbados aboard the *Tartar* on 28 March 1764 and gave an error of less than 40 seconds (the trip lasted 47 days). These two results were judged as excellent, largely enough to win the prize. This was nevertheless not enough for the Board of Longitude to award the prize. Harrison was asked to disclose his entire design to the Astronomer Royal in order to allow him to make and test such a timekeeper.

Figure 2.1 Harrison's H4 clock.

---

8 It is nevertheless not so clear to find out the way the comparison was carried out and the accuracy of the reference clock.

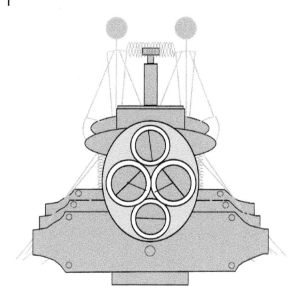

Figure 2.2 Harrison's H1 clock.

This was the condition to obtain half the prize, the second half being potentially awarded when the other timekeepers had exhibited similar performance at sea.

Harrison finally accepted to disclose the inner mechanism of H4 and received the first half of the longitude prize. A lot of stories concerning innovation are very similar to this, requiring huge efforts from the inventor to prove his truth against the well-established " authorities", but that is another story.

H1 (1730–1735) is a portable version of the previous wooden clocks. It was based on the use of springs in order to allow the effects of gravity to be dealt with, unlike a pendulum clock (note that overcoming the effects of gravity is the second difficulty at sea, together with the accuracy problem).

H2 and H3 (1737–1759) were not successful in resolving the longitude problem. As a matter of fact, the accuracy required to win the prize was not reached, despite the numerous innovations. Harrison requested additional funding from the Board to continue his efforts. At the end of this period, Harrison was convinced that the design should be entirely new: this led to H4.

Indoor positioning has similarities with the longitude problem in the way many technologies are available and have been tested, and performances evaluated. The main point is that no solution has so far met the expectations (although the expectations themselves are far from clear).

2.3 Quick Summary of the Indoor Problem

There are a lot of technologies that are able to solve part of the problem, but none able to solve the whole problem as it is perceived today. This is mainly due to the very large diversity of constraints we would like to overcome at once. Thus, if this first analysis is correct, the problem is indeed not a technological one, but rather linked to the difficulty in classifying the relative importance of the constraints.

An additional difficulty lies in the fact that indoor positioning is not seen as a must. It does not exist today on a representative scale, but there are not so many activities that are then not possible. It is indeed seen as a "comfort" nobody is really ready to pay for. In addition, nobody tries to define "minimal sets of constraints," which would lead to a real industrial (and hence commercial) development and business. Such an exercise would nevertheless be quite useful for the various communities of the indoor positioning domain but has not been carried out so far.

The purpose of the book is not to establish such sets and to organize the community, mainly because this should be a collegiate action including technologists, industrialists, users, building managers, civil engineers, services operators, etc.

Bibliography

1 Boorstin, D.J. (1983). *The Discoverers*. New York: Random House.
2 Bostrom, C.O. and Williams, D.J. (1998). The Space Environment. *Johns Hopkins APL Technical Digest* 19 (1): 43–52.
3 Gardner, A.C. (1958). *Navigation*. Bungay: Hodder and Stoughton Ltd.
4 Ifland, P. (1998). *Taking the Stars, Celestial Navigation from Argonauts to Astronauts*. Newport News, VA/Malabar, FL: The Mariners' Museum/The Krieger Publishing Company.
5 Parkinson, B.W. and Spilker, J.J. Jr. (1996). *Global Positioning System: Theory and Applications*. American Institute of Aeronautics and Astronautics.
6 Sobel, D. (1996). *Longitude*. London: Fourth Estate Limited.
7 Eissfeller, B., Gänsch, D., Müller, S. and Teuber, A. (2004). Indoor positioning using wireless LAN radio signals. *ION GNSS 17th International Technical Meeting of the Satellite Division*, Long Beach, CA (21–24 September 2004).
8 Gezici, S., Tian, Z., Giannakis, G.B. et al. (2005). Localization via ultra-wideband radios – a look at positioning aspects of future sensor networks. *IEEE Signal Processing Magazine* 22 (4): 70–84.
9 Gilliéron, P.-Y. and Merminod, B. (2003). Personal navigation system for indoor applications. *11th IAIN World Congress*, Berlin, Germany.
10 Hightower, J. and Borriello, G. (2001). Location systems for ubiquitous computing. *Computer* 34 (8): 57–66.
11 Koshima, H. and Hoshen, J. (2000). Personal locator services emerge. *IEEE Spectrum* 37 (2): 41–48.

12 Krishnan, P., Krishnakumar, A. S., Ju, W.-H. et al. (2004). A system for LEASE: location estimation assisted by stationary emitters for indoor RF wireless networks. *IEEE INFOCOM 2004*.

13 Mattos, P.G. (2003). "Assisted GPS without network cooperation using GPRS and the internet." *ION GPS/GNSS 2003*, Portland, OR (September 2003).

14 Pateli, A., Fouskas, K., Kourouthanassis, P., and Tsamakos, A. (2002). On the potential use of mobile positioning technologies in indoor environments. *15th Bled Electronic Commerce Conference*, Bled, Slovenia (June 2002).

15 Zagami, J.M., Parl, S.A., Bussgang, J.J., and Devereaux Melillo, K. (1998). Providing universal location services using a wireless E911 location network. *IEEE Communications Magazine* 36 (2): 66–71.

16 Angrisani, L., Arpaia, P., and Gatti, D. (2017). Analysis of localization technologies for indoor environment. In: *2017 IEEE International Workshop on Measurement and Networking (M&N)*, Naples, 1–5. IEEE.

17 Kırkağac, Y. and Doğruel, M. (2018). Performance criteria based comparative analysis of indoor localization technologies. In: *2018 26th Signal Processing and Communications Applications Conference (SIU)*, Izmir, 1–4. IEEE.

18 Melamed, R. (2016). Indoor localization: challenges and opportunities. In: *2016 IEEE/ACM International Conference on Mobile Software Engineering and Systems (MOBILESoft)*, Austin, TX, 1–2. IEEE.

3

General Introduction to Positioning Techniques and Their Associated Difficulties

Abstract

In this book, we make a semantic difference between techniques and technologies. The latter are assigned to real implementations, whereas the former describe the way positioning is achieved, i.e. with which measurements and with which calculations. Indeed, techniques are the basics of technologies. They were often designed long time ago and have not evolved much, whereas technologies are continuing to evolve together with measurement capabilities.

Keywords *Techniques; technologies; physical measurements; position calculations; difficulties*

This chapter is devoted to positioning techniques. In addition, we shall mention the main difficulties associated with these techniques. Usually, they are due to the measurements, or to the error linked to the measurements, and to the resulting positioning accuracy. Successively, we shall describe techniques associated with angle measurements, distance measurements, Doppler measurements, physical measurements, and finally image measurements. A specific paragraph concerns techniques implementing simultaneous multiple measurements, at the end of the chapter. As you will see, a given technique often leads to various approaches, i.e. a given type of measurement can be split into different approaches.

3.1 Angle-Based Positioning Technique

3.1.1 Pure Angle-Based Positioning Technique

The first navigation systems were certainly the stars, followed by the sun when the Portuguese realized that the pole star was not available when south.

In early navigation, lighthouses had the role of warning sailors of the emergence of a part of the ground. This was somehow the characterization of the

Indoor Positioning: Technologies and Performance, First Edition. Nel Samama.
© 2019 The Institute of Electrical and Electronics Engineers, Inc. Published 2019 by John Wiley & Sons, Inc.

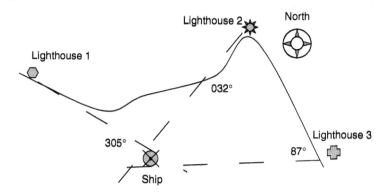

Figure 3.1 The compass bearing technique.

limit between the sea and the land. Some lighthouses were identifiable and hence were able to give additional information. Furthermore, these seamarks were used for years, and are still used, for positioning purposes, using the intersection of base lines. The idea is simply to measure the compass bearing of two lighthouses and to use these angles in order to draw the corresponding lines on a map to evaluate the location of the observation point. Figure 3.1 gives an illustration of this technique. Of course, the accuracy of the positioning greatly depends on the accuracy of the compass bearing (and of the map).

3.1.2 Triangulation-Based Positioning Technique

The property that is used is the following:

A triangle is fully defined by one side and the two adjacent angles.[1]

The idea of triangulation was first to define a base line,[2] thus defining two points of the triangle and a segment length (precisely the one corresponding to the base line, of course). The next step is to carry out two angle measurements (see Figure 3.2 for illustration).

- α from the first point (A) of the base line to the measured point M
- β from the second point (B) of the base line to the measured point M

Applying the rule described above, knowing the base line and the two angles allows the complete determination of the triangle AMB, thus leading to the

---

1 There are many other similar properties, hence many other ways to achieve "triangulation."
2 The approach often implemented is to use an invar cable of 17 m as the first base line. The material used, invar, has the particularity of being very stable and it presents an extremely low expansion coefficient, thus leading to an accurate base line length value in all weather, pressure, or temperature conditions.

Figure 3.2 The triangulation technique.

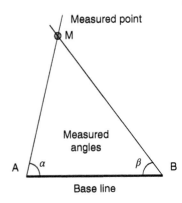

location of M. This method can be applied further considering, for example, as the new base line, the MA segment. Gradually, it is then possible to triangulate a whole country.

Note that this technique is quite different from the compass bearing as it consists of sighting a targeted point (the measured one, M in our example) from two known points, when the compass bearing is almost the opposite approach consisting of "receiving" the signal at the measuring location from two lighthouses.

3.2 Distance-Based Positioning Technique

3.2.1 Distances to Known Environment-Based Positioning Technique

In some specific environments, knowing the location of obstacles can enable complex laser-based systems to carry out self-positioning. Let us imagine a system composed of three laser beams as described in Figure 3.3 in a closed environment indicated by the polygon. The laser telemetry system allows d_1, d_2, and d_3 to be obtained with an accuracy that can reach a few millimeters,

Figure 3.3 Possible distances positioning system and environment.

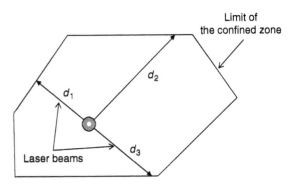

even when the rays are not perpendicular to the reflected surfaces.[3] Knowing the shape of the confined zone, it is possible to carry out computations in order to define the location of the laser system and its orientation. Of course, this example is only 2D, but a similar approach can be taken with additional measurements in order to achieve 3D positioning and orientation.

The main difficulty is of course that potential obstacles can certainly lead to wrong distance measurements. This is also the case with open doors in indoor environments or with windows (although not always true in this latter case). A solution could be to have the system pointing at the "sky," i.e. up to the ceiling, for example, like the radio satellite-based systems. In such a case, more measurements are required because the "sky" is a perfect plane. This approach could present some interest in static environments where the positions of objects and the structures are well defined.

3.2.2 Radar Method

The principle of the RAdio Detection And Ranging (radar) was found at the beginning of the twentieth century. The typical measurements are identical to those carried out by the modern global navigation satellite system (GNSS), namely time measurements in order to provide distance and frequency[4] measurements in order to provide velocity. Although there are many different types of radars, let us just deal with a simple implementation that should allow us to understand the basic principles.

A transmitter transmits characteristic waves at a given power level. The wave travels in free space and is reflected by any kind of target (a plane or a missile in the first applications) in various directions in space. The function that characterizes the way the target reradiates the incident wave is specified by the so-called radar effective area of the target. It corresponds roughly to the directions and intensity of the radiated reflected power. As shown in Figure 3.4, a

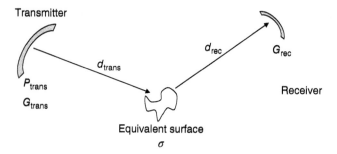

Figure 3.4 The bistatic radar principle.

---

3 Note that this is a major difference from ultrasound telemetry systems.
4 Doppler shift frequency indeed.

receiver can be placed in any location and can measure the power level, time shift, and Doppler shift of the reflected signal at the receiver's end. Knowing the main characteristics of the transmitted signal, the distance and velocity of the target can be computed.

When the transmitter and the receiver are not located in the same place, the radar is called "bi-static," as opposed to the "monostatic" radar. Let us now deal with this most common latter configuration. Many technical points are quite easy to deal with, for instance, the time reference. A typical wave form is given in Figure 3.5: the signal is a repetition of pulses equally time spaced. Each pulse is made up of a microwave carrier frequency in order to use the good propagation characteristics of high frequencies and properties linked to the geometry of the antenna. By measuring the time needed for the transmitted wave to be received back by the receiver (considered to be in the same location as the transmitter), one can easily obtain twice the distance from the radar to the target. Of course, the signal received is much lower than the transmitted one. The propagation equation[5] shows that the power decrease is proportional to the power of four of the distance, meaning that this power decrease is very rapid with distance. Typical distance attenuations of more than 150 dB are usual.

Furthermore, the received frequency can be shifted by the amount of the Doppler frequency if the target (or the radar) is moving. Thus, by measuring the shift between the transmitted frequency and the received one, the radar can calculate the radial projection of the relative velocity of the target with respect to the radar. It is interesting to note that modern GNSS use exactly the same approach in order to define both the location and velocity of a typical receiver. The time determination is quite different but the principle is the same: time measurements to deduce distance. As a matter of fact, in the case of the monostatic radar, things are simplified by the colocation of the transmitter and the receiver (which is not really the case with GNSS). Both time and frequency generators are available for comparison of the received signal. In such a case,

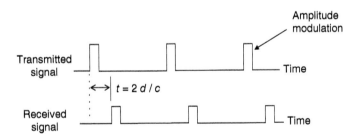

Figure 3.5 The bistatic radar principle.

---

5 The equation is indeed $P_r = \frac{P_e G^2 \lambda^2 \sigma}{(4\pi)^3 d^4 L}$, where P_e is the transmitted power, P_r the received power level, G the antenna gain, λ the wavelength, σ the radar equivalent surface, d the distance separating the radar to the target, and L all the possible loss along the path.

it is possible to provide a very accurate clock and very accurate frequency shift measurements. Distance and velocity measurements of radars are thus of excellent quality.

Another radar technique was to use nonmodulated pulses of very short duration: this was called the ultrawide band (UWB) radar. The name is associated with the fact that very short pulses have a very large equivalent frequency spectrum. However, one has to remember that the UWB radar is rather based on a time principle instead of a frequency one. The very short pulse allows a very accurate distance measurement as long as the receiver is able to detect the reflected pulse. One of the main advantages is that when considering the frequency spectrum, the bandwidth is so large that effects related to certain frequency bands can be ignored if they affect only a limited part of the bandwidth. Thus, attenuation when traveling through walls is less than for particular frequencies and multipath can be reduced because of the time-pulsed approach (the width of the pulse gives the maximum nonvisible multipath-induced delay).

A similar concept is bound to be implemented in the UWB wireless local area network (WLAN) approach. Unfortunately, although the name is identical, the principle is quite different: the time reference is not the same as for radar. This means that in order to define flight times, one will require additional infrastructure for synchronization purposes (see Chapter 7 for details).

3.2.3 Hyperbolic Method

It is a typical so-called hyperbolic system as it is based on the time difference between the receptions of two signals from two radio stations. Considering one knows the exact location of both transmitters and that they are synchronized, then the time difference gives the possible locations on a hyperbola whose focal points are the two stations. Thus, the receiver knows, from this first time difference, that it lies on a hyperbola: this is of course not enough to define a fix and a second measurement is required. This is achieved through the use of a third station, synchronized with the two others. A new time difference is measured, taking one of the preceding two stations as the reference. This second measurement gives a new hyperbola and the location is determined by the intersection of the two hyperbolae obtained from the two time differences (refer to Figure 1.2 for details).

3.2.4 Mobile Telecommunication Networks

In order to forward communications, the GSM/UMTS/4/5G mobile network needs to have access to a database that keeps track of the mobile locations. Indeed, the location is simply the identification of the base station (BS) providing the greatest power level to the receiver. Of course, as base stations are

Figure 3.6 The Cell-Id concept (BS stands for base station).

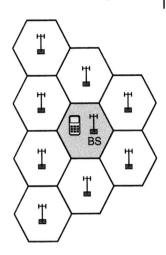

rather large installations, their locations are known and they are not movable. Thus, the so-called Cell-Id (see Figure 3.6), for identification of the telecommunication cell where the terminal is, is embedded in the network. Furthermore, it is implemented for all the deployed networks, as it is compulsory. Therefore, localization features exist in all large-area networks. The advantage of this positioning is that it gives a location in all places where the network is available, i.e. where the reception is possible, including indoors. The main disadvantage is the low level of accuracy it provides: the typical size of the result is the cell itself. Therefore, it goes from 100 m in densely populated urban areas to 20 or 30 km in rural zones. Another difficulty, where the density of base stations is high, is that the cell associated with the mobile is the one whose reception is the most powerful, i.e. not necessarily the one that is the nearest. Thus, the accuracy can easily drop to a few hundred meters even though there are base stations in the vicinity of the mobile terminal.

This method is quite comparable to radio signal strength (RSS) measurements as it is based on power level estimation, but with only one base station. The main advantage of this approach is that it works with only one base station in radio visibility.

As it is difficult to deal with very different accuracy figures depending on the network deployment, the Cell-Id positioning has not been used for years, except for telecommunication purposes.

For positioning purposes, this Timing Advanced can of course be used to give a first idea of the distance the terminal is from the base station. It is not really very accurate as its resolution is typically a bit length, hence 3.6 μs. However, this is far better than with the cell identification method. The graphical representation of this combined Cell-Id plus Timing Advance approach is given in Figure 3.7. The diagram on the left-hand side shows that the resulting position of the terminal is reduced compared with the Cell-Id technique. The graph

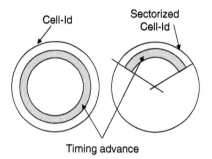

Figure 3.7 The Cell-Id + Timing Advance technique (left) alone and (right) with sectorized antennas.

on the right-hand side is the one obtained when considering, in addition, a sectorized approach where sectorized antennas are used. Using such antennas is quite normal in mobile networks, once again in order to improve the network's capabilities.

Many of the above-listed techniques can either be initiated and location calculated at the terminal end or at the base station end.

3.3 Doppler-Based Positioning Approach

3.3.1 Doppler Radar Method

The principle of the radar was found at the beginning of the twentieth century. The typical measurements are identical to those carried out by the modern GNSS, namely time measurements in order to provide distance and frequency[6] measurements in order to provide velocity. Although there are many different types of radars, let us just deal with a simple implementation that should allow us to understand the basic principles.

Furthermore, the received frequency can be shifted by the amount of the Doppler frequency if the target (or the radar) is moving. Thus, by measuring the shift between the transmitted frequency and the received one, the radar can calculate the radial projection of the relative velocity of the target with respect to the radar. It is interesting to note that modern GNSS use exactly the same approach in order to define both the location and velocity of a typical receiver. The time determination is quite different but the principle is the same: time measurements to deduce distance. As a matter of fact, in the case of the monostatic radar, things are simplified by the colocation of the transmitter and the receiver (which is not really the case with GNSS). Both time and frequency generators are available for comparison of the received signal. In such a case, it is possible to provide a very accurate clock and very accurate frequency shift measurements. Distance and velocity measurements of radars are thus of excellent quality.

---

6 Doppler shift frequency indeed.

3.3.2 Doppler Positioning Approach

The Argos system, such as the COSPAS–SARSAT and the DORIS systems, is based on Doppler measurement positioning, illustrated in Figure 3.8. One has to remember that the first observations of Sputnik by the members of the Department of Applied Physics of the John Hopkins Laboratory concerned the Doppler shift of the transmitted signal. Also remember that the first navigation satellite systems, such as TRANSIT, PARUS, and TSIKADA, were also based on Doppler shift. In the present case, the situation is quite different: the transmission is achieved from the Argos beacon and the satellite is the receiving element (see Figure 3.9 for illustration).

The Doppler shift of the received frequency is zero when the satellite passes the closest point to the transmitter. In such conditions, the beacon is located on a circle of unknown radius, perpendicular to the satellite's orbit. The intersection between this circle and the Earth's surface gives a possible line of locations, as illustrated in Figure 3.12. Note that this line is perpendicular to the satellite's Earth track at the closest point.

To be able to provide a more accurate location of the beacon, the slope of the typical Doppler versus time curve is used. Indeed, this slope allows the determination of the angle between the satellite's orbit and the beacon's location. Thus, from both the closest location of the satellite and the slope of the Doppler curve, one is able to provide the user with two points (one on each side of the Earth track of the satellite, perpendicular to the track). The only way to eliminate one point is to wait for another satellite, with another Earth track.

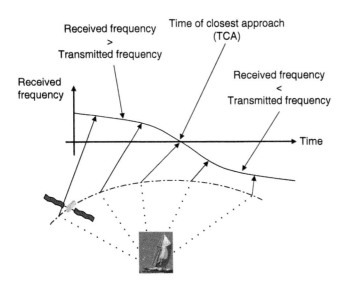

Figure 3.8 The Doppler-based positioning technique – I.

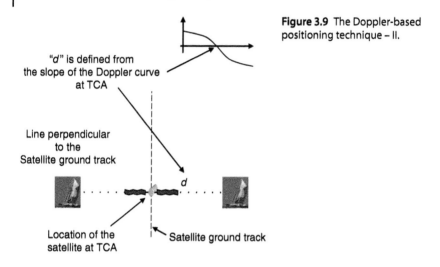

Figure 3.9 The Doppler-based positioning technique – II.

3.4 Physical Quantity-Based Positioning Approaches

3.4.1 Luminosity Measurements

As the problem of indoor positioning is a real challenge (see Chapters 9 and 10 for details), many original solutions have been investigated. Among others, the measurement of the level of the surrounding light level is an unusual way of applying fingerprinting. This technique relies on the observation, under certain conditions, of the variation of a given parameter (here the light levels). Following a calibration phase, where measurements are carried out throughout the whole place, a database is established. Then, an instantaneous light level measurement allows pattern matching recognition in the database and possibly location determination. This general approach can easily be extended to a lot of physical parameters, like the temperature or radio power levels, either locally generated (WLAN) or regionally generated (TV signals, for example).

3.4.2 Local Networks

The WLANs and wireless personal area networks (WPANs) are bound to be widely deployed in the near future. These networks, designed for local telecommunication purposes, are usually installed in indoor environments. As long as global positioning systems (GPS and GLONASS, for instance) either do not work indoors or are not very accurate, there is a need for indoor positioning solutions. The major point is that future applications will certainly be demanding in terms of continuity of service. There is currently no such continuity, mainly because of the poor indoor coverage and performances of localization systems.

The usual methods used for localization purposes are described in Chapter 8. The one that is usually chosen for WLAN positioning is based on the RSS, which we will focus on. Nevertheless, some WLAN-based systems have been proposed using time of flight of radio signal[7]: the AirLocation® from Hitachi is a network-based positioning system that includes a very precise clock in the base station network, allowing accurate time measurements.

The idea of signal strength measurement is not really a new one as it was thought of in the early eighteenth century for the terrestrial magnetic field in order to solve the problem of longitude. The principal difficulty of this approach lies in the fact that there are few locations that show an identical value of power level (in the current case): so the radio visibility of more than one transmitter (base station) is required if an "accurate" location is required. Nevertheless, the idea is basically to draw a map of radio signal strength received. A typical result of such a process is represented in Figure 3.10: the values are the power levels in dB over a characteristic value in this case (Bluetooth system using the RSS indicator). From this figure, it is possible to see that many different locations are characterized by, say, an received signal strength indicator (RSSI) value of

5	9	5	5	6	10,5							
6	10	9	7	5,5	5							
9,5	10	6,5	8,5	6	7	9,5	8	11,5	12,5	7,5	9,5	7,5
7,5	10	9,5	10,5	5	7,5	10	10	10,5	7,5	9,5	9	10,5
5,5	11,5	7	9,5	8	6	11,5	9	7,5	10	12	12,5	8
8	10,5	10,5	8	8	10,5	10,5	8,5	13	11,5	12,5	10	14,5
9,5	6,5	8,5	9,5	10,5	14,5	14,5	13,5	13	10	13,5	12	12,5
11	9,5	12,5	10	11,5	12,5	14	2	19,5	16,5	12,5	10	11
9,5	9	10,5	12	10	13	13	17,5	14,5	13,5	14,5	9	10,5
12	9	8	10	11,5	5,5	11	16	16,5	15,5	13	12	11
9,5	8	10,5	6,5	8	14,5	11	13,5	13	13	14	9,5	12,5
3,5	0,5	1,5	1,5	9,5	11,5	13,5	13,5	14,5	13,5	9	10	9
1,5	3	2	4	0,5	16,5	10,5	9,5	10,5	12	11,5	10,5	7
0,5	3	1,5	5,5	2,5	7,5	11,5	7,5					
0	2,5	1	3,5	2	10	12,5	11					
0	0	1,5	0	0	0,5	7,5	10,5					
0	0	0	0	0	6,5	9,5	11,5					
0	0	0	0	0	2,5	0,5	6,5					
					12	5	2,5					
					0	2,5	1,5					

Figure 3.10 A typical RSS map (1 m step in both north and east directions – Bluetooth technology).

---

7 It is indeed time difference of arrival by comparing the time of arrival at base stations.

10. Thus, with only one base station, the accuracy obtained is poor, or indeed the number of possible locations is high. Furthermore, these locations can be scattered throughout the place (due to the nontrivial propagation scheme indoors). If accuracy is not required, this can be a simple positioning method.

When accuracy is required,[8] there is the need for more than one base station. When a few base stations are considered, the principle is to find the nearest location within the database containing all the values of all the base stations. It will be the most probable location of the mobile. As a matter of fact, many different algorithms have been tried and proposed, all taken from the first phase of establishing the database: this involves a campaign of measurements. Then, two methods are possible: either a pattern matching approach consisting of finding the nearest element in the database or a propagation-based approach consisting of extracting a model of indoor radio wave propagation, i.e. some mathematical relationship between the power level received and the distance from the mobile to the base station in consideration. More sophisticated methods, like defining possible tracks, have been evaluated, with good results. Indeed, if one accepts an increase in the constraints on the "resulting locations," the accuracy can be very good: the main difficulty remains to be able to easily apply the method to a new place.

From a few representative works, it is possible to outline the major trends of a WLAN-based positioning system, as it appears today. The following remarks can be made:

✓ Pattern matching approaches give quite good results (in terms of accuracy).
✓ Positioning based on propagation models appears to be less accurate.
✓ There is quite a large range of accuracy values.
✓ The distribution of measurements exhibits a rather large error margin (often around 10 dB).
✓ The orientation of the mobile terminal, relative to the base stations, is a parameter of concern.
✓ The direct impact of the number of access points that are used in the system is of great influence.
✓ The use of sophisticated algorithms for a "nearest neighbor" search leads to no improvement.
✓ Use of privileged tracks provides a tremendous increase in accuracy, when staying on these tracks.
✓ More complex infrastructures are likely to provide for both simplification of the calibration phase (almost always required) and efficiency of the location finding process.

---

8 In general, as indoor places are smaller than outdoor ones, the typical accuracy required is somewhere between one-third and one-half the floor height. This should then allow the floor level to be determined as well. A typical value of 1 m can then be the tracked value. Note that very few works deal with the full three-dimensional WLAN approach at the moment.

From this first analysis, another approach could be conceived that deals with a "symbolic" positioning, i.e. in terms of rooms and corridors instead of accuracy in meters. The principle of a symbolic positioning was used in the infrared system and could be once again implemented in the present case.

3.4.3 Attitude and Heading Reference System

Inertial systems include all the techniques that take advantage of inertial properties of any movement. For instance, if a straight-line displacement at a constant speed is curved, a force F appears and thus a corresponding acceleration γ: both are linked by the simple formula $F = m\gamma$, where m is the mass of an object. The same appears if the movement is simply accelerated. As the acceleration is also the first derivative of the velocity, it can also be used in order to define the evolution of the velocity over time. A second integration could then be carried out to obtain the displacement. The gyroscopic effect is also a result of acceleration but exhibits different sensitivity errors than accelerometers, thus allowing a combined use in order to reduce measurement errors. By extension, other physical measurements have been included in the inertial systems, such as barometers and magnetometers: the first ones could help in defining the floor level indoors and the second ones the absolute orientation of the mobile terminal.

One has to note that inertial systems are those that allowed the first navigation systems to be installed in automobiles. A few months before the official availability of GPS, the first navigation system was available in a car. To achieve such a goal without GPS,[9] there was the need for autonomous sensors that were able to "follow" the car's movement: inertial sensors. Nowadays, GNSS receivers are widely used in order to make the best possible system, thanks to the very good accuracy provided, but for a while, GPS was mainly used to calibrate the inertial system dynamically. The main sensors used in car navigation are accelerometers, gyroscopes, and odometers. Others can be used indoors or for pedestrian purposes, such as barometers or magnetometers.

It is interesting to note that inertial systems are the modern implementation of "dead reckoning."[10] In ancient times, the "*lochs*" were used in order to allow more continuous navigation than astronomical positioning. Nowadays, when GNSS signals are not available, i.e. in a few environments and especially indoors, there is still a demand for continuous positioning. The basic needs are indeed to determine the velocity vector, i.e. both its amplitude and its direction, in real time. As physical measurements are achieved at a given rate, this introduces errors: current systems can typically achieve 100 Hz. The integration

---

9 GPS was declared operational in 1995.
10 "Dead" stands for "where positioning is no longer possible with the current technique in use."

(and thus the sum) of errors finally lead to increasing positioning inaccuracy with time.

Inertial personal devices for pedestrians are complex to implement, in comparison with those for automobiles, mainly because of the much larger range of physical measurements required and the nonconstant attitude of the mobile terminal.[11] For instance, let us imagine a mobile phone equipped with a GNSS receiver and also with an inertial system in order to allow dead reckoning when GNSS signals are no longer available (indoors for instance). As a mobile terminal, the phone is subject to many small but violent movements such as rotation, hand shaking, or even falls. All these movements must be analyzed by the inertial system and should not lead to the accumulation of errors. Unfortunately, as the basic principle is integration, all the errors are added up: as these kinds of motions are frequent, the resulting error can be significant. Moreover, the other types of movements that are bound to be difficult for pedestrian mobile terminals are also very slow and hesitant. For example, when one is moving from one leg to the other very gently, the resulting signals are almost identical whether there is a real displacement or not. In this second case, the errors are not the major concern, but the interpretation of the motion is. Thus, although some remarkable implementations have been achieved, the use of inertial systems for pedestrian mobile terminals has not yet been widely deployed.

3.4.3.1 Accelerometers

Acceleration is the rate of change of velocity. Accelerometers are used for both measuring vibration and shock (which can be considered as sudden acceleration) and acceleration of bodies. In this latter case, it is thus possible to use the sensor in order to provide velocity determination (by integrating the signal), positioning (by a double integration), or distance traveled (by successive integrations). Accelerometers are also used to define the attitude of a body, with reference to the horizontal plane, taking the gravity of the Earth (which can be considered as acceleration) as the reference.

The most popular accelerometers use the piezoelectric effect[12]: it consists of a mass attached to the piezoelectric element, as shown in Figure 3.11. When acceleration occurs, the mass applies a force on the piezoelectric crystal, leading to the appearance of a charge across the crystal, thus polarizing its metal faces and producing a voltage. As the electric output signal is a function of the force, it is then possible to make a measurement of the acceleration.

The main error in accelerometer-based positioning is due to the integration process (the measurement errors are added up).

---

11 Within the car, the horizontal plane should roughly remain unchanged.
12 The piezoelectric effect was discovered by Pierre and Jacques Curie in 1880. Some crystals, called piezoelectric crystals, when subject to a mechanical constraint, exhibit electric charges of opposite polarity to appear on their sides.

Figure 3.11 A piezoelectric accelerometer (modern implementations use nanotechnology).

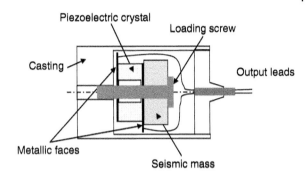

3.4.3.2 Gyrometers

The output of a gyrometer is the rate of change of the angle of its axis. This is used in order to measure the variation of the direction of the mobile terminal. A three-dimensional accelerometer could also be used to achieve such a goal, but a gyrometer exhibits slightly different error bias: the use of accelerometers for velocity determination and gyroscopes for angle determination also allows both sensors' biases to be compensated.

Although many different technologies are available, let us describe briefly the way it works. A gyrometer is made of a rotor, rotating at a high speed around its axis, which could have 1 or 2 degrees of freedom. The fundamental principle of the operation of a gyros gyrometer cope is that a moment[13] applied perpendicularly to the rotation axis leads to a displacement that is perpendicular to both the rotation axis and the orientation of the moment. Here again, to be able to define precisely the relative orientation changes in the movement of the mobile, the need for a three-dimensional gyrometer is important. Figure 3.12 shows the principle of a one axis gyroscope.

3.4.3.3 Odometers

Although it is possible to use accelerometers in order to obtain the relative distance traveled, this means calculating a double integration from the measurements. This is certainly not a very efficient way as errors and biases are going to accumulate. When possible, in an automobile system, for instance, it is preferable to use direct displacement measurement through the use of odometers. In a car, this is simply a sensor that can count the rotation of the wheels and convert the value into a linear distance. When required, it is also possible to implement a differential approach on the wheels of a given axle in order to define the direction of the displacement.

---

13 A moment is the product of a force applied to the gyroscope with the distance from the force to the center of the gyroscope.

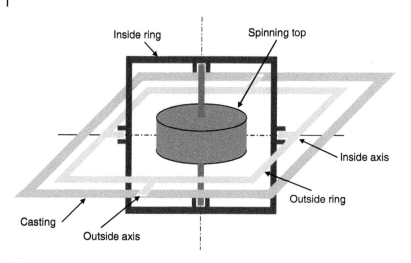

Figure 3.12 Principle of a gyroscope.

For indoor purposes, the same idea applies for rolling objects. For pedestrians, it is different as it is not convenient to use wheels in any way. Once again, accelerometers can be used to count the number of footsteps and then convert them into a distance.

3.4.3.4 Magnetometers

As accelerometers, gyroscopes, and odometers are primarily relative sensors, there could be the need for absolute ones in order to allow an absolute positioning, such as GNSS fixes. We have seen that with accelerometers designed in order to achieve an inclinometer, it is possible to define the horizontality of the mobile. This is a first approach of an absolute sensor since this is achieved without the need for any former attitude. Another important parameter is the absolute orientation of the mobile[14]: in applications where the discovery of the environmental world is required, this feature is a must. For instance, in a Museum, the electronic guide should certainly take advantage of the fact that it knows what the visitor is looking at. This is also important when one wants to be oriented when taking a first step. With current GNSS receivers, one needs to start moving before this information is relevant.

Magnetometers are sensitive to the Earth's magnetic field and are thus available all around the world, without any calibration required. The main direction is the magnetic north, which is slightly different from the geographical north (this must be taken into account, at least by staying in the same referential, either magnetic or geographical). The difference is the declination, experimentally discovered by Christopher Columbus during his travels to "India."

---

14 Note that GNSS signals do not provide this information, unless in dynamic mode.

3.5 Image-Based Positioning Approach

The camera is a passive optical system in the sense it transmits no signal. The scene is captured and forms an image that can then be processed. Therefore, such a system cannot be saturated and is also silent, unlike telecommunication systems. Nevertheless, obstacles could be present between the camera and the object one would like to locate and in this case, only a system using multiple cameras or multiple photographs taken at different instants could be acceptable.

For positioning purposes, the idea is to carry out specific form recognition within the image. This can be either markers or "natural" forms such as buildings, roads, doors, or characteristic monuments (churches, stadium, transportation systems, etc.). With a certain level of knowledge of the environment, the location of the camera can be extracted from one or several views. Note that for characteristic environments such as indoors (with windows, doors, ceilings, and angles of walls), location calculation can be quite efficient. The counterpart is the fact that the positioning is "highly" relative, in comparison with an absolute positioning such as the GNSS-based one.

A few other approaches have been developed, such as the simultaneous localization and mapping one, which consists in defining the relative displacement from one image to the next, giving identical patterns detected in the successive images. Calibration is generally required and is a source of positioning error. Note that blurred images are also a potential source of error.

Another interesting method consists in using the so-called markers, visible in the image, which helps in georeferencing the images. Relative as well as absolute positioning can be achieved, depending on the way the georeferencing of the markers has been carried out. Mathematical transformations have to be applied to the image in order to be able to determine the location of any pixel. Note that depending on the complexity of the system, the accuracy and the positioning mode will be different (one or several cameras, one or several images, etc.). A more detailed description is given in Chapter 8.

3.6 ILS, MLS, VOR, and DME

In some specific cases, such as civil aviation, the need rapidly appeared for dedicated systems allowing guidance and landing approach assistance to planes. The terrestrial positioning systems were not accurate and reliable enough. The main systems were

- The VHF Omni-Directional Radio Range (VOR)
- The distance measuring equipment (DME)
- The instrument landing system (ILS)
- The microwave landing system (MLS)

The VOR system is a rotating radio lighthouse. The transmitted frequency is in the range of 108–118 kHz and the signal is modulated by two 30 Hz signals whose phase difference gives the azimuth with reference to a characteristic direction, which is usually the magnetic north. If this direction is changed to the direction of the VOR station, then the phase difference is zero while the plane stays on the route of the transmitter. With three such VOR stations, it is theoretically possible to calculate a location, but VOR is usually used for alignment purposes.

In order to define the distance of the plane from the station, the DME system is used. It works in the 962–1213 MHz band and is based on a plane interrogating the DME station. Then, the station answers all the interrogations: the plane should find the response that corresponds to its request. Note that each transmitter is characterized by a specific pulse sequence.

The VOR and DME systems are coupled in order to provide a location of the plane in polar coordinates. The "positioning" accuracy is typically a few hundred meters.

This accuracy is not enough for landing phase approach. Thus, the ILS has been developed. It defines a light slope rectilinear trajectory for the landing by the way of the intersection of two surfaces. It then requires two radio lighthouses: the first one defines an alignment on the runway (the "localizer") and the second one is for the descent (the "glide-slope"). The system is also completed with two or three vertically radiating radio markers that play the role of spot locations in front of the runway. The runway alignment radio lighthouse uses a frequency in the 108–112 kHz band and the descent one in the 328–335 MHz band.

All these systems of so-called "goniometry[15]" are limited in accuracy because the frequencies that are used are not free from multipath disturbances. Nevertheless, the MLS also uses the rotating radio lighthouse principle but at microwave frequencies, in the 5 GHz band. The running mode is always a radio lighthouse one, but the radiating pattern is rather narrow (1–3°) and the free space is successively scanned from right to left and back again. Knowing the scanning pattern, the receiver can determine its angular location by analyzing the time delay separating two successive beams of the MLS. The angles are thus provided in a continuous mode, unlike ILS. The landing path is then evaluated in comparison to a predefined optimal path that can have any desired form (and not necessarily a rectilinear one, see Figure 3.13 for illustration). This component is not widely deployed because of the number of sites to be equipped and the corresponding costs.[16]

---

15 The fact of measuring angles.
16 And the advent of satellite-based navigation systems.

Figure 3.13 The MLS concept.

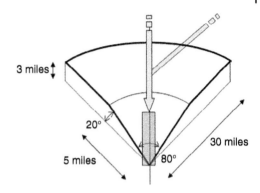

3.7 Summary

Table 3.1 is a summary of the techniques with their associated limitations. It is clearly not exhaustive but gives you a global view of the difficulties one can expect in the implementations of these techniques.

Table 3.1 Specification by main domains.

Technique	Main characteristics	Foreseen possible difficulties
Angle based		
Pure angle	Compass bearing approach	Must be line of sight
Triangulation	Use of triangle properties	Reflected path must be avoided
Distance based		
To a known environment	Accurate measurements	Obstacles
Radar	Return time of flight	Line of sight
Hyperbolic	No need for receiver synchronization	Positioning accuracy sensitivity to noise
Mobile networks	Indoor availability	Very poor accuracy
Doppler based		
Doppler radar	Very precise measurements	Line of sight
Doppler positioning	Geometrical approach	Reflected path to be avoided
Physical quantities		
Luminosity	Simple to measure	Complex to calibrate
Local telecom networks	Large availability indoors	Limited reliability

(Continued)

Table 3.1 (Continued)

Technique	Main characteristics	Foreseen possible difficulties
Inertial		
Accelerometer	Environment independent	Need to be calibrated often
Gyroscopes	Accurate sensor	Calibration
Odometers	Simple measurement	Associated noise
Magnetometers	Positioning and orientation	Reliability
Image based		
Pattern matching	Image sensor available	Need to refer to huge databases
Image processing	Very powerful tool	Obstacles and calibration

Bibliography

1 Cafferi, J. (2000). *Wireless Location in CDMA Cellular Radio Systems.* Kluwer Academic Publishers.

2 Cafferi, J.J. and Stüber, G.L. (1998). Overview of radiolocation in CDMA cellular systems. *IEEE Communications Magazine* 36 (4): 38–45.

3 Duffett-Smith, P.J. and Tarlow, B. (2005). E-GPS: indoor mobile phone positioning on GSM and W-CDMA. *ION GNSS 2005*, Long Beach, CA (September 2005).

4 Kaplan, E.D. and Hegarty, C. (2017). *Understanding GPS: Principles and Applications*, 3e. Artech House.

5 Küpper, A. (2005). *Location Based Services – Fundamentals and Operation.* Chichester: Wiley.

6 Ladetto, Q. and Merminod, B. (2003). Digital magnetic compass and gyroscope integration for pedestrian navigation. *GPS/GNSS 2003*, Portland, USA (September 2003).

7 Parkinson, B.W. and Spilker, J.J. Jr. (1996). *Global Positioning System: Theory and Applications.* American Institute of Aeronautics and Astronautics.

8 Patwari, N., Ash, J.N., Kyperountas, S. et al. (2005). Locating the nodes – cooperative localization in wireless sensor networks. *IEEE Signal Processing Magazine* 22 (4): 54–69.

9 Shankar, P.M. (2002). *Introduction to Wireless Systems.* Wiley.

10 SnapTrack. Location technologies for GSM, GPRS and UMTS networks. White Paper http://www.snaptrack.com.

11 Yazdi, N., Ayazi, F., and Najafi, K. (1998). Micromachined inertial sensors. *Proceedings of the IEEE* 86 (8): 1640–1659.

12 Wirola, L., Laine, T.A., and Syrjärinne, J. (2010). Mass-market requirements for indoor positioning and indoor navigation. In: *2010 International Conference on Indoor Positioning and Indoor Navigation*, Zurich, 1–7. IEEE.

13 Deng, Z., Yu, Y., Yuan, X. et al. (2013). Situation and development tendency of indoor positioning. *China Communications* 10 (3): 42–55.

14 Bozkurt, S., Yazici, A., Günal, S., and Yayan, U. (2015). A survey on RF mapping for indoor positioning. In: *2015 23nd Signal Processing and Communications Applications Conference (SIU)*, Malatya, 2066–2069. IEEE.

15 Al-Ammar, M.A., Alhadhrami, S., and Al-Salman, A. (2014). Comparative survey of indoor positioning technologies, techniques, and algorithms. In: *2014 International Conference on Cyberworlds*, Santander, 245–252. IEEE.

16 Mainetti, L., Patrono, L., and Sergi, I. (2014). A survey on indoor positioning systems. In: *2014 22nd International Conference on Software, Telecommunications and Computer Networks (SoftCOM)*, Split, 111–120. IEEE.

17 Liu, H., Darabi, H., Banerjee, P., and Liu, J. (2007). Survey of wireless indoor positioning techniques and systems. *IEEE Transactions on Systems, Man, and Cybernetics, Part C (Applications and Reviews)* 37 (6): 1067–1080.

18 Macagnano, D., Destino, G., and Abreu, G. (2014). Indoor positioning: a key enabling technology for IoT applications. In: *2014 IEEE World Forum on Internet of Things (WF-IoT)*, Seoul, 117–118. IEEE.

19 Abdat, M., Wan, T., and Supramaniam, S. (2010). Survey on indoor wireless positioning techniques: Towards adaptive systems. In: *2010 International Conference on Distributed Frameworks for Multimedia Applications*, Yogyakarta, 1–5. IEEE.

20 Witrisal, K., Hinteregger, S., Kulmer, J. et al. (2016). High-accuracy positioning for indoor applications: RFID, UWB, 5G, and beyond. In: *2016 IEEE International Conference on RFID (RFID)*, Orlando, FL, 1–7. IEEE.

21 Zhang, D., Xia, F., Yang, Z. et al. (2010). Localization technologies for indoor human tracking. In: *2010 5th International Conference on Future Information Technology*, Busan, 1–6. IEEE.

22 Yassin, A., Nasser, Y., Awad, M. et al. (2017). Recent advances in indoor localization: a survey on theoretical approaches and applications. *IEEE Communications Surveys & Tutorials* 19 (2): 1327–1346.

23 Yan, J., Tiberius, C.C.J.M., Janssen, G.J.M. et al. (2013). Review of range-based positioning algorithms. *IEEE Aerospace and Electronic Systems Magazine* 28 (8): 2–27.

24 Harle, R. (2013). A survey of indoor inertial positioning systems for pedestrians. *IEEE Communications Surveys & Tutorials* 15 (3): 1281–1293.

4

Various Possible Classifications of Indoor Technologies

Abstract

A real difficulty while dealing with indoor positioning relates to the multiple parameters that are involved in the classification of technologies and in the evaluation of their performances. From purely usage-related ones to purely technical ones, they all combine in many constraints that are almost impossible to de-interleave. The purpose of this chapter is to carry out a tentative redefinition of the most important parameters in order to propose a reading of this complexity in an "as simple as possible" way. As you will see, the problem is indeed complex as the respective importance of the numerous parameters highly depends on the use that will be made of the positioning system.

Keywords *Indoor parameters; classifications; technologies; performances*

4.1 Introduction

The first part of the chapter deals with the various parameters one should take into account when characterizing an indoor positioning system. A quick description of each of them is also provided, but a more complete discussion is left for the second part. The third one gives initial details concerning the various technologies selected and produces some references to the following chapters where more complete descriptions are given. The fourth paragraph is then a succession of tables in accordance with the four major categories used for the classification of the main parameters. Finally, the fifth paragraph gives a few possible classifications of the technologies depending on only one or two parameters. The last paragraph describes the classification finally held for the rest of the book, based on the so-called "range" parameter.

Indoor Positioning: Technologies and Performance, First Edition. Nel Samama.
© 2019 The Institute of Electrical and Electronics Engineers, Inc. Published 2019 by John Wiley & Sons, Inc.

4.2 Parameters to Be Considered

It is possible to consider a large number of parameters in order to characterize a given technology, but the 20 parameters mentioned below seem to be the most important. Furthermore, they can be classified into four categories depending on the aspect they are dealing with.

- Parameters related to the hardware of the system:
  - Infrastructure complexity
  - Infrastructure maturity
  - Infrastructure-estimated cost
  - Terminal complexity
  - Terminal maturity
  - Terminal cost
- Parameters related to the type and performances of the system:
  - Positioning type
  - Accuracy
  - Reliability
  - Range
  - Positioning mode
  - Feasibility of the in–out transition of the positioning
- Parameters related to the real implementation of the system:
  - Availability on smartphones
  - Sensitivity to environmental conditions
  - Need for calibration
  - Complexity of the calibration, if any
- Parameters related to the physical aspects of the system:
  - Type of physical measurements carried out
  - Signal processing algorithm(s) used
  - The way the position is calculated
  - Physical quantity associated with the measurement

Such a list seems to be plethoric but is indeed linked to the complexity of the real indoor world. In addition, the next paragraph shows that for a given parameter, it is necessary to go into even more detail to cope with any real situation. The description should be even more precise.

In addition, another important parameter is energy consumption, which is required by the infrastructure and also necessary for the terminal. However, this parameter is not directly related to the positioning approach, evolves very quickly with technological advances, and is also extremely difficult to estimate properly. Therefore, without forgetting it in our discussions, we will not deal with it directly.

4.3   Discussion About These Parameters

All these parameters have potentially different meanings. In this paragraph, the "standard" definition used is described, together with the possible options. Where applicable, a few practical examples are provided to the reader in order to help the understanding. Note that the order of the different levels (the possible choices indeed) considered for each parameter progresses from the less constraining to the most constraining.

4.3.1   Parameters Related to the Hardware of the System

→ *Infrastructure complexity*: As already discussed in previous chapters, the need for a locally deployed infrastructure is an important parameter. Of course, any "infrastructure-free" technology will be seen advantageously, but until now, these kinds of solutions have not shown sufficiently "good" global performances. We have used six levels for this parameter: none, low, medium, high, very high, and new. The first five are easily understandable. The last one means that research works are required to completely define the infrastructure. In this case, one does not consider that the complexity is at a high level but just that it is not yet possible to give it a rank.

→ *Infrastructure maturity*: The second parameter relating to infrastructure concerns the maturity. The levels considered are none, existing, integration, development, research, and new. The first two are quite clear. The third one, integration, means that there is a limited effort required to integrate the existing components or functions. "Development" means the effort is slightly more important. The last two, "research" and "new," parameters mean the infrastructure does not yet exist at an industrial level. Once again, it does not mean that it would be difficult to get it to that level but just that it is not yet available. "New" also means that it has to be designed from almost zero.

→ *Infrastructure-estimated Cost*: This is really an **estimated** cost. The levels considered are simple and the values assigned to the technologies are highly debatable: they reflect the perception of the author.

→ *Terminal complexity*: This parameter is linked to that of the "smartphone," in particular when the "existing" level is selected. It is nevertheless more accurate and directly linked to the technology.

→ *Terminal maturity*: This parameter is similar to that of the infrastructure.

→ *Terminal cost*: This parameter has the same comments that apply to infrastructure.

4.3.2 Parameters Related to the Type and Performances of the System

→ *Positioning type*: This is a fundamental parameter. Some technologies are very efficient but limited to a specific aspect of positioning or navigation. This is typically the case of a magnetometer for orientation purposes, or a microbarometer for altitude determination. Thus, five levels are possible: "orientation" or "altitude," for example, for the two technologies just mentioned and also "relative" for those systems providing a location with respect to an initial one, as opposed to "absolute" for those, like Global Positioning System (GPS), for instance, that provide the user with a complete set of coordinates in an absolute reference frame. The last level is called "symbolic": it is associated with those technologies that propose a positioning on the basis of areas rather than coordinates. Such a system could be the one that gives you your location by just defining the room or the corridor you are supposed to be in, but nothing more precise.

→ *Accuracy*: This parameter is dangerous and must be linked to the following one: the "reliability." Let us take an example. Consider a GPS receiver: what is its accuracy? Some people will answer "a few meters" for a typical receiver embedded in a smartphone. However, what happens when it is located indoors, for example, in an underground car park? As a matter of fact, it no longer provides you with any location. Thus, accuracy alone is not a sufficient indicator and it must be coupled with another parameter, either concerning the availability or the reliability of the positioning (I have chosen the second one in fact). Nevertheless, this parameter exists and is of course of major importance. The values available are numerous, ranging from centimeters to hundreds of meters. Note that these values characterize the corresponding technology when it is implemented indoors, thus leading, for instance, to poor values when dealing with GPS.

→ *Reliability*: It is probably one of the most sought after characteristics. As a matter of fact, it is often forgotten that positioning should be accompanied with such a reliability-related indicator. Without this information, it is almost impossible to design a useful system as one never knows whether the positioning is correct or not. This parameter goes from very high to low through four levels. Surprisingly, almost only GNSS has implemented such an indicator. This parameter is also very valuable when one wants to carry out fusion between systems: without the reliability specified, how can the algorithms be chosen properly?

→ *Range*: This parameter is mainly here in order to explain the classification used in the present book, i.e. the order of the following chapters. When somebody envisages deploying a positioning system (not a researcher who wants to imagine a new approach but someone who really wants to install a system), one of the first questions is to determine the range of the system, i.e. the size of the area that should be covered. Indoors, this size can vary greatly from a few square meters (for very localized robot applications, for instance) to tens of thousands of square meters (in the case of very large malls, for example). Thus, knowing about the range of a given technology is of utmost importance in order to judge the efforts that will be required, the cost of the solution, etc.

→ *Positioning mode*: This parameter is related to the actual way the positioning is achieved. Many different aspects can be considered: in the case of GPS, the mode is set to "continuous" because the signals are supposed to be available almost everywhere. The positioning is thus potentially continuous both in time and in space (although we know this is not the case in indoors). At the other side of the spectrum, a technology based on near-field communication (NFC) tags will require a specific action from the user who wants to locate her(him)self. Once again, the role of the tables presented in this chapter, and hence of the parameters, is not to say one technology is better than another but to specify each technology with pertinent parameters in order to give useful guidelines to the deployment of a system. The two intermediate levels are "discrete" when the positioning is only available on demand or from time to time and almost continuous when service interruptions are inherent to the system, but for a reduced part of either time or space (or both).

→ *Feasibility of the in–out transition of the positioning*: The question is: is it possible, simple, or impossible to use this technology outdoors, as well as indoors? Five levels are possible, from "already exists" to "impossible," through respectively, "easy," "moderate," and "difficult."

4.3.3 Parameters Related to the Real Implementation of the System

At this stage, the apparent simplicity of the problem seems less evident than at the beginning of the chapter. Let us go one step further by introducing the potential difficulties of real implementation.

→ *Availability on smartphones*: Smartphones are today absolutely unavoidable when it comes to discussing the terminal any mass market (and even professional now) system should be deployed on. The problem is clearly that for a sensor to be eligible to be included in the standard package, it should be, at least, of much reduced size (because the game is to propose slimmer and slimmer telephones). Thanks to nanotechnology, microelectromechanical systems (MEMS, such as magnetometers or accelerometers) are small enough and are thus present in all smartphones. The case of GPS is specific because the integration required to develop some dedicated antennas. The pressure of the market is at such a level that unless it gives a real commercial advantage, the introduction of new sensors is not easy. Seven levels are considered for this parameter, from "not applicable" because there is no sense considering it to "almost impossible" taking into account the technology. The five other levels are, respectively, "existing," "near future" if we think it should be available soon, "easy" in case it would not be a problem if useful, "future" if we think it could be done, and "difficult" because the integration would require a significant amount of work in order to be miniaturized or to make it available on smartphones.

→ *Sensitivity to environmental conditions*: Many solutions show quite acceptable performance (in terms of accuracy, for example) when the environment is well mastered, but are almost unusable in "normal"[1] situations. This is in particular the case with some optical-based approaches where the laser beam must have a completely clear line of sight (LOS) to its target: the presence of anything (a person for instance) means no correct measurement. Thus, the sensitivity of the technology to the environment in which it is deployed is important. The five levels are "no impact," "low," "moderate," "high," and "very high."

→ *Need for calibration*: A few technologies require calibrating, either the system itself or the environment, sometimes both. Such calibrations can be time-consuming, as in the case of a fingerprint approach with typical 1-m steps in both X and Y directions. An additional difficulty occurs when the calibration also depends on environmental changes, leading to the need for recurring calibrations, and sometimes continuously. The five explicit reserved levels are "none," "once," "several times," "often," and "permanently."

→ *Complexity of the calibration, if any*: The notion of complexity of the calibration also matters. System designers try to find ways to carry out the calibration automatically, without the need for human actions. This is not always possible. We retain four levels, from "none" to "heavy" through "light" and "medium." Many different cases occur: sometimes, one needs to calibrate the infrastructure by knowing the power emitted by different transmitters in order, for example, to calibrate a propagation model. In such a case, it can be imagined to collect measurements in order to calibrate automatically the changes of these transmitted powers. Sometimes, it is a lot more complex, for accelerometers, for instance, and the calibration must be carried out regularly. The availability of other "sensors" could then be used in order to achieve the calibration automatically. Thus, once again, this parameter cannot be used alone. This is the complexity of the indoor positioning problem, which increases rapidly indeed.

4.3.4 Parameters Related to the Physical Aspects of the System

This last series of parameters is much more technical and probably not the first one to be used in order to specify a positioning system. Nevertheless, they are typically used for classification purposes, although this would not be our choice.

---

1 "Normal" is clearly not a good adjective. As a matter of fact, two indoor environments are never identical, although classes of environments can be imagined.

→ *Type of physical measurements carried out:* This is a classical way to characterize a technology: the type of measurements. Indeed, it has a tremendous effect on the performances and can directly link the "accuracy," "reliability," or "sensitivity to environment" parameters to the hardware. For instance, when a system is based on time measurements, one knows that hypotheses are proposed concerning the propagation model used, hence a certain dependency on environmental conditions, such as walls, for instance. The eight levels considered are "time," "distance," "angle," phase," "frequency," "physical," "image," and "fusion." It is important to understand that we are dealing with the actual physical measurement and not with the way the positioning is calculated (this is another parameter!). For instance, the GPS carries out time measurements and then converts them into distances for positioning, but physical measurements are truly time-based ones. The parameters "angle" and "phase" are indeed quite different as although angles can be determined by phase measurements (of electromagnetic signals for example), they can also be determined by angular sensors such as code wheels, hence mechanical sensors. We choose to use the "frequency" parameter instead of "Doppler" in order to remain at the physical level for this parameter. In addition, some pure frequency differences are used in a few radars in order to achieve distance estimations (but it remains frequency measurements). "Image" is used when the system uses image or images, regardless of the approach (image processing or image analysis, which is slightly different). "Physical" means a physical quantity is measured, either inertial values, power values, or even pressure values. The distinction between the quantities has not been considered valuable in the scope of the present discussion, and hence, they have been grouped into the generic term "physical." The last parameter, "fusion," means there is no possibility to achieve positioning without additional data coming from another source. This is typical in the case of wired networks, such as the Internet.

→ *Signal processing algorithm(s) used:* When one goes one step further, the way the measurements are dealt with is the next parameter. Here again, seven levels are taken into account: "propagation modeling," "correlation," "classification," "detection," "pattern recognition," "pattern matching," and "a combination of." The great diversity of approaches is then highlighted, ranging from purely physical techniques (relating to propagation, for example) to data science ones (such as classification or correlation), passing through data analysis, such as pattern matching approaches. The very interesting point regarding this diversity of domains involved reflects the attractiveness of positioning. A special mention should be given to "classification," as it indicates that some techniques are oriented toward the extraction of "classes" in order to achieve the positioning. These approaches are particularly interesting when the fusion between various sensors is investigated. The level "a combination of" means that several techniques, among those quoted, are required together.

→ *The way the position is calculated*: Once the measurements have been obtained and the signal is processed, it is time to calculate the position. Depending on the type of the data available, various kinds of mathematical calculations can be carried out. Geometry is the major domain implemented, either in two dimensions or in three dimensions. Sometimes, the analytical solution is possible, sometimes not, but as measurement and processing error must be considered, the current techniques almost always implement numerical resolution methods. The first six levels relate to geometry: "∩ spheres," "∩ hyperbolae," "∩ circles," "∩ straight lines," "∩ lines + distance(s)," and "∩ plans + distance(s)." For simple intersections of geometrical forms, no specific explanations are required (although not always simple to calculate, i.e. for intersection of hyperbolae). However sometimes, measurements are not of the same type: this can be the case for local radar, for instance, where both phase measurements are carried out in order to define the angle of arrival and distance measurements. In such a case, the calculation mixes intersection of straight lines (if two angles are used) and a distance (i.e. the intersection of the two lines with a sphere). The next level is "spot location" and applied when the positioning is provided to the user as a reduced circular area: this is often the case with very low range technologies, such as NFC-based approaches, which show discrete positioning, both in time and space. More complex mathematical functions are needed when one wants to calculate the trajectory of a terminal using inertial data. In the case of acceleration measurements, one needs to integrate once in order to obtain the speed and twice in order to obtain the distance traveled. It corresponds to the level "math functions $(\int, \int\int, \int\int\int, ...)$." The level "matrices calculus" is common for all the technologies based on image processing or analysis. Images are dealt with as matrices and numerous transformations are available for positioning purposes. The last level, "zone determination," is used when the position calculation is indeed the determination of an area of highest probability of presence: this is the case of so-called "symbolic" methods.

→ *Physical quantity associated with the measurement*: The last parameter concerns the physical domain involved in the measurements. Many systems use electromagnetic waves, either in the radio, microwave, or optical spectrums. Others take advantage of mechanical waves, such as in the case of ultrasound-based systems. Thus, "electromagnetic waves" and "mechanical waves" are the first two levels. Then, the third one is "image sensor," while the fourth is "physical sensor." The latter level refers to the equivalent in the "*Type of physical measurements realized*" parameter list. The last two levels are, respectively, "electronics" and "optoelectronics," used only for two very special technologies: theodolite and wired networks. It just means that this is a piece of electronics or optoelectronics that allows the measurement.

4.4   Technologies Considered

Forty technologies are taken into account, ranging from classical GNSS-based approaches to quite specific theodolites, going through WiFi or LiFi. Quick explanations are given below for each of them.

- *Accelerometer*: Such sensors are nowadays present in almost all smartphones. Nevertheless, the performances of those embedded in smartphones are not good enough, today, for longtime positioning. Some exist, which allow positioning but at a quite higher cost.
- *Bar codes*: These tags are associated with many products today for logistics purposes. As the readers are usually at a fixed location, the scan of a bar code means it was at that location when scanned: this is also positioning!
- *BLE*: The Bluetooth technology has evolved these in the past few years toward a much simpler transmission scheme allowing for fast connectivity and hence reduced energy consumption, leading to the so-called Bluetooth low energy (BLE). Although many techniques could be implemented with BLE, the most common one is based on the establishment of a geographical map of the covered area. The measurements available are mainly received signal strength, hence received power levels. The main difficulty lies with the high dependency of these measurements on the environment, both human and architectural, of the receiver.
- *Contactless cards*: Here is another way to achieve what can be called "spot location," meaning the location is carried out punctually in time and in space. The card was at the location of the reader, at the time of reading, generally very accurate and reliable, but not continuous.
- *Cospas Sarsat, Argos*: This system is usually deployed for search and rescue, or for cohort following purposes. It uses Doppler measurements and provides the user with a location in a nonpermanent manner.
- *Credit cards*: In the same way as for contactless cards, the credit (or debit) cards can be used for positioning. This is often considered as evidence in trials. Once again, the reader is either wired or linked to a base that is nearby and can deliver a rough location of the card at a precise time.
- *GNSS*: The satellite-based positioning systems are widely used outdoors and probably have no competitors in this environment today. The problem is that they need at least three or four satellites in view with a sufficient power level, together with so-called LOS signals, meaning a direct path from the satellite to the receiver. Indoors, this situation is not likely to happen often. Moreover, it predominantly happens when the receiver is located near a window. In such a case, the vision of the sky is "partial" such that that only part of it is seen, leading to a very poor geometry between the satellites and the receiver for the positioning and hence a location that could be very poor (and typically outside the building indeed). In the tables, the line "GNSS" is dedicated to

current GNSS, available almost everywhere in the mass market (details are given in Chapter 11).

- *Global system for mobile, GSM/3/4/5G*: Many works were carried out in the early 2000s, leading to potential approaches allowing typically 100 m accuracy of positioning, both outdoors and indoors. These kinds of performances can be reached when the signal propagation conditions are very good, which is almost never the case in urban areas, for example. Unfortunately, it means that this technology offers good performance where GNSS is excellent, thus reducing its interest. On the other hand, GSM/3/4/5G signals are available indoors, whereas GNSS ones are not. Current works are still under progress and mainly oriented toward the use of statistical analysis in order to propose a best estimate of the probable position. Research works have shown, in research environments, quite good performance in terms of accuracy, down to typically less than 10 m. As usual, from research to industry, a lot can happen. In the following tables, the case of the most used positioning aspect of the mobile telecommunication networks is considered: the so-called cell-id. The positioning is determined by considering the area covered by the "base station." The previous discussion concerning the low interest of mobile networks in urban areas is thus inverted, as a lot of "microcells" are deployed in dense urban areas in order to satisfy the data rates and the high number of users, leading to a reduced size of the above-mentioned areas, thus increasing the accuracy. A complete discussion on this point is provided in Chapter 10.
- *Gyrometer*: The same approach applies for accelerometers. Some are excellent, but the purpose of the book is the continuity of the positioning service for the general public, hence concentrating on gyrometers available on a large scale (today it means on smartphones). Thus, the current performances are not at the top level.
- *High-accuracy GNSS*: Almost all current GNSS receivers are so-called high sensitivity or assisted GNSS. They implement either signal processing techniques designed in order to increase the sensitivity or get additional aiding data from the mobile network to decrease the time required to obtain a first position. Receivers can also implement both high sensitivity and assisted GNSS. The complete aspects are described in Chapter 11.
- *Image markers*: Image processing or analysis are quite efficient ways for positioning. Indeed, it is even possible to determine the location of the camera in a completely unknown environment (in such a case in a relative manner). Another approach consists in distributing so-called markers in the environment that will be visible on the images. The fact that these markers have some specificities (size, location, form, etc.) helps in determining the position of the camera or of any pixel in the image. This approach is possible with a single image, but restrictions appear (see Chapter 8), or multiple images, either simultaneous or sequential.

- *Image recognition (site, people)*: Another simple principle that relies on the comparison of one image with a database of images and in analyzing the level of resemblance. Huge databases are required, but they are now available, and the performance also depends on the specificity of the object photographed. For the Eiffel Tower, the recognition algorithm is bound to be quite rapid and accurate. It would be more difficult in the case of a "standard" house in the suburbs of London. The location given is the one characterizing the object in the database. Thus, it is mainly applied to fixed objects.
- *Image relative displacement*: With or without markers, successive images taken when the camera is moving can be used in order to determine the trajectory of the camera. In this case, either there are markers and the trajectory can be given in an absolute manner or remarkable forms are visible on successive images allowing the camera to reconstruct its trajectory, in a relative manner. Indeed, in the second case, there is the need to find a scaling factor, corresponding roughly to the real sizes of the remarkable forms, which are unknown of the problem.
- *Image SLAM (simultaneous localization and mapping)*: The principle of the image-relative displacement is applied, but made more complex by the idea to reconstruct the environment while moving and determining the location of the camera, all of which has to happen at the same time. Thus, once you have moved along a track, you also get the map of this track. The same problem as for image relative displacement occur: there is a scaling factor and identical specific forms must be present in successive images. This second constraint is nevertheless not really a problem indoors as the displacements are usually quite slow.
- *Indoor GNSS*: The idea of having your GNSS receiver working indoors is not new. The idea is based on local transmitters, called pseudolites (contraction of Pseudo and Satellites) and usually referred to as PL, which transmit equivalent signals to those transmitted by the satellites. The pseudolites could be located on the roof of the building. The main technical difficulties are associated with the synchronization needed between the pseudolites, the so-called near-far effect which, in some cases, makes it impossible to detect a given pseudolite because another one "blurs" it, and the need to transmit at very reduced power in order not to disturb outdoor signals transmitted by the satellites. For this last reason, regulations exist in many countries in order to forbid or limit the deployment of PLs. Many approaches of indoor GNSS exist and will be detailed in Chapter 9. The positioning performance can be quite good, but once again mainly when the propagation environment is not too bad, i.e. in large halls, in exhibition centers, or in warehouses. The other disadvantage is the need for a local infrastructure, of course.
- *Infrared*: Optical waves have many ways to achieve positioning. The one commonly described in the literature with infrared signals is based on a receiver mounted on the ceiling in such a way that it can detect any signal in

the room. Once a transmitter enters the room, it is detected. As long as each transmitter has its own identifier, it is located in the room. Note that in this case, we call the positioning "symbolic," meaning the location is provided as an area and no longer as coordinates.

- *Laser*: A laser can be used in order to obtain a distance with a very good accuracy in the range of a few millimeters over distances of a few hundreds of meters. Such an instrument is called a telemeter. Achieving a few such measurements and knowing the shape of your environment allows you to determine your location with a very good accuracy. Such measurements are carried out in the next technology, the theodolite. This "instrument" is used by topographers and allows them to define the relative positions of any location accurately. The combination of a laser telemeter and some electronic coding wheels allow us to determine, quite accurately, two angles (elevation and azimuth) as well as a distance from the so-called theodolite and any objects in its LOS.

- *Lidar*: Lidar stands for light detection and ranging, the optical equivalent of radar. There are many applications to lidar, but in our case, the usual way to use it concerns the three-dimensional imaging of an environment (see Chapter 6). For indoor positioning, the idea is in fact to invert the problem: from the lidar measurement and knowing about the environment, i.e. the shape of the building, it is possible to go back to the location where the lidar is. This is of course quite a complex and expensive approach, not intended for the general public (as it should be in this book), but we believe it is worth mentioning it.

- *LiFi*: The equivalent of WiFi for optical waves. The main idea is to use light bulbs that transmit a modulated signal, together with the light it is supposed to transmit. This latter is not visible to the human eye but can be detected and processed by an adequate receiver. Current implementations also allow smartphone cameras to detect the LiFi. Note that if one wants to have LiFi available during the day, lights should be permanently "on." This could sound strange, but this is exactly the same for WiFi: if one wants it, it has to be on.

- *Light opportunity*: Let us imagine you know about the building: the location of the lights and of the windows. An analysis of the luminosity could help in defining the area where you are within a room: close to a window, just at the door, in the corridor, etc. Of course, additional data would be required in order to define a more precise location, but it could give useful data.

- *LoRa*: The LoRa network is deployed for Internet of Things purposes. Nevertheless, as radio signals are transmitted from known locations, it is possible to extract data. Depending on the density of the network and of the power levels used, the positioning will largely differ, and the same applies to the environment in which the receiver is. As for numerous radio systems, different approaches are possible, from cell-id to time-based measurements, but require more or less effort. LoRa reports positioning better than 100 m, which depends on the number of antennas deployed.

- *Magnetometer*: This is the third sensor in the so-called "inertial" systems. A sort of electronic compass that measures the strength of the magnetic field. There are several techniques, but mass market positioning approaches are mainly based on two techniques: the first one is based on the use of the terrestrial magnetic field in order to determine the absolute orientation of a displacement, and the second one is based on the cartography of the magnetic field over a given area. This latter approach is called "fingerprint" (similar to WiFi techniques) and is often used for positioning when measurements are not of a very high quality. See Chapter 11 for details.
- *NFC*: Near-field communication is based on a magnetic field coupling between a transmitter, usually called the "reader," and a receiver called the "tag." The latter can be passive (without embedded energy) or active, usually with batteries. When the reader is close to the tag, a data exchange takes place. In the case of positioning, this is typically the location of the fixed part (which can be either the reader or the tag), which is transmitted. As the range of the transmission is very short (a few centimeters for passive tags and up to 1 m for active ones), the location is quite good, although not continuous in time and space.
- *Pressure*: Barometers are sometimes available on mobile terminals. They allow two types of measurements: variations in the time of atmospheric pressure and a rough estimation of your altitude. In the first case, it requires a long time as atmospheric pressure is slow to change, typically a few hours in "normal" conditions. Thus, if you are using your barometer for a few minutes or even dozens of minutes, then you are able to notice a change in altitude as atmospheric pressure depends on the altitude. Current microbarometers are able to typically detect a 1 m change in altitude. It allows you, under initialization process, to determine the floor level you have reached as you entered the building (supposing a calibration at the entrance).
- *QR codes*: These two-dimensional bar codes are developing rapidly (QR stands for quick response). Nevertheless, they are also an interesting way for positioning. In this case, as for NFC or bar codes, the positioning will be discontinuous both in time and in space, but it is an accurate and very reliable mean, as long as the code provides you with location information. The main differences between QR and NFC codes are the need to "open" your camera and the possibility to make the codes unreadable by painting them (pen, marker, paint, etc.): this is also possible with NFC, but it is slightly more difficult.
- *Radar*: Radar stands for radio detection and ranging. It is a remarkable sensor also allowing the identification of targets. The ranging capabilities are used for positioning, and one great advantage is that, as it carries out return travels of the signal with (usually but not always) no cooperation of the target, this is the same clock that manages the transmission and the reception, hence leading to a huge decrease in the complexity of the synchronization process. Chapter 7 gives details, but there are different ways to implement radar for

positioning: distance measurements are the basis, but angles of arrival are also possible, and the possible signal characteristics are very numerous, from pulsed to chirp in frequency or in Doppler, depending on the goals.

- *Radio 433/868/... MHz*: The so-called ISM (for industrial, scientific, and medical) radio bands are free to **use in many countries**. This leads to a large number of devices using these bands. Unlike WiFi transmitters, a complete database of those transmissions is not available, probably because of the rapid change of the systems. Thus, it is not easy to convert such a signal into a location. Nevertheless, in conjunction with a digital map, one could conclude that the terminal is in the right zone or not. Indeed, the radio range of these devices is not so great and they are mainly deployed in inhabited areas: receiving such signals means you are not far away from civilization.

- *Radio AM (amplitude modulated)/FM (frequency modulated)*: The main difference with ISM bands comes from the fact that the locations of the transmitters are known. Moreover, the transmitters are of high power, leading to a greater range. This is not always an advantage when one wants to carry out positioning, but at least you are sure you will receive it indoors. Usually, the signal incorporates a digital signature telling you the name of the station. Thus, having a correspondence table between the frequency of the signal and the names of the stations, there is a possibility to have a rough idea of your global location. Of course, this is clearly not guaranteed to provide you with an accurate indoor positioning.

- *RFID*: The name stands for radio-frequency identification and has been designed for short-range wireless data exchanges. In such a way, this is another technology allowing for local and reduced range cell-id positioning. Nevertheless, a few other approaches have been proposed for improving the range and for carrying out distance measurements (see Chapter 7 for details).

- *Sigfox*: The same as for LoRa applies, the two networks being quite close in principle. The 868 MHz frequency band generates a few constraints, but the network manager states that it is possible to achieve positioning with accuracy better than 1000 m. We shall discuss this assertion in Chapter 10, but this is an interesting idea to achieve positioning with a low-power system.

- *Opportunity radio signals*: With this technology, we are once again changing the philosophy of the positioning. It is no longer a system dedicated to positioning but rather the use of all radio signals available (this is "radio signal" here, but could be optical or sound signals). The process is based on deductions rather than on a direct use of measurements. We are entering the "software sensor" world where the principle is to use any detected signals to achieve positioning. Chapter 11 is dealing with these kinds of approach.

- *Sonar*: The sonar is the sound-based equivalent of radar or lidar. The main interest of sound waves that are in fact mechanical waves, unlike radar or lidar that are based on electromagnetic waves, is the reduced speed of the waves (typically 340 m s^{-1} in the air and around 1500 m s^{-1} in water). This low speed dramatically reduces the synchronization-related constraints: 1 ms is now about the same as 1 ns for radio waves, but 1 ms is much easier to obtain with a low drift. Distance measurements are then achievable with an accuracy of a few centimeters quite easily. The main problem remains the very high sensitivity to the environment as the distance provided is the one from the transmitter to the first obstacle found in the direction of propagation. Unfortunately, the directivity of the beam is larger than radio or light signals.

- *Sound*: What has been said about sound for sonar remains correct here, but the way we mean positioning is slightly different. The idea is to use sounds of opportunity indeed: if you can hear the sound from a loudspeaker, it means you are in its vicinity. The higher the sound, the closer you are (this assertion is not so obvious but let us consider it for now). If you know the locations of such speakers in the mall, you begin to have an idea of your possible location. Other sounds can also be used, such as specific advertisement messages from the flower shop, for instance. In addition, of course, this can be just part of the aggregation of other types of opportunity (or not) signals.

- *Theodolites*: These measurement instruments are used by civil engineers or by geodesists. They are made of two very precise angular measurement instruments in order to determine the direction of viewing, from the theodolite to any given point. The distance between the theodolite and the point is obtained, thanks to a laser telemeter. Thus, from the known location of the theodolite, it is possible to calculate a very accurate positioning of the viewing point (indeed, two angles and a distance are enough for a three-dimensional positioning). This is the way civil engineers can obtain a location very accurately (down to a centimeter) of any specific indoor point, for instance. We mention the theodolite in this review because this is often the instrument that is used to verify the accuracy of an indoor positioning system.

- *TV*: The use of television signals in order to achieve positioning is of the same idea as for other opportunity signals. For indoor positioning, the efficiency has not really been proved and is probably of poor performance. We just mention it, although some companies proposed it a few years ago.

- *Ultrasound*: Based on the physical aspects of sound, it presents the same advantages and disadvantages such as low speed of propagation or inability to pass through obstacles. However, the interesting point concerns the fact that speakers of current smartphones are able to detect such signals, allowing, for example, proximity detection systems to use ultrasounds. Moreover, the distance measurement performance can reach a very good accuracy, as long as the environmental conditions are the right ones.

- *UWB*: The ultra-wideband technology was invented in the 1960s by radar operators in order to be able to "see" behind obstacles. The idea is to generate a signal with a very large frequency bandwidth because the hypothesis made is based on the fact that any obstacles correspond to a given frequency band rejection. If the UWB signal has a significantly higher bandwidth, then only part of this signal will be removed, hopefully still allowing the detection of the target. In addition, current UWB technological manufacturing is based on low-cost nanotechnologies. The telecommunication domain has thus considered UWB as a potential approach for low-cost, low-power local area system, but with very high data rates. As already mentioned, radar is a good way to carry out distance measurements; thus, UWB can be seen as a potential candidate for indoor positioning, hoping that the signal will pass through the walls, for instance. The main difficulties are indeed the synchronization, as always, between the transmitters and the receivers (but some clever techniques have been proposed, see Chapter 7 for details), and the limited transmitted power level allowed, which is often not enough to actually pass through the walls indoors.

- *WiFi*: Certainly, this is the most known and deployed system intended for positioning purposes, both indoors and outdoors in some cases. Many techniques have been proposed, experimented, and deployed, ranging from the current most used fingerprint to the synchronized network of access points in order to achieve time-of-flight measurements. Fingerprint remains the most efficient approach today, mainly because the only real measurement available in such wireless local area networks is the received power level. This quantity is moreover very dependent on the local environment and reduced changes can produce high differences in the received power because of propagation conditions. Thus, besides the need for a first calibration of the environment in order to "produce" the initial fingerprints, it is necessary, when the environment is likely to change, to update this calibration from time to time. The recurrence not only depends on the reliability sought but also on the variability of the environment. In addition, as with all radio-based technologies, the presence of people creates interferences by blocking the signals to certain extend. This should be taken into account and increases the complexity of implementation. The resulting positioning accuracy also depends on the number of deployed access points: one can consider that the need for positioning is slightly higher than that for telecommunication purposes.

- *Wired networks*: Everybody has realized that after a few minutes spent on the web searching for something, some advertisements pop up corresponding to your searches but linked to your physical address. This is possible, thanks (or not) to the geo localization of the Internet Protocol (IP) address of your computer. Even though there is not a direct link between your address and your location, gradually the various traces you leave (by entering your postal address when you create an account, for example) are used and the matching is carried out.

- *WLAN (wireless local area telecommunication networks) symbolic*: There is an opportunity between cell-id approaches and purely distance or angle measurements associated with trilateration or triangulation ones. The "WLAN symbolic" is a sort of improved cell-id, preserving its simplicity, but increasing the area determination without any decrease in reliability. The main idea, fully described in Chapter 8, is based on a simple assertion: if the received power level is high, it means that you are close to the transmitter. All other assertions are likely to be untrue: for example, it happens when you are close with a low level because of obstacles. Thus, by using thresholds in power received, one can easily define successive areas of probable presence. In the case of several transmitters received, the intersection of these areas is carried out in order to supply an estimated location. Note that the areas must not be exclusive of each other, at the risk of ruining the efficiency of the method. The symbolic aspect lies in the fact that the positioning is now available under the form of an area and no longer provided to the user with coordinates. Of course, this symbolic approach can be implemented with technologies other than WLAN.

From all the above descriptions, one can understand that a synthesis is not obvious. Nevertheless, we have provided such a summary, but please note that it must be discussed and not considered as definitive. In addition, when multiple choices were possible for a given technology, we chose the one we consider is the most frequently implemented (but once again, this is debatable).

4.5 Complete Tables

As previously said, the term "technology" means indeed "a practical implementation" of a technique. Hence, technologies have to be seen as "systems" used in order to achieve indoor positioning. In the following tables, about 40 such technologies have been considered, from very local ones such as NFC based on already deployed (but not so efficient) GNSS ones.

Please note that all the values given to all the parameters are debatable and reflect the point of view of the author. Nevertheless, a few are almost always true, whereas in some cases, a real discussion would be welcome. Completing the boxes also depends on the objective assigned to the system.

It is not always easy to allocate a value to a parameter, but in order to be able to run the tool developed to help identify the best technologies that meet a given list of deployment constraints, it is necessary to allocate a value to all the parameters. A few examples of the use of the above-mentioned tool are given in the next paragraph.

In the tables, one should consider the following:

- The values concern the characteristics of the indoor performance of the technology, as considered at the time of writing.

- These values are provided as an average value (more details are given in the chapters dealing specifically with the technology).
- Note that almost each statement in the table must be qualified (this is part of the "indoor problem," which is more complex than it seems at first sight).
- Other values could have been assigned quite easily. Thus, one has to consider the box as just one value within a possible list.

In order to ease the reading of the tables, the classification of Section 4.2 is used, respectively, for Tables 4.1–4.4. Table 4.1 is related to the links between technologies and the infrastructure and terminal characteristics. Table 4.2 relates to performance parameters. Table 4.3 relates to real implementation parameters and finally Table 4.4 to physical aspects.

The list of technologies is organized in alphabetical order.

Table 4.1 Hardware table of indoor positioning technologies.

Technology	Infrastructure complexity	Infrastructure maturity	Infrastructure cost	Terminal complexity	Terminal maturity	Terminal cost
Accelerometer	None	None	Zero	Medium	Hardware development	Medium
Bar codes	None	Existing	Low	Low	Software development	Low
BLE	Low	Existing	Low	None	Software development	Zero
Contactless cards	None	Existing	Zero	None	Existing	Low
Cospas Sarsat – Argos	None	None	Zero	High	Existing	Medium
Credit cards	None	Existing	Zero	None	Existing	Low
GNSS	None	None	Zero	None	Existing	Zero
GSM/3/4/5G	None	Existing	Zero	None	Existing	Low
Gyrometer	None	None	Zero	Medium	Hardware development	Medium
High-accuracy GNSS	High	None	Zero	High	Existing	Very high
Image markers	Low	None	Zero	Low	Software development	Zero
Image recognition (site, people)	None	None	Zero	Low	Software development	Zero
Image-relative displacement	None	None	Zero	Low	Software development	Zero
Image SLAM	None	None	Zero	Low	Software development	Zero

Table 4.1 (Continued)

Technology	Infrastructure complexity	Infrastructure maturity	Infrastructure cost	Terminal complexity	Terminal maturity	Terminal cost
Indoor GNSS	Medium	Research	Medium	Low	Software development	Zero
Infrared	Medium	Development	Medium	Medium	Integration	Mediu
Laser	None	None	Zero	Medium	Existing	Mediu
Lidar	None	None	Zero	High	Existing	Very high
LiFi	Medium	Development	Medium	Medium	Integration	Mediu
Light opp	None	Research	Zero	Low	Software development	Low
LoRa	None	Existing	Zero	Low	Integration	Low
Magnetometer	None	None	Zero	Low	Existing	Low
NFC	Low	Existing	Low	Low	Software development	Low
Pressure	None	None	Zero	Low	Existing	Low
QR codes	None	Existing	Low	Low	Software development	Low
Radar	Medium	Research	Medium	High	Research	Mediu
Radio 433/ 868/... MHz	None	None	Zero	Low	Integration	Low
Radio AM/FM	None	None	Zero	Low	Integration	Low
RFID	Low	Existing	Low	Low	Software development	Low
Sigfox	None	Existing	Zero	Low	Integration	Low
Opportunity radio signals	None	Existing	Zero	Medium	Integration	Mediu
Sonar	None	None	Zero	Low	Integration	Low
Sound	None	None	Zero	low	Integration	Low
Theodolites	None	None	Zero	Very high	Existing	very high
TV	None	Existing	Zero	Medium	Integration	Mediu
Ultrasound	High	Existing	Medium	Medium	Integration	Low
UWB	Low	Development	High	Medium	Integration	mediu
WiFi	Low	Existing	Low	None	Software development	Zero
Wired networks	None	Existing	Zero	None	Existing	Zero
WLAN Symbolic	None	Existing	Zero	None	Software development	Zero

le 4.2 Type and performances table.

chnology	Positioning type	Accuracy	Reliability	Range	Positioning mode	In/out transition
ccelerometer	Relative	$f(t)$	Medium	Block	Continuous	Already exist
ar codes	Absolute	Decimeter	Very high	Proximity	User action needed	Easy
.E	Absolute	A few meters	Medium	Building	Almost continuous	EASY
ontactless cards	Absolute	A few centimeters	Very high	Proximity	Discrete	Impossible
ospas arsat – Argos	Absolute	>100 m	Medium	World	Continuous	Easy
redit cards	Absolute	A few centimeters	Very high	Proximity	Discrete	Impossible
NSS	Absolute	100 m	Low	World	Continuous	Easy
SM/3/4/5G	Absolute	>100 m	Low	City	Continuous	Moderate
yrometer	Relative	$f(t)$	Medium	Building	Continuous	Already exist
igh-accuracy NSS	Absolute	100 m	Low	City	Continuous	Difficult
nage markers	Absolute	<1 m	Medium	Proximity	Almost continuous	Easy
nage recognition ite, people)	Absolute	A few Dm	Medium	Proximity	Almost continuous	Easy
nage-relative splacement	Relative	<1 m	Medium	Building	Almost continuous	Moderate
nage SLAM	Relative	<1 m	Medium	Building	Almost continuous	Moderate
door GNSS	Absolute	A few decimeters	Medium	Building	Continuous	Easy
frared	Symbolic	A few m	High	Room	Almost continuous	Easy
aser	Absolute	Less than centimeter	Very high	Room	Almost continuous	Difficult
dar	Absolute	Less than centimeter	Very high	Room	Almost continuous	Difficult
Fi	Symbolic	A few meters	Low	Room	Almost continuous	Easy

Table 4.2 (Continued)

Technology	Positioning type	Accuracy	Reliability	Range	Positioning mode	In/out transition
Light opp	Relative	100 m	Low	Room	Almost continuous	Moderate
LoRa	Absolute	>100 m	Low	City	Continuous	Easy
Magnetometer	Orientation	A few degrees	Medium	World	Continuous	Already exist
NFC	Absolute	A few centimeters	Very high	Proximity	User action needed	Easy
Pressure	Relative	1 m	High	World	Continuous	Easy
QR Codes	Absolute	Decimeter	Very high	Proximity	User action needed	Easy
Radar	Absolute	A few centimeters	Medium	A few rooms	Continuous	Easy
Radio 433/ 868/… MHz	Absolute	>100 m	Low	County	Continuous	Moderate
Radio AM/FM	Absolute	>100 m	Low	County	Continuous	Moderate
RFID	Absolute	Decimeter	High	Proximity	Discrete	Easy
Sigfox	Absolute	>100 m	Low	City	Continuous	Easy
Opportunity radio signals	Absolute	>100 m	Low	World	Almost continuous	Moderate
Sonar	Relative	A few centimeters	Medium	Room	Continuous	Easy
Sound	Relative	>100 m	Low	Building	Continuous	Difficult
Theodolites	Absolute	A few centimeters	Very high	Building	Continuous	Difficult
TV	Absolute	>100 m	Low	County	Continuous	Moderate
Ultrasound	Absolute	A few decimeters	Low	Room	Continuous	Easy
UWB	Absolute	A few centimeters	Medium	A few rooms	Continuous	Easy
WiFi	Absolute	A few meters	Medium	Building	Continuous	Easy
Wired networks	Absolute	An address	Medium	World	Discrete	Impossibl
WLAN symbolic	Symbolic	Dm	Very high	Building	Continuous	Easy

Table 4.3 Real implementation parameters table.

Technology	Smartphone	Sensitivity to environment	Calibration needed	Calibration complexity
Accelerometer	Existing	No impact	Often	Medium
Bar codes	Existing	Low	None	None
BLE	Existing	High	Several times	Medium
Contactless cards	Near future	No impact	None	None
Cospas Sarsat – Argos	Almost impossible	High	None	None
Credit cards	Not applicable	No impact	None	None
GNSS	Existing	Very high	None	None
GSM/3/4/5G	Existing	High	None	None
Gyrometer	Existing	No impact	Often	Medium
High-accuracy GNSS	Future	Very high	Once	Light
Image markers	Existing	Very high	Once	Medium
Image recognition (site, people)	Existing	Very high	None	None
Image-relative displacement	Existing	High	Several times	Medium
Image SLAM	Existing	High	Several times	Medium
Indoor GNSS	Existing	High	None	None
Infrared	Near future	Very high	None	None
Laser	Future	Very high	None	None
Lidar	Almost impossible	Very high	Once	Medium
LiFi	Near future	Very high	None	None
Light opp	Near future	Very high	Often	Light
LoRa	Easy	High	None	None
Magnetometer	Existing	Moderate	Several times	Light
NFC	Existing	No impact	None	None
Pressure	Easy	No impact	Several times	Light
QR codes	Existing	Low	None	None
Radar	Difficult	High	none	None
Radio 433/868/… MHz	Easy	High	None	None
Radio AM/FM	Easy	High	None	None
RFID	Easy	Low	None	None
Sigfox	Easy	High	None	None
Opportunity radio signals	Near future	High	None	None

Table 4.3 (Continued)

Technology	Smartphone	Sensitivity to environment	Calibration needed	Calibration complexity
Sonar	Difficult	Very high	None	None
Sound	Existing	High	Often	Light
Theodolites	Almost impossible	Very high	Once	Medium
TV	Future	High	None	None
Ultrasound	Easy	Very high	None	None
UWB	Near future	High	None	None
WiFi	Existing	High	Several times	Medium
Wired networks	Not applicable	No impact	None	None
WLAN symbolic	Existing	Low	None	None

Table 4.4 Physical aspects table of indoor positioning technologies.

Technology	Technique	Signal processing	Position calculation	Physics used
Accelerometer	Physical	Detection	Math functions $(\int, \int\int, \int\int\int, \ldots)$	Physical sensor
Bar codes	Image(s)	Pattern recognition	Spot location	Image sensor
BLE	Physical	Pattern matching	Math functions $(\int, \int\int, \int\int\int, \ldots)$	EM waves
Contactless cards	Physical	Detection	Spot location	EM waves
Cospas Sarsat – Argos	Frequency(ies)	A combination of	∩ Straight lines	EM waves
Credit cards	Physical	Detection	Spot location	Electronics
GNSS	Time(s)	A combination of	∩ Spheres	EM waves
GSM/3/4/5G	Distance(s)	Propagation modeling	∩ Circles	EM waves
Gyrometer	Physical	Detection	Math functions $(\int, \int\int, \int\int\int, \ldots)$	Physical sensor
High-accuracy GNSS	Phase(s)	A combination of	∩ Spheres	EM waves
Image markers	Image(s)	A combination of	Matrices calculus	Image sensor
Image recognition (site, people)	Image(s)	Pattern recognition	Spot location	Image sensor
Image-relative displacement	Image(s)	A combination of	Math functions $(\int, \int\int, \int\int\int, \ldots)$	Image sensor
Image SLAM	Image(s)	A combination of	Math functions $(\int, \int\int, \int\int\int, \ldots)$	Image sensor
Indoor GNSS	Phase(s)	Correlation	∩ Hyperbolae	EM waves

(Continued)

Table 4.4 (Continued)

Technology	Technique	Signal processing	Position calculation	Physics used
Infrared	Physical	Detection	Zone determination	EM waves
Laser	Phase(s)	Propagation modeling	∩ Spheres	EM waves
Lidar	Time(s)	Correlation	∩ Plans + distance(s)	EM waves
LiFi	Physical	Detection	Spot location	EM waves
Light opp	Physical	Classification	Zone determination	EM waves
LoRa	Distance(s)	Propagation modeling	∩ circles	EM waves
Magnetometer	Physical	Detection	Math functions $(\int, \int\int, \int\int\int, ...)$	Physical sensor
NFC	Physical	Detection	Spot location	EM waves
Pressure	Physical	Detection	Zone determination	Physical sensor
QR Codes	Image(s)	Pattern recognition	Spot location	Image sensor
Radar	Phase(s)	A combination of	∩ Plans + distance(s)	EM waves
Radio 433/ 868/... MHz	Physical	Propagation modeling	∩ Circles	EM waves
Radio AM/FM	Physical	Propagation modeling	∩ Circles	EM waves
RFID	Physical	Detection	Spot location	EM waves
Sigfox	Distance(s)	Propagation modeling	∩ Circles	EM waves
Opportunity radio signals	Physical	Propagation modeling	∩ Circles	EM waves
Sonar	Time(s)	Detection	∩ Plans + distance(s)	Physical sensor
Sound	Physical	Detection	∩ Circles	Physical sensor
Theodolites	Angle(s)	A combination of	∩ Plans + distance(s)	Optoelectronics
TV	Physical	Propagation modeling	∩ Circles	EM waves
Ultrasound	Time(s)	Propagation modeling	∩ Spheres	Mechanic waves
UWB	Time(s)	A combination of	∩ Spheres	EM waves
WiFi	Physical	Pattern matching	Math functions $(\int, \int\int, \int\int\int, ...)$	EM waves
Wired networks	Fusion	Correlation	Zone determination	Electronics
WLAN symbolic	Physical	Propagation modeling	Zone determination	EM waves

EM, electro-magnetic.

4.6 Playing with the Complete Table

A very simple tool was developed by our research group at the Mines-Telecom Institute, in order to have a quick selection of suitable technologies, given a certain number of constraints. Indeed, this is probably the sort of tool useful for many positioning engineers who want to find the best direction to choose to implement a system, or a researcher who wants to determine what to do in order to improve the "science." However, of course, it does not give you the solution to the global indoor positioning problem: this is just an aid to help you given your own constraints and your own way to classify them.

We present three tables with all the technologies classified according to a specific criterion, respectively: accuracy (Table 4.5), positioning type (Table 4.6),

Table 4.5 Technologies classified according to the "Accuracy" parameter.

Technology	Positioning type	Accuracy	Reliability	Range	Sensitivity to environment	Positioning mode
Laser	Absolute	Less than centimeter	Very high	Room	Very high	Almost continuous
Lidar	Absolute	Less than centimeter	Very high	Room	Very high	Almost continuous
Contactless cards	Absolute	A few centimeters	Very high	Proximity	No impact	Discrete
Credit cards	Absolute	A few centimeters	Very high	Proximity	No impact	Discrete
NFC	Absolute	A few centimeters	Very high	Proximity	No impact	User action needed
Radar	Absolute	A few centimeters	Medium	A few rooms	High	Continuous
Sonar	Relative	A few centimeters	Medium	Room	Very high	Continuous
Theodolites	Absolute	A few centimeters	Very high	Building	Very high	Continuous
UWB	Absolute	A few centimeters	Medium	A few rooms	High	Continuous
Bar codes	Absolute	Decimeter	Very high	Proximity	Low	User action needed
QR codes	Absolute	Decimeter	Very high	Proximity	Low	User action needed
RFID	Absolute	Decimeter	High	Proximity	Low	Discrete
Indoor GNSS	Absolute	A few decimeters	Medium	Building	High	Continuous
Ultrasound	Absolute	A few decimeters	Low	Room	Very high	Continuous
Image markers	Absolute	<1 m	Medium	Proximity	Very high	Almost continuous

(Continued)

Table 4.5 (Continued)

Technology	Positioning type	Accuracy	Reliability	Range	Sensitivity to environment	Positioning mode
Image-relative displacement	Relative	<1 m	Medium	Building	High	Almost continuous
Image SLAM	Relative	<1 m	Medium	Building	High	Almost continuous
Pressure	Relative	1 m	High	World	No impact	Continuous
BLE	Absolute	A few meters	Medium	Building	High	Almost continuous
Infrared	Symbolic	A few meters	High	Room	Very high	Almost continuous
LiFi	Symbolic	A few meters	Low	Room	Very high	Almost continuous
WiFi	Absolute	A few meters	Medium	Building	High	Continuous
WLAN symbolic	Symbolic	Dm	Very high	Building	Low	Continuous
Image recognition (site, people)	Absolute	A few Dm	Medium	Proximity	Very high	ALMOST continuous
GNSS	Absolute	100 m	Low	World	Very high	Continuous
High-accuracy GNSS	Absolute	100 m		City	Very high	Continuous
Light opp	Relative	100 m	Low	Room	Very high	Almost continuous
Cospas Sarsat – Argos	Absolute	>100 m	Medium	World	High	Continuous
GSM/3/4/5G	Absolute	>100 m	Low	City	High	Continuous
LoRa	Absolute	>100 m	Low	City	High	Continuous
Radio 433/868/... MHz	Absolute	>100 m	Low	County	High	Continuous
Radio AM/FM	Absolute	>100 m	Low	County	High	Continuous
Sigfox	Absolute	>100 m	Low	City	High	Continuous
Signaux radio opp	Absolute	>100 m	Low	World	High	Almost continuous
Sound	Relative	>100 m	Low	Building	High	Continuous
TV	Absolute	>100 m	Low	County	High	Cositinuous
Accelerometer	Relative	$f(t)$	Medium	Block	No impact	Continuous
Gyrometer	Relative	$f(t)$	Medium	Building	No impact	Continuous
Magnetometer	Orientation	A few degrees	Medium	World	Moderate	Continuous
Wired networks	Absolute	An address	Medium	World	No impact	Discrete

Table 4.6 Technologies classified according to the "positioning mode" parameter.

Technology	Positioning type	Accuracy	Reliability	Range	Sensitivity to environment	Positioning mode
Bar codes	Absolute	Decimeter	Very high	Proximity	Low	User action needed
NFC	Absolute	A few centimeters	Very high	Proximity	No impact	User action needed
QR codes	Absolute	Decimeter	Very high	Proximity	Low	User action needed
Contactless cards	Absolute	A few centimeters	Very high	Proximity	No impact	Discrete
Credit cards	Absolute	A few centimeters	Very high	Proximity	No impact	Discrete
RFID	Absolute	Decimeter	High	Proximity	Low	Discrete
Wired networks	Absolute	An address	Medium	World	No impact	Discrete
BLE	Absolute	A few meters	Medium	Building	High	Almost continuous
Image markers	Absolute	<1 m	Medium	Proximity	Very high	Almost continuous
Image recognition (site, people)	Absolute	A few Dm	Medium	Proximity	Very high	Almost continuous
Image-relative displacement	Relative	<1 m	Medium	Building	High	Almost continuous
Image SLAM	Relative	<1 m	Medium	Building	High	Almost continuous
Infrared	Symbolic	A few meters	High	Room	Very high	Almost continuous
Laser	Absolute	<1 cm	Very high	Room	Very high	Almost continuous
Lidar	Absolute	<1 cm	Very high	Room	Very high	Almost continuous
LiFi	Symbolic	A few meters	Low	Room	Very high	Almost continuous
Light opp	Relative	100 m	Low	Room	Very high	Almost continuous
Signaux radio opp	Absolute	>100 m	Low	World	High	Almost continuous
Accelerometer	Relative	$f(t)$	Medium	Block	No impact	Continuous
Cospas Sarsat – Argos	Absolute	>100 m	Medium	World	High	Continuous
GNSS	Absolute	100 m	Low	World	Very high	Continuous

(Continued)

Table 4.6 (Continued)

Technology	Positioning type	Accuracy	Reliability	Range	Sensitivity to environment	Positioning mode
GSM/3/4/5G	Absolute	>100 m	Low	City	High	Continuous
Gyrometer	Relative	$f(t)$	Medium	Building	No impact	Continuous
High-accuracy GNSS	Absolute	100 m		City	Very high	Continuous
Indoor GNSS	Absolute	A few dm	Medium	Building	High	Continuous
LoRa	Absolute	>100 m	Low	City	High	Continuous
Magnetometer	Orientation	A few degrees	Medium	World	Moderate	Continuous
Pressure	Relative	1 m	High	World	No impact	Continuous
Radar	Absolute	A few cm	Medium	A few rooms	High	Continuous
Radio 433/ 868/… MHz	Absolute	>100 m	Low	County	High	Continuous
Radio AM/FM	Absolute	>100 m	Low	County	High	Continuous
Sigfox	Absolute	>100 m	Low	City	High	Continuous
Sonar	Relative	A few centimeters	Medium	Room	Very high	Continuous
Sound	Relative	>100 m	Low	Building	High	Continuous
Theodolites	Absolute	A few centimeters	Very high	Building	Very high	Continuous
TV	Absolute	>100 m	Low	County	High	Continuous
Ultrasound	Absolute	A few decimeters	Low	Room	Very high	Continuous
UWB	Absolute	A few centimeters	Medium	A few rooms	High	Continuous
WiFi	Absolute	A few meters	Medium	Building	High	Continuous
WLAN Symbolic	Symbolic	Dm	Very high	Building	Low	Continuous

and range (Table 4.7). For each table, the technologies are then classified in the increasing order of the parameter considered. The values that are special, for instance, the accuracy of the accelerometer that depends on time, are at the end of the tables.

We decided to keep six parameters in each table (positioning type, accuracy, reliability, range, sensitivity to environment, and positioning mode). The three tables show the resulting order of the technologies obtained when considering, respectively, the three parameters. Note that in case of a similar value of the parameter, the alphabetical order is chosen.

Table 4.7 Technologies classified according to the "Range" parameter.

Technology	Positioning type	Accuracy	Reliability	Range	Sensitivity to environment	Positioning mode
Cospas Sarsat – Argos	Absolute	>100 m	Medium	World	High	Continuous
GNSS	Absolute	100 m	Low	World	Very high	Continuous
Magnetometer	Orientation	A few degrees	Medium	World	Moderate	Continuous
Pressure	Relative	1 m	High	World	No impact	Continuous
Signaux radio opp	Absolute	> 100 m	Low	World	High	Almost continuous
Wired networks	Absolute	An address	Medium	World	No impact	Discrete
Radio 433/ 868/… MHz	Absolute	>100 m	Low	County	High	Continuous
Radio AM/FM	Absolute	>100 m	Low	County	High	Continuous
TV	Absolute	>100 m	Low	County	High	Continuous
GSM/3/4/5G	Absolute	>100 m	Low	City	High	Continuous
High-accuracy GNSS	Absolute	100 m	low	City	Very high	Continuous
LoRa	Absolute	>100 m	Low	City	High	Continuous
Sigfox	Absolute	>100 m	Low	City	High	Continuous
Accelerometer	Relative	$f(t)$	Medium	Block	No impact	Continuous
BLE	Absolute	A few meters	Medium	Building	High	Almost continuous
Gyrometer	Relative	$f(t)$	Medium	Building	No impact	Continuous
Image relative displacement	Relative	<1 m	Medium	Building	High	Almost continuous
Image SLAM	Relative	<1 m	Medium	Building	High	Almost continuous
Indoor GNSS	Absolute	A few decimeters	Medium	Building	High	Continuous
Sound	Relative	>100 m	Low	Building	High	Continuous
Theodolites	Absolute	A few centimeters	Very high	Building	Very high	Continuous
WiFi	Absolute	A few meters	Medium	Building	High	Continuous
WLAN Symbolic	Symbolic	Dm	Very high	Building	Low	Continuous
Radar	Absolute	A few centimeters	Medium	A few rooms	High	Continuous

(Continued)

Table 4.7 (Continued)

Technology	Positioning type	Accuracy	Reliability	Range	Sensitivity to environment	Positioning mode
UWB	Absolute	A few centimeters	Medium	A few rooms	High	Continuous
Infrared	Symbolic	A few meters	High	Room	Very high	Almost continuous
Laser	Absolute	Less than centimeter	Very high	Room	Very high	Almost continuous
Lidar	Absolute	Less than centimeter	Very high	Room	Very high	Almost continuous
LiFi	Symbolic	A few meters	Low	Room	Very high	Almost continuous
Light opp	Relative	100 m	Low	Room	Very high	Almost continuous
Sonar	Relative	A few centimeters	Medium	Room	Very high	Continuous
Ultrasound	Absolute	A few decimeters	Low	Room	Very high	Continuous
Bar codes	Absolute	Decimeter	Very high	Proximity	Low	User action needed
Contactless cards	Absolute	A few centimeters	Very high	Proximity	No impact	Discrete
Credit cards	Absolute	A few centimeters	Very high	Proximity	No impact	Discrete
Image markers	Absolute	<1 m	Medium	Proximity	Very high	Almost continuous
Image recognition (site, people)	Absolute	A few Dm	Medium	Proximity	Very high	Almost continuous
NFC	Absolute	A few centimeters	Very high	Proximity	No impact	User action needed
QR Codes	Absolute	Decimeter	Very high	Proximity	Low	User action needed
RFID	Absolute	Decimeter	High	Proximity	Low	Discrete

Unsurprisingly, the order is completely modified depending on the parameter chosen. Moreover, all the techniques are also completely mixed, meaning there is not a "best technique" that appears clearly. We believe that these tables are a first explanation concerning the difficulty to find a good way toward indoor positioning: everything blends together. This is also an additional argument toward the "fusion" or "statistical" approaches, which will either combine various technologies (at different levels) or use historical data and results in order to estimate the current positioning. All these techniques are discussed in a specific chapter (Chapter 12).

Our selection tool also allows us to specify either a constraint or a combination of constraints. For example, one could have chosen to list all the technologies leading to accuracy better than 1 m. In order to move forward in our description of the impact of multiple constraints, we choose to combine the three parameters: accuracy, positioning mode, and range and to modify them with currently sought values, i.e. "a few meters," "continuous," and "building," respectively. It means that we are interested in those technologies leading to positioning accuracies better than a few meters, or running in a continuous way (with a continuous positioning), or with a coverage better than a building. Tables 4.8–4.11 are the results for given combinations.

When looking attentively at this first table, it can be seen that not all the resulting technologies are of the same category. If a few can effectively be positioning systems, others are not real candidates alone. Further analyses are thus necessary: inertial sensors need to be calibrated (more or less often depending

Table 4.8 Technologies obtained for accuracy better than a few meters and a continuous positioning mode.

Technology	Positioning type	Accuracy	Reliability	Range	Sensitivity to environment	Positioning mode
Accelerometer	Relative	$f(t)$	Medium	Block	No impact	Continuous
Gyrometer	Relative	$f(t)$	Medium	Building	No impact	Continuous
Magnetometer	Orientation	A few degrees	Medium	World	Moderate	Continuous
Radar	Absolute	A few centimeters	Medium	A few rooms	High	Continuous
Sonar	Relative	A few centimeters	Medium	Room	Very high	Continuous
Theodolites	Absolute	A few centimeters	Very high	Building	Very high	Continuous
UWB	Absolute	A few centimeters	Medium	A few rooms	High	Continuous
Indoor GNSS	Absolute	A few decimeters	Medium	Building	High	Continuous
Ultrasound	Absolute	A few decimeters	Low	Room	Very high	Continuous
Pressure	Relative	1 m	High	World	No impact	Continuous
WiFi	Absolute	A few meters	Medium	Building	High	Continuous

Table 4.9 Technologies obtained for accuracy better than a few meters and a "building" range.

Technology	Positioning type	Accuracy	Reliability	Range	Sensitivity to environment	Positioning mode
Magnetometer	Orientation	A few degrees	Medium	World	Moderate	Continuous
Wired networks	Absolute	An address	Medium	World	No impact	Discrete
Pressure	Relative	1 m	High	World	No impact	Continuous
Accelerometer	Relative	$f(t)$	Medium	Block	No impact	Continuous
Gyrometer	Relative	$f(t)$	Medium	Building	No impact	Continuous
Theodolites	Absolute	A few centimeters	Very high	Building	Very high	Continuous
Indoor GNSS	Absolute	A few decimeters	Medium	Building	High	Continuous
Image relative displacement	Relative	<1 m	Medium	Building	High	Almost continuous
Image SLAM	Relative	<1 m	Medium	Building	High	Almost continuous
BLE	Absolute	A few meters	Medium	Building	High	Almost continuous
WiFi	Absolute	A few meters	Medium	Building	High	Continuous

Table 4.10 Technologies obtained for a "building" range and a continuous positioning mode.

Technology	Positioning type	Accuracy	Reliability	Range	Sensitivity to environment	Positioning mode
Cospas Sarsat – Argos	Absolute	>100 m	Medium	World	High	Continuous
GNSS	Absolute	100 m	Low	World	Very high	Continuous
Magnetometer	Orientation	A few degrees	Medium	World	Moderate	Continuous
Pressure	Relative	1 m	High	World	No impact	Continuous
Radio 433/ 868/… MHz	Absolute	>100 m	Low	County	High	Continuous
Radio AM/FM	Absolute	>100 m	Low	County	High	Continuous
TV	Absolute	>100 m	Low	County	High	Continuous
GSM/3/4/5G	Absolute	>100 m	Low	City	High	Continuous
High-accuracy GNSS	Absolute	100 m		City	Very high	Continuous
LoRa	Absolute	>100 m	Low	City	High	Continuous

Table 4.10 (Continued)

Technology	Positioning type	Accuracy	Reliability	Range	Sensitivity to environment	Positioning mode
Sigfox	Absolute	>100 m	Low	City	High	Continuous
Accelerometer	Relative	$f(t)$	Medium	Block	No impact	Continuous
Gyrometer	Relative	$f(t)$	Medium	Building	No impact	Continuous
Indoor GNSS	Absolute	A few decimeters	Medium	Building	High	Continuous
Sound	Relative	>100 m	Low	Building	High	Continuous
Theodolites	Absolute	A few centimeters	Very high	Building	Very high	Continuous
WiFi	Absolute	A few meters	Medium	Building	High	Continuous
WLAN Symbolic	Symbolic	Dm	Very high	Building	Low	Continuous

Table 4.11 Technologies obtained for accuracy better than a few meters, a continuous positioning mode, and a "building" range.

Technology	Positioning type	Accuracy	Reliability	Range	Sensitivity to environment	Positioning mode
Magnetometer	Orientation	A few degrees	Medium	World	Moderate	Continuous
Pressure	Relative	1 m	High	World	No impact	Continuous
Accelerometer	Relative	$f(t)$	Medium	Block	No impact	Continuous
Gyrometer	Relative	f(t)	Medium	Building	No impact	Continuous
Theodolites	Absolute	A few centimeters	Very high	Building	Very high	Continuous
Indoor GNSS	Absolute	A few decimeters	Medium	Building	High	Continuous
WiFi	Absolute	A few meters	Medium	Building	High	Continuous

on their quality). The same applies to pressure sensors that can only be used for elevation purposes, thus as a complement to another system.

The same remarks as for Table 4.8 apply, leading indeed to a reduced set of potential candidates, namely indoor GNSS, image-based technologies, and wireless local area networks, when not considering fusion approaches. One can see that the addition of high reliability of the positioning would have automatically led to "no real acceptable solution."

The choice is enlarged when accuracy is not forced, but the majority of the resulting technologies are of really poor accuracy. Although we agree with the relative real importance of this parameter (preferring the reliability one), more than 100 m indoors is of reduced usefulness (this could however be discussed).

From these tables, one can observe that even when taking into account only technical criteria, the range of real possibilities reduces drastically very quickly. When adding technological aspects such as maturity of technologies, or even more binding, cost aspects or availability on current smartphones, the choice is reduced to only one or two possibilities that are furthermore not totally satisfactory and that usually need to reduce the level of requirement on the other parameters.

Thus, it seems that there is no issue indeed, and the current directions of work, toward fusion and statistics, are the right thing to do. As a matter of fact, as will be developed in Chapter 12, it is not as simple as that: new constraints arise when dealing with these approaches that are as complex (and may be even more complex) as that the problem we want to overcome. Thus, adding a new level of complexity does not compulsorily lead to better systems.

The alternative could be somewhere else. My opinion is that there are no sufficient discussions between users (in the broad sense) and technicians. Thus, the really essential needs have not been identified. Everything is left with the faith that "science" will provide us a solution. The problem is indeed that it was the case with GPS: nobody had asked for anything but it arrived nevertheless. For indoor positioning, everybody would like to have it, but it is not yet here. Indoors, propagation problems (among others) are dramatic, often leading to reduced reliability or reduced accuracy. However, what do we need exactly? As already mentioned in the previous chapters, knowing about the room in which we are is quite enough: from a technical point of view, the approach is however completely different than if we seek an accuracy of one meter. Quite often, the "needs" are indeed a translation from users of what they think should be technically possible. This may be part of the current problem: specifications are not really based on needs, but already on technical requirements (without the first phase of technical translation between needs and specifications). Thus, the result is that when "technical" staff get involved in the domain and finally face a blocking point, they do not really know if it is important to confront it or bypass it. Usually, the approach adopted is to choose the easiest way to progress, which has not yet shown to supply the best answer.

4.7 Selected Approach for the Rest of the Book

We have to decide how to present the various technologies and their performance in the rest of the book. A classification of chapters based on technical considerations, such as the signal processing approach, followed by the way positioning is calculated is a possibility, but risks being a complex structure for one who wants to select a potential candidate for a specific application or environment.

When considering the range parameter (see Table 4.12), one should keep in mind that the values entered in the table are not absolute values for several reasons: first of all, because the border between two categories is not so clear and also because different embodiments of a given technology are often available. In the table, we choose the most commonly implemented one, thus leading to possible discussions (these discussions are carried out in the following chapters). As an example, let us consider the case of LiFi in Table 4.12. The way it works is based on the detection of the light modulation produced by a given light bulb, which the terminal can "see." Thus, it is necessary to be in the close vicinity of the bulb in order to achieve the positioning. The range should then be "room," at best. However, we put "building," because the implementation of such a system is clearly not intended to be restricted to a single room (although it could be), but to a complete building. However, in the case of UWB, we chose the opposite way round and put "a few rooms" in the corresponding box considering the local deployment of an elementary system, although it is probably also intended to cover the entire building. The reason, which can once again be discussed, is the following: in the case of LiFi, the room coverage is achieved with a single light bulb. Thus, the "room" coverage is not significant of the system itself, but only of an elementary part of the system, unlike UWB, where the characteristics provided (accuracy, positioning mode, etc.) are associated with a system composed of several transmitters and a global management approach. As one can see, even just filling in the tables is not at all simple.

In a similar approach, we consider two other potential candidates for the next chapters: the accuracy and the reliability. When drawing up the table with the parameter "accuracy" as a sorting criterion, another list appears. It is also important to pay some attention to the interpretation of parameter values. Not all are at the same level. For example, the "a few centimeters" allotted to NFC is a real one as there is no possibility to detect it farther away (in its classical implementation). This is not the same for the "Sonar" technology, where propagation and obstacles are bound to produce noise signals and the figure is thus considered as the optimal value. Moreover, for "accuracy" as it was for "range," the limits are not so clear and should be considered blurred. Table 4.13 gives the classification obtained.

The "reliability" parameter is probably one of the most important, although not to be used alone. Table 4.14 is the result of such a choice as the primary classification parameter. Its advantage is that we have only four possibilities, reducing the number of corresponding potential chapters. Unfortunately, the counterpart is that the corresponding chapters would include a large diversity of techniques and technologies, too many for a comprehensive discussion. Once again, the values assigned to the various technologies are debatable. The "low" value given to "high-accuracy GNSS" is due to the fact the signals are almost never available indoors, whereas the "low" value assigned to the "ultrasound"

Table 4.12 The classification on the "range" parameter.

Technology	Positioning type	Accuracy	Reliability	Range	Sensitivity to environment	Calibration needed	Positioning mode	Technique	Signal processing	Position calculation
Cospas Sarsat – Argos	Absolute	>100 m	Medium	World	High	None	Continuous	Frequency(ies)	A Combination of	∩ Straight lines
GNSS	Absolute	100 m	Low	World	Very high	None	Continuous	Time(s)	A combination of	∩ Spheres
High-accuracy GNSS	Absolute	100 m	Low	World	Very high	Once	Continuous	Phase(s)	A combination of	∩ Spheres
Magnetometer	Orientation	A few degrees	Medium	World	Moderate	Several times	Continuous	Physical	Detection	Math functions $(f, f', f/f', ...)$
Pressure	Relative	1 m	High	World	No impact	Several times	Continuous	Physical	Detection	Zone determination
Signaux radio opp	Absolute	>100 m	Low	World	High	None	Almost continuous	Physical	Propagation modeling	∩ Circles
Wired networks	Absolute	An address	Medium	World	No impact	None	Discrete	Fusion	Correlation	Zone determination
Radio AM/FM	Absolute	>100 m	Low	County	High	None	Continuous	Physical	Propagation modeling	∩ Circles
TV	Absolute	>100 m	Low	County	High	None	Continuous	Physical	Propagation modeling	∩ Circles
GSM/3/4/5G	Absolute	>100 m	Low	City	High	None	Continuous	Distance(s)	Propagation modeling	∩ Circles
LoRa	Absolute	>100 m	Low	City	High	None	Continuous	Distance(s)	Propagation modeling	∩ Circles
Sigfox	Absolute	>100 m	Low	City	High	None	Continuous	Distance(s)	Propagation modeling	∩ Circles
Radio 433/ 868/... MHz	Absolute	>100 m	Low	Block	High	None	Continuous	Physical	Propagation modeling	∩ Circles

Accelerometer	Relative	$f(t)$	Medium	Block	No impact	Often	Continuous	Physical	Detection	Math functions $(\int f, \int\int f, \ldots)$
BLE	Absolute	A few meters	Medium	Building	High	Several times	Almost continuous	Physical	Pattern matching	Math functions $(\int f, \int\int f, \ldots)$
Gyrometer	Relative	$f(t)$	Medium	Building	No impact	Often	Continuous	Physical	Detection	Math functions $(\int f, \int\int f, \ldots)$
Image relative displacement	Relative	<1 m	Medium	Building	High	Several times	Almost continuous	Image(s)	A combination of	Math functions $(\int f, \int\int f, \ldots)$
Image SLAM	Relative	<1 m	Medium	Building	High	Several times	Almost continuous	Image(s)	A combination of	Math functions $(\int f, \int\int f, \ldots)$
Indoor GNSS	Absolute	A few decimeters	Medium	Building	High	None	Continuous	Phase(s)	Correlation	∩ Hyperbolae
LiFi	Symbolic	A few meters	Low	Building	Very high	None	Almost continuous	Physical	Detection	Spot location
Light opp	Relative	100 m	Low	Building	Very high	Often	Almost continuous	Physical	Classification	Zone determination
Sound	Relative	>100 m	Medium	Building	High	Often	Continuous	Physical	Detection	∩ Circles
Theodolites	Absolute	A few centimeters	Very high	Building	Very high	Once	Continuous	Angle(s)	A combination of	∩ Plans + distance(s)
WiFi	Absolute	A few meters	Medium	Building	High	Several times	Continuous	Physical	Pattern matching	Math functions $(\int f, \int\int f, \ldots)$
WLAN symbolic	Symbolic	Dm	Very high	Building	Low	None	Continuous	Physical	Propagation modeling	Zone determination

(Continued)

Table 4.12 (Continued)

Technology	Positioning type	Accuracy	Reliability	Range	Sensitivity to environment	Calibration needed	Positioning mode	Technique	Signal processing	Position calculation
Radar	Absolute	A few centimeters	Medium	A few rooms	High	None	Continuous	Phase(s)	A combination of	∩ Plans + distance(s)
RFID	Absolute	Decimeter	High	A few rooms	Low	None	Discrete	Physical	Detection	Spot location
UWB	Absolute	A few centimeters	Medium	A few rooms	High	None	Continuous	Time(s)	A combination of	∩ Spheres
Image markers	Absolute	<1 m	Medium	Room	Very high	Once	Almost continuous	Image(s)	A combination of	Matrices calculus
Infrared	Symbolic	A few meters	High	Room	Very high	None	Almost continuous	Physical	Detection	Zone determination
Laser	Absolute	Less than centimeter	Very high	Room	Very high	None	Almost continuous	Phase(s)	Propagation modeling	∩ Spheres
Lidar	Absolute	Less than centimeter	Very high	Room	Very high	Once	Almost continuous	Time(s)	Correlation	Spot location
Sonar	Relative	A few centimeters	Medium	Room	Very high	None	Continuous	Time(s)	Detection	∩ Plans + distance(s)
Ultrasound	Absolute	A few decimeters	Low	Room	Very high	None	Continuous	Time(s)	Propagation modeling	∩ Spheres
Bar codes	Absolute	Decimeters	Very high	Proximity	Low	None	User action needed	Image(s)	Pattern recognition	Spot location
Contactless cards	Absolute	A few centimeters	Very high	Proximity	No impact	None	Discrete	Physical	Detection	Spot location
Credit cards	Absolute	A few centimeters	Very high	Proximity	No impact	None	Discrete	Physical	Detection	Spot location
Image recognition (site, people)	Absolute	A few Dm	Medium	Proximity	Very high	None	Almost continuous	Image(s)	Pattern recognition	Spot location
NFC	Absolute	A few centimeters	Very high	Proximity	No impact	None	User action needed	Physical	Detection	Spot location
QR codes	Absolute	Decimeter	Very high	Proximity	Low	None	User action needed	Image(s)	Pattern recognition	Spot location

Table 4.13 The classification on the "accuracy" parameter.

Technology	Positioning type	Accuracy	Reliability	Range	Sensitivity to environment	Calibration needed	Positioning mode	Technique	Signal processing	Position calculation
Accelerometer	Relative	$f(t)$	Medium	Block	No impact	Often	Continuous	Physical	Detection	Math functions ($f, ff, fff, \ldots$)
Gyrometer	Relative	$f(t)$	Medium	Building	No impact	Often	Continuous	Physical	Detection	Math functions ($f, ff, fff, \ldots$)
Magnetometer	Orientation	A few degrees	Medium	World	Moderate	Several times	Continuous	Physical	Detection	Math functions ($f, ff, fff, \ldots$)
Wired networks	Absolute	An address	Medium	World	No impact	None	Discrete	Fusion	Correlation	Zone determination
Laser	Absolute	Less than centimeter	Very high	Room	Very high	None	Almost continuous	Phase(s)	Propagation modeling	∩ Spheres
Lidar	ABSOLUTE	Less than centimeter	very high	Room	Very high	Once	Almost continuous	Time(s)	Correlation	∩ Plans + distance(s)
Contactless cards	Absolute	A few centimeter	Very high	Proximity	No impact	None	Discrete	Physical	Detection	Spot location
Credit cards	Absolute	A few centimeter	Very high	Proximity	No impact	None	Discrete	Physical	Detection	Spot location
NFC	Absolute	A few centimeter	Very high	Proximity	No impact	None	User action needed	Physical	Detection	Spot location
Radar	Absolute	A few centimeter	Medium	A few rooms	High	None	Continuous	Phase(s)	A combination of	∩ Plans + distance(s)
Sonar	Relative	A few centimeter	Medium	Room	Very high	None	Continuous	Time(s)	Detection	∩ Plans + distance(s)
Theodolites	Absolute	A few centimeter	Very high	Building	Very high	Once	Continuous	Angle(s)	A combination of	∩ Plans + distance(s)
UWB	Absolute	A few centimeter	Medium	A few Rooms	High	None	Continuous	Time(s)	A combination of	∩ Spheres
Bar codes	Absolute	Decimeter	Very high	Proximity	Low	None	User action needed	Image(s)	Pattern recognition	Spot location

(Continued)

Table 4.13 (Continued)

Technology	Positioning type	Accuracy	Reliability	Range	Sensitivity to environment	Calibration needed	Positioning mode	Technique	Signal processing	Position calculation
QR codes	Absolute	Decimeter	Very high	Proximity	Low	None	User action needed	Image(s)	Pattern recognition	Spot location
RFID	Absolute	Decimeter	High	A few rooms	Low	None	Discrete	Physical	Detection	Spot location
Indoor GNSS	Absolute	A few decimeters	Medium	Building	High	None	Continuous	Phase(s)	Correlation	∩ Hyperbolae
Ultrasound	Absolute	A few decimeters	Low	Room	Very high	None	Continuous	Time(s)	Propagation modeling	∩ Spheres
Image markers	Absolute	<1 m	Medium	Room	Very high	Once	Almost continuous	Image(s)	A combination of	Matrices calculus
Image-relative displacement	Relative	<1 m	Medium	Building	High	Several times	Almost continuous	Image(s)	A combination of	Math functions $(f, ff, fff, \ldots)$
Image SLAM	Relative	<1 m	Medium	Building	High	Several times	Almost continuous	Image(s)	A combination of	Math functions $(f, ff, fff, \ldots)$
Pressure	Relative	1 m	High	World	No impact	Several times	Continuous	Physical	Detection	Zone determination
BLE	Absolute	A few m	Medium	Building	High	Several times	Almost continuous	Physical	Pattern matching	Math functions $(f, ff, fff, \ldots)$
Infrared	Symbolic	A few meters	High	Room	Very high	None	Almost continuous	Physical	Detection	Zone determination
Lifi	Symbolic	A few meters	Low	Building	Very high	None	Almost continuous	Physical	Detection	Spot location
WiFi	Absolute	A few meters	Medium	Building	High	Several times	Continuous	Physical	Pattern matching	Math functions $(f, ff, fff, \ldots)$
WLAN Symbolic	Symbolic	Dm	Very high	Building	Low	None	Continuous	Physical	Propagation modeling	Zone determination
Image recognition (site, people)	Absolute	A few Dm	Medium	Proximity	Very high	None	Almost continuous	Image(s)	Pattern recognition	Spot location

GNSS	Absolute	100 m	Low	World	Very high	None	Continuous	Time(s)	A combination of ∩ Spheres
High-accuracy GNSS	Absolute	100 m	Low	World	Very high	Once	Continuous	Phase(s)	A combination of ∩ Spheres
Light opp	Relative	100 m	Low	Building	Very high	Often	Almost continuous	Physical	Classification Zone determination
Cospas Sarsat – Argos	Absolute	>100 m	Medium	World	High	None	Continuous	Frequency(ies)	A combination of ∩ Straight lines
GSM/3/4/5G	Absolute	>100 m	Low	City	High	None	Continuous	Distance(s)	Propagation modeling ∩ Circles
LoRa	Absolute	>100 m	Low	City	High	None	Continuous	Distance(s)	Propagation modeling ∩ Circles
Radio 433/868/... MHz	Absolute	>100 m	Low	Block	High	None	Continuous	Physical	Propagation modeling ∩ Circles
Radio AM/FM	Absolute	>100 m	Low	County	High	None	Continuous	Physical	Propagation modeling ∩ Circles
Sigfox	Absolute	>100 m	Low	City	High	None	Continuous	Distance(s)	Propagation modeling ∩ Circles
Signaux radio opp	Absolute	>100 m	Low	World	High	None	Almost continuous	Physical	Propagation modeling ∩ Circles
Sound	Relative	>100 m	Low	Building	High	Often	Continuous	Physical	Detection ∩ Circles
TV	Absolute	>100 m	Low	County	High	None	Continuous	Physical	Propagation modeling ∩ Circles

Table 4.14 The classification on the "reliability" parameter.

Technology	Positioning type	Accuracy	Reliability	Range	Sensitivity to environment	Calibration needed	Positioning mode	Technique	Signal processing	Position calculation
Bar codes	Absolute	Decimeter	Very high	Proximity	Low	None	User action needed	Image(s)	Pattern recognition	Spot location
Contactless cards	Absolute	A few centimeter	Very high	Proximity	No impact	None	Discrete	Physical	Detection	Spot location
Credit cards	Absolute	A few centimeter	Very high	Proximity	No impact	None	Discrete	Physical	Detection	Spot location
Laser	Absolute	Less than centimeter	Very high	Room	Very high	None	Almost continuous	Phase(s)	Propagation modeling	Spot location
Lidar	Absolute	Less than centimeter	Very high	Room	Very high	Once	Almost continuous	Time(s)	Correlation	$\cap$ Plans + distance(s)
NFC	Absolute	A few centimeters	Very high	Proximity	No impact	None	User action needed	Physical	Detection	Spot location
QR codes	Absolute	Decimeter	Very high	Proximity	Low	None	User action needed	Image(s)	Pattern recognition	Spot location
Theodolites	Absolute	A few centimeters	Very high	Building	Very high	Once	Continuous	Angle(s)	A combination of	$\cap$ Plans + distance(s)
WLAN symbolic	Symbolic	Dm	Very high	Building	Low	None	Continuous	Physical	Propagation modeling	Zone determination
Infrared	Symbolic	A few meters	High	Room	Very high	None	Almost continuous	Physical	Detection	Zone determination
Pressure	Relative	1 m	High	World	No impact	Several times	Continuous	Physical	Detection	Zone determination
RFID	Absolute	Decimeters	High	A few rooms	Low	None	Discrete	Physical	Detection	Spot location
Accelerometer	Relative	$f(t)$	Medium	Block	No impact	Often	Continuous	Physical	Detection	Math functions $(\int, \int\int, \int\int\int, ...)$

BLE	Absolute	A few meters	Medium	Building	High	Several times	Almost continuous	Physical	Pattern matching	Math functions $(\int, \int\int, \int\int\int, \ldots)$
Cospas Sarsat – Argos	Absolute	>100 m	Medium	World	High	None	Continuous	Frequency(ies)	A combination of	∩ Straight lines
Gyrometer	Relative	$f(t)$	Medium	Building	No impact	Often	Continuous	Physical	Detection	Math functions $(\int, \int\int, \int\int\int, \ldots)$
Image markers	Absolute	<1 m	Medium	Room	Very high	Once	Almost continuous	Image(s)	A combination of	Matrices calculus
Image recognition (site, people)	Absolute	A few Dm	Medium	Proximity	Very high	None	Almost continuous	Image(s)	Pattern recognition	Spot location
Image-relative displacement	Relative	<1 m	Medium	Building	High	Several times	Almost continuous	Image(s)	A combination of	Math functions $(\int, \int\int, \int\int\int, \ldots)$
Image SLAM	Relative	<1 m	Medium	Building	high	Several times	Almost continuous	Image(s)	A combination of	Math functions $(\int, \int\int, \int\int\int, \ldots)$
Indoor GNSS	Absolute	A few decimeters	Medium	Building	High	None	Continuous	Phase(s)	Correlation	∩ Hyperbolae
Magnetometer	Orientation	A few degrees	Medium	World	Moderate	Several times	Continuous	Physical	Detection	Math functions $(\int, \int\int, \int\int\int, \ldots)$
Radar	Absolute	A few centimeters	Medium	A few rooms	High	None	Continuous	Phase(s)	A combination of	∩ Plans + distance(s)
Sonar	Relative	A few centimeters	Medium	Room	Very high	None	Continuous	Time(s)	Detection	∩ Plans + distance(s)
UWB	Absolute	A few centimeters	Medium	A few rooms	High	None	Continuous	Time(s)	A combination of	∩ Spheres
WiFi	Absolute	A few meters	Medium	Building	High	Several times	Continuous	Physical	Pattern matching	Math functions $(\int, \int\int, \int\int\int, \ldots)$
Wired networks	Absolute	An address	Medium	World	No impact	None	Discrete	Fusion	Correlation	Zone determination
GNSS	absolute	100 m	Low	World	Very high	None	Continuous	Time(s)	A combination of	∩ Spheres

(Continued)

Table 4.14 (Continued)

Technology	Positioning type	Accuracy	Reliability	Range	Sensitivity to environment	Calibration needed	Positioning mode	Technique	Signal processing	Position calculation
GSM/3/4/5G	Absolute	>100 m	Low	City	High	None	Continuous	Distance(s)	Propagation modeling	∩ Circles
High-accuracy GNSS	Absolute	100 m	Low	World	Very high	Once	Continuous	Phase(s)	A combination of	∩ Spheres
LiFi	Symbolic	A few meters	Low	Building	Very high	None	Almost continuous	Physical	Detection	Spot location
Light opp	Relative	100 m	Low	Building	Very high	Often	Almost continuous	Physical	Classification	Zone determination
LoRa	Absolute	>100 m	Low	City	High	None	Continuous	Distance(s)	Propagation modeling	∩ Circles
Radio 433/868/... MHz	Absolute	>100 m	low	Block	High	None	Continuous	Physical	Propagation modeling	∩ Circles
Radio AM/FM	Absolute	>100 m	Low	County	High	None	Continuous	Physical	Propagation modeling	∩ Circles
Sigfox	Absolute	>100 m	Low	City	High	None	Continuous	Distance(s)	Propagation modeling	∩ Circles
Signaux radio opp	Absolute	>100 m	Low	World	High	None	Almost continuous	Physical	Propagation modeling	∩ Circles
Sound	Relative	>100 m	Low	Building	High	Often	Continuous	Physical	Detection	∩ Circles
TV	Absolute	>100 m	Low	County	High	None	Continuous	Physical	Propagation modeling	∩ Circles
Ultrasound	Absolute	A few decimeters	Low	Room	Very high	None	Continuous	Time(s)	Propagation modeling	∩ Spheres

technology is mainly due to the potential presence of obstacles in the environment. The two "low values" are thus not exactly of the same type.

Our choice for the rest of the book is finally to retain the range parameter as the first classifier. Moreover, we shall start with the proximity ranges and in reverse order to technologies with "world" coverage. The second classifier considered, i.e. the subchapter groupings, is related to the techniques implemented.

Thus, the organization of the following chapters is as follows:

- Chapter 5: "proximity" associated technologies.
- Chapter 6: "room" associated technologies.
- Chapter 7: "a few rooms" associated technologies.
- Chapter 8: "building" associated technologies.
- Chapter 9: the specific case of indoor GNSS "building" associated technologies.
- Chapter 10: "block," "city," and "county" associated technologies.
- Chapter 11: "world" associated technologies.

Then, although discussed throughout the chapters, a specific chapter (i.e. Chapter 12) will deal with fusion and statistical possible approaches designed in order to overcome the difficulties that are going to be described in these seven chapters.

Bibliography

1 Blaunstein, N. and Christodoulou, C.G. (2014). Indoor radio propagation. In: *Radio Propagation and Adaptive Antennas for Wireless Communication Networks* (ed. N. Blaunstein and C. Christodoulou), 302–334. Wiley.

2 Frattasi, S. and Rosa, F.D. (2016). Indoor positioning in WLAN. In: *Mobile Positioning and Tracking: From Conventional to Cooperative Techniques* (ed. S. Frattasi and F. Della Rosa), 261–282. Wiley.

3 Zekavat, R. and Michael Buehrer, R. (2012). Smart antennas for direction-of-arrival indoor positioning applications. In: *Handbook of Position Location: Theory, Practice and Advances* (ed. S.A.R. Zekavat and R.M. Buehrer), 319–358. Wiley.

4 Kavehrad, M., Sakib Chowdhury, M.I., and Zhou, Z. (2015). Indoor positioning methods using VLC LEDs. In: *Short Range Optical Wireless: Theory and Applications* (ed. M. Kavehrad, S. Chowdhury, and Z. Zhou), 225–262. Wiley.

5 Frattasi, S. and Rosa, F.D. (2016). Ultra-wideband positioning and tracking. In: *Mobile Positioning and Tracking: From Conventional to Cooperative Techniques* (ed. S. Frattasi and F. Della Rosa), 225–260. Wiley.

6 Zekavat, R. and Michael Buehrer, R. (2012). Remote sensing technologies for indoor applications. In: *Handbook of Position Location: Theory, Practice and Advances* (ed. S.A.R. Zekavat and R.M. Buehrer), Wiley.

7 Zekavat, R. and Michael Buehrer, R. (2012). Kernel methods for RSS-based indoor localization. In: *Handbook of Position Location: Theory, Practice and Advances* (ed. S.A.R. Zekavat and R.M. Buehrer), 457–486. Wiley.

8 Zekavat, R. and Michael Buehrer, R. (2012). On the performance of wireless indoor localization using received signal strength. In: *Handbook of Position Location: Theory, Practice and Advances* (ed. S.A.R. Zekavat and R.M. Buehrer), 425–456. Wiley.

9 Frattasi, S. and Rosa, F.D. (2016). Error mitigation techniques. In: *Mobile Positioning and Tracking: From Conventional to Cooperative Techniques* (ed. S. Frattasi and F. Della Rosa), 163–188. Wiley.

10 Kavehrad, M., Sakib Chowdhury, M.I., and Zhou, Z. (2015). Analyses of indoor optical wireless channels based on channel impulse responses. In: *Short Range Optical Wireless: Theory and Applications* (ed. M. Kavehrad, S. Chowdhury, and Z. Zhou), 67–110. Wiley. Edited by Mohsen Kavehrad, Sakib Chowdhury and Zhou Zhou.

11 Geng, H. (2017). Beacon technology with IoT and big data. In: *Internet of Things and Data Analytics Handbook* (ed. H. Geng), 267–282. Wiley.

12 Song, H., Srinivasan, R., Sookoor, T., and Jeschke, S. (2017). Smart lighting. In: *Smart Cities: Foundations, Principles, and Applications* (ed. H. Song, R. Srinivasan, T. Sookoor, and S. Jeschke), 697–724. Wiley.

13 Blaunstein, N. and Christodoulou, C.G. (2014). Adaptive antennas for wireless networks. In: *Radio Propagation and Adaptive Antennas for Wireless Communication Networks* (ed. N. Blaunstein and C. Christodoulou), 216–279. Wiley.

14 Harle, R. (2013). A survey of indoor inertial positioning systems for pedestrians. *IEEE Communications Surveys and Tutorials* 15 (3): 1281–1293.

15 He, S. and Gary Chan, S.-H. (2016). Wi-Fi fingerprint-based indoor positioning: recent advances and comparisons. *IEEE Communications Surveys and Tutorials* 18 (1): 466–490.

16 Jimenez Ruiz, A.R., Seco Granja, F., Prieto Honorato, J.C., and Guevara Rosas, J.I. (2012). Accurate pedestrian indoor navigation by tightly coupling foot-mounted IMU and RFID measurements. *IEEE Transactions on Instrumentation and Measurement* 61 (1): 178–189.

17 Kim, H., Kim, D., Yang, S. et al. (2013). An indoor visible light communication positioning system using a RF carrier allocation technique. *Journal of Lightwave Technology* 31 (1): 134–144.

18 Conti, A., Guerra, M., Dardari, D. et al. (2012). Network experimentation for cooperative localization. *IEEE Journal on Selected Areas in Communications* 30 (2): 467–475.

19 Faragher, R. and Harle, R. (2015). Location fingerprinting with bluetooth low energy beacons. *IEEE Journal on Selected Areas in Communications* 33 (11): 2418–2428.

20 Yang, C. and Shao, H. (2015). WiFi-based indoor positioning. *IEEE Communications Magazine* 53 (3): 150–157.

21 Zhang, C., Kuhn, M.J., Merkl, B.C. et al. (2010). Real-time noncoherent UWB positioning radar with millimeter range accuracy: theory and experiment. *IEEE Transactions on Microwave Theory and Techniques* 58 (1): 9–20.

22 Yang, S., Dessai, P., Verma, M., and Gerla, M. (2013). FreeLoc: calibration-free crowd sourced indoor localization. In: *2013 Proceedings IEEE INFOCOM*, 2481–2489. Turin: IEEE.

23 Wang, G., Gu, C., Inoue, T., and Li, C. (2014). A hybrid FMCW-interferometry radar for indoor precise positioning and versatile life activity monitoring. *IEEE Transactions on Microwave Theory and Techniques* 62 (11): 2812–2822.

24 Yassin, A., Nasser, Y., Awad, M. et al. (2017). Recent advances in indoor localization: a survey on theoretical approaches and applications. *IEEE Communications Surveys and Tutorials* 19 (2): 1327–1346.

25 Lee, S., Kim, B., Kim, H. et al. (2011). Inertial sensor-based indoor pedestrian localization with minimum 802.15.4a configuration. *IEEE Transactions on Industrial Informatics* 7 (3): 455–466.

26 Sheinker, A., Ginzburg, B., Salomonski, N. et al. (2013). Localization in 3-D Using beacons of low frequency magnetic field. *IEEE Transactions on Instrumentation and Measurement* 62 (12): 3194–3201.

27 Moghtadaiee, V., Dempster, A.G., and Lim, S. (2011). Indoor localization using FM radio signals: a fingerprinting approach. In: *2011 International Conference on Indoor Positioning and Indoor Navigation*, 1–7. Guimaraes: IEEE.

28 Angermann, M., Frassl, M., Doniec, M. et al. (2012). Characterization of the indoor magnetic field for applications in localization and mapping. In: *2012 International Conference on Indoor Positioning and Indoor Navigation (IPIN)*, 1–9. Sydney, NSW: IEEE.

29 Panta, K. and Armstrong, J. (2012). Indoor localisation using white LEDs. *Electronics Letters* 48 (4): 228–230.

30 Matic, A., Papliatseyeu, A., Osmani, V., and Mayora-Ibarra, O. (2010). Tuning to your position: FM radio based indoor localization with spontaneous recalibration. In: *2010 IEEE International Conference on Pervasive Computing and Communications (PerCom)*, 153–161. Mannheim: IEEE.

5

Proximity Technologies: Approaches, Performance, and Limitations

Abstract

The so-called proximity technologies have the typical advantages of their disadvantages. For instance, the very short range involves the use of a large number of sensors or actuators in order to provide sufficient coverage. However, correlatively, the positioning accuracy is bound to be quite good. The main problem that remains is linked to the fact that the positioning is also often discontinuous in time or space. Nevertheless, one can easily see very interesting implementations, for example, in situations where one needs to be autonomous and does not want to be "tracked" by any infrastructure continuously.

Keywords *Proximity; very short range; coverage; accuracy; limitations*

The classification described in Chapter 4 led to the following table concerning these technologies (Table 5.1). Please note that the last paragraph of the chapter shows a slightly different point of view and proposes a discussion concerning other technologies that could be considered as "proximity" ones but which are dealt with in other chapters.

5.1   Bar Codes

A barcode is a one-dimensional code consisting of a series of vertical lines and spaces of different widths and allowing a string of numeric or alphanumeric characters to be translated. Different coding techniques are available, leading to different formulations of the same chain.

For example, it is possible to code the string "Indoor Positioning" (randomly chosen) by the following bar codes (Figure 5.1): using a Code-128, or in this other form using a GS1–128 encoding (UCC/EAN-128). They look similar but are not (Figure 5.2).

Many other forms are possible and allow a very wide adaptation to the various domains in which the bar codes are used. A well-known form is that found

Indoor Positioning: Technologies and Performance, First Edition. Nel Samama.
© 2019 The Institute of Electrical and Electronics Engineers, Inc. Published 2019 by John Wiley & Sons, Inc.

Table 5.1 Main proximity technologies.

Technology	Positioning type	Accuracy	Reliability	Range	Sensitivity to environment	Calibration needed	Positioning mode	Technique	Signal processing	Position calculation
Bar codes	Absolute	Decimeters	Very high	Proximity	Low	None	User action needed	Image(s)	Pattern recognition	Spot location
Contactless cards	Absolute	A few centimeters	Very high	Proximity	No impact	None	Discrete	Physical	Detection	Spot location
Credit cards	Absolute	A few centimeters	Very high	Proximity	No impact	None	Discrete	Physical	Detection	Spot location
Image recognition (site, people)	Absolute	A few decameters	Medium	Proximity	Very high	None	Almost continuous	Image(s)	Pattern recognition	Spot location
NFC	Absolute	A few centimeters	Very high	Proximity	No impact	None	User action needed	Physical	Detection	Spot location
QR codes	Absolute	Decimeters	Very high	Proximity	Low	None	User action needed	Image(s)	Pattern recognition	Spot location

Figure 5.1 "Indoor Positioning" coded under a Code-128 format.

Indoor Positioning

Figure 5.2 "Indoor Positioning" coded under a GS1–128 encoding format.

Indoor Positioning

Figure 5.3 "9781234567897" coded under an ISBN 13 format.

on food products (ISBN 13): we have all been confronted with such a code (Figure 5.3).

The use of barcodes for positioning has been imagined in several ways. The first, obvious but rarely actually implemented, is simply coding a position. By simply reading the code, we deduce our position. This requires an action (that of reading the code), and the position is obtained only when one performs the reading.

A second approach has been developed in an industrial setting of positioning a code reader on an assembly line. In such a case, it is reading the part of a code that is in front of the reader that provides the position of the latter. Such codes can reach several kilometers in length. The bar code is then a band presenting a succession of lines and vertical spaces that the reader will read (see Figure 5.4).

Figure 5.4 The BPS 300i system from Leuze electronic. Source: Courtesy of Leuze electronic.

In the context of a consumer positioning system, a conventional approach is to place barcodes in the environment at locations that are simply accessed by a user. All he has to do is to scan the barcode using his smartphone to retrieve the code and convert it into a position. Two typical cases can then arise: either the code makes it possible to make a correspondence with the position by analysis of a database or the code is directly translatable into a position (the code could, for example, be directly the position in longitude, latitude, and possibly altitude, in any geographical format). A simplified representation of such an approach is visible in Figure 5.5.

The various parameters considered for this technology are given in Chapter 4. However, a few could be discussed. Let us explain those that are quite questionable (in the positive sense of the word) (Table 5.2).

Infrastructure complexity: This parameter is set to "none" because the deployment for mass market positioning purposes is quite simple and easy. Nevertheless, it does not mean that it is free, but the infrastructure cost is low.

Terminal maturity: If we talk about the implementation on smartphone, then it is necessary to imagine an application that provides the position according to the code read. This presents no difficulty.

Figure 5.5 A typical deployment of a bar code positioning system in a building.

Table 5.2 Summary of the main parameters for bar codes.

Infrastructure complexity	Infrastructure maturity	Infrastructure cost	Terminal complexity	Terminal maturity	Terminal cost	Smart-phone	Calibration complexity
None	Existing	Low	Low	Software development	Low	Existing	None

Positioning type	Accuracy	Reliability	Range	Sensitivity to environment	Positioning mode	In/out transition	Calibration needed
Absolute	Decimeters	Very high	Proximity	Low	User action needed	Easy	None

Terminal cost: Bar codes reading applications already exist on smartphones, using the always available cameras.

Calibration complexity: The only real constraint concerns the fact that once a bar code is installed, it must not be removed to another place. The problem is identical for the infrastructure components of the majority of positioning systems. The main problem in the present case is that these components must be located at places accessible by everyone.

Accuracy: The decimeter-reported value is typically the range the readers are able to detect the code. Of course, this value also depends on the size of the code: if it is a 1-m high code, it will be detected much further away.

Reliability: It is indeed excellent as the detection is almost perfect once the reader can "see" the code, unless it is damaged or optically modified (for example, by adding black lines where there were none).

Range: Nonline of sight is obviously not possible. Nevertheless, the range should here be associated with an individual bar code as a large number of such codes could be distributed throughout a large area in order to provide a positioning system (as shown in Figure 5.4). Thus, this "range" parameter should always be considered carefully.

Sensitivity to environment: The "low" value could be discussed as nonline of sight is not possible. Indeed, it is insensitive to the main electromagnetic disturbances, except those relative to visible light (mainly obstacles to light propagation).

In/out transition: There is absolutely no difficulty to install bar codes indoors or outdoors.

5.2 Contactless Cards and Credit Cards

Contactless cards are smart cards, such as credit cards, including a wireless communications system that allows you to avoid having to insert the card into a reader, but only to approach it (see Figure 5.6). In order to secure the

Figure 5.6 Principle of a contactless payment.

transactions and not to trigger a nonvoluntary payment, the distance of action is extremely low (of the order of a few centimeters). This makes the voluntary act indispensable.

Credit cards, and now contactless cards, are part of everyday life for many people. For purposes of securing and traceability of transactions, these are time stamped but also geo-located. Indeed, each reading terminal must be connected in order to verify and validate the transaction (which allows it to be dated) but is also located geographically. This correspondence between the location of the reading terminal and the location of the user makes it possible to "follow" a user, in space and in time. However, here again, the positioning is discontinuous both in time and in space (unless the buyer is on a spending spree).

The various parameters considered for this technology are given in Chapter 4. However, let us explain those that are potentially questionable. The top lines are relative to contactless cards, whereas the bottom ones are relative to credit cards. As one can see, they are very similar. The only difference relates to the availability on smartphones, which is already the reality for contactless cards in the form of near field communication (NFC) payment in a few countries (Table 5.3).

Infrastructure and terminal parameters: It is not complex in that it already exists for a completely different purpose and it is not likely to disappear quickly in the near future. At worst, it will be replaced by a new system that will have the same characteristics in terms of securing transactions, therefore in terms of positioning (time and place). In addition, note that the system is always up-to-date as it must be updated regularly in order to satisfy the customers.

Table 5.3 Summary of the main parameters for contactless and credit cards.

Infrastructure complexity	Infrastructure maturity	Infrastructure cost	Terminal complexity	Terminal maturity	Terminal cost	Smart-phone	Calibration complexity
None	Existing	Zero	None	Existing	Low	Near future	None
None	Existing	Zero	None	Existing	Low	Not applicable	None

Positioning type	Accuracy	Reliability	Range	Sensitivity to environment	Positioning mode	In/out transition	Calibration needed
Absolute	A few centimeters	Very high	Proximity	No impact	Discrete	Impossible	None
Absolute	A few centimeters	Very high	Proximity	No impact	Discrete	Impossible	None

Smartphone: Current developments are going in the direction of making seamless contactless payments with smartphones. However, the preferred technology seems to be NFC (the majority of contactless cards use this technology).

Calibration: It is achieved through the database of terminal locations and the link to the network for time synchronization purposes.

Accuracy and reliability: The same as for bar codes apply. The range is even reduced in order to respect the required level of confidentiality and security.

5.3 Image Recognition

The basic idea is to use the image obtained with any kind of camera and to compare it with a large database of recorded images. The database could be obtained easily from professional photographers, journalists, artists, or of course from everybody leaving photos on the Internet. The important point is of course to associate with each photo a location, whose accuracy can be anything (but it would be better if it is specified, but this is not compulsory). Then, the location process is as follows:

- One takes a photo
- Launches the image recognition location processing tool
- The tool searches the database in order to identify the best match (or the best matches), based on image processing techniques
- The best estimate of one's location is provided

Image processing techniques are not intended to be dealt with in this book, but it appears quite obvious that the acquisition of the image is a fundamental process that could have a real impact on the quality of the resulting positioning. For instance, parameters such as the resolution and the method used for the coding of the image during the digitization process are important. The same applies to the optical settings of the camera, the lighting conditions, or the noise of the signal.

A few typical characteristics that can be obtained from a digital image are average value if one copes with the global image. For example, one can cite the luminance, the color, the sharpness, or the contrast. A few specific operations can be achieved on an image, either directly on each pixel or on a group of pixels, or on a much larger part of the image, or even on the whole image. It is, for example, possible to change the dynamics of the image by modifying, through binary operators, for instance, the pixels (adding, combining, differencing, etc.). It is then possible to detect contours by analyzing the boundary of an object or a scene in the image (looking at the contrast changes). At the image level, it is possible to analyze the distribution of any parameter: brightness, levels of gray, various histograms, etc., and then to extract some classifications or to carry out segmentation (technique consisting of extracting primitives from an image) or skeletonization (reduction of the dimension of an object without losing its topological or geometrical information). Classification, for instance, could be a first approach to image recognition, reducing the size of the database used to look for images. A few other techniques related to image analysis are described in Chapters 5 and 6 and allow distances between pixels to be calculated. Filtering is also a major function in image processing: noise reduction, smoothing, detection, etc., are all likely to benefit from filtering.

This positioning technique is very efficient when the image shows a well-known landmark, like the Eiffel Tower, for instance (see Figure 5.7a) but is much more complex when remarkable features are not identified (see Figure 5.7b).

The various parameters considered for this technology are given in Chapter 4. However, we shall comment on a few of them (Table 5.4).

(a)   (b)

Figure 5.7 (a) The Eiffel Tower. Source: Photo by Anthony Delanoix on Unsplash. (b) A country road somewhere. Source: Photo by Thomas Lefebvre on Unsplash.

Camera #2 / Location #2

Camera #3 / Location #3   Camera #1 / Location #1

Figure 5.8 Simple representation of the position uncertainty.

Table 5.4 Summary of the main parameters for image recognition.

Infrastructure complexity	Infrastructure maturity	Infrastructure cost	Terminal complexity	Terminal maturity	Terminal cost	Smart-phone	Calibration complexity
None	None	Zero	Low	Software development	Zero	Existing	None

Positioning type	Accuracy	Reliability	Range	Sensitivity to environment	Positioning mode	In/out transition	Calibration needed
Absolute	A few decameters	Medium	Proximity	Very high	Almost continuous	Easy	None

Infrastructure and terminal parameters: No complexity at all for all these parameters as cameras are easily available. The complexity is indeed transferred to the servers and the associated databases and matching algorithms. This technology requires access, in one way or another, to the databases mentioned.

Smartphone: No problem today.

Calibration complexity: There is actually no calibration, but the need for large databases, which could be considered as a sort of calibration.

Positioning type: Since the image of reference (the one in the database) is probably associated with an absolute location in the database, the positioning will be of the same type.

Accuracy: Once again, it is completely dependent on the parameters entered in the database, associated with the images. However, it is slightly more complex: unless the characteristics of the camera one is using are incorporated into the positioning process, as well as the orientation of the camera, it is not possible to define the real location of the camera (see Figure 5.8). Usually, accuracy is not the parameter of importance when one is using this kind of technology.

Reliability: It is associated with the efficiency of the recognition algorithms. It goes from "excellent" for the Eiffel Tower to "zero" when in the middle of a desert or at sea with no visibility to the shore.

Range: "Proximity" has been chosen because this is the classical way it is used, in front of a monument or a street sign.

Sensitivity to environment: As always with the optical technologies, obstacles are the main restriction. If only a part of the monument is available in the image, the efficiency of the recognition process will decrease. In addition, this approach will not be able to provide any location (or at least not an acceptable one) where no reference image has been recorded.

In/out transition: No impact whether indoors or outdoors.

Positioning type: When used for tourists in order to provide them with an idea of their location, and thus in "touristic" areas, it could be considered as "almost continuous." In other environments, it would probably be considered as "discrete."

5.4   Near-Field Communication – NFC

This technology is, in many aspects, quite similar to bar codes or contactless cards, at least in the way it is used. Note that contactless cards are usually implementing an NFC technology. NFC is based on the association of a tag and a reader. The tag can be active (including a power supply) or passive (without any power supply). The difference will refer to the range of detection of the tag by the reader. Passive tags are the most used ones. The transmission between the tag and the reader is achieved through magnetic coupling at a very short distance, as shown in Figure 5.9. The radiated electromagnetic field is captured by the antenna of the tag, which in turn provides the sufficient power supply to activate the tag, which transmits its identifier and eventually an additional message. The tag is powered as long as the reader is in the right range.

When thinking of a positioning system based on NFC, the main idea is to associate one component, the tag or the reader, to a fixed location (on the walls, on the ground, or on the doors of a building). Usually, we prefer to install the passive tags on these locations for energy-saving reasons (see Figure 5.10). Then, a user needs only to approach a reader (a few current smartphones are equipped, in particular, due to the development of contactless payment with smartphones) in order to get the unique identifier of the tag. This

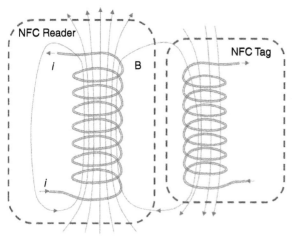

Figure 5.9 A typical electromagnetic passive NFC coupling between the tag and the reader.

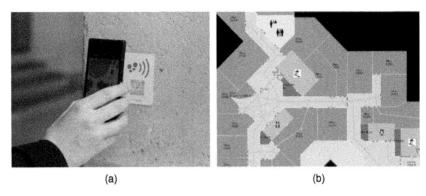

(a)   (b)

Figure 5.10 A passive tag fixed on a wall, read by a smartphone (a), and the associated indoor map (b). Source: Courtesy of LittleThumb.

identifier is advantageously directly convertible into a location. This location should be provided in the most appropriate format, i.e. in coordinates that allow the user to somehow "understand" his or her location in the building. Outdoors, the most used format is an address or a global positioning system (GPS)-compatible format.

Note that a complete positioning system, including guidance and map aspects, requires a whole process that is much more complex that just the positioning described above. In particular, indoor maps are not so easily available, especially when talking in terms of navigation. These aspects are dealt with in Chapter 13.

The various parameters considered for this technology are given in Chapter 4. Nevertheless, additional details are needed in order to explain the assigned values (Table 5.5).

Infrastructure complexity: It is quite simple and consists only in NFC tags scattered throughout the area one wants to cover. Even the number of tags has

Table 5.5 Summary of the main parameters for NFC.

Infrastructure complexity	Infrastructure maturity	Infrastructure cost	Terminal complexity	Terminal maturity	Terminal cost	Smart-phone	Calibration complexity
Low	Existing	Low	Low	Software development	Low	Existing	None

Positioning type	Accuracy	Reliability	Range	Sensitivity to environment	Positioning mode	In/out transition	Calibration needed
Absolute	A few centimeters	Very high	Proximity	No impact	User action needed	Easy	None

no real importance and should be adapted to the real needs of navigation. Nevertheless, tags still have to be deployed.

Infrastructure maturity: NFC passive tags have the advantage, in addition to requiring no power supply, i.e. no wire, of being a mature technology. It means, for example, that the lifetime of tag is typically 15–20 years (with no maintenance).

Infrastructure cost: There are a lot of different tags depending on the size of the memory included. It goes from a few cents to fractions of a dollar.

Terminal complexity, maturity, and cost: A lot of current smartphones are "NFC enabled" and one can imagine that in the near future, almost all of them will be equipped as the integration is very cheap and that "mobile payment" is developing rapidly.

Calibration complexity: As often, the calibration is not required, but the right location of the tag in the associated map is a fundamental aspect.

Positioning type, accuracy, and reliability: This very simple approach of positioning shows some very interesting characteristics. The positioning is given under any type of format: absolute with respect to a global coordinate system, relative with respect to a given building, symbolic, semantic, or whatever. In addition, as long as the tag has been correctly referenced in the map, the accuracy is very good. Furthermore, the reliability of such a system is at a very high level.

Range: The potential ambiguity of the term "range" appears clearly here. If one thinks in terms of the elementary sensor range, it is really "proximity," and one should be very close to the tag indeed. However, the complete system is obtained through the distribution of as many tags as one wants in the environment, leading to a potentially very large coverage.

Sensitivity to environment: The reading of the tag is achieved through a specific action of the user who needs to get very close to the tag. Thus, in addition to the conscious movement of the user, it cannot be achieved at the same location by several users at the same time.

In/out transition: As with bar codes, there is no limitation to outdoors or indoors deployment.

5.5 QR Codes

A Quick Response Code (QR Code) is typically a two-dimensional bar code. Here again, there are a large number of different formats as shown in Figure 5.11.

Such a code has the advantage, compared to the bar code, of being more compact, and so allows a very fast reading. Similarly, it makes it possible to store, for the same surface, a larger amount of data. In addition, many applications can

(a)   (b)   (c)

Figure 5.11 A few examples of two-dimensional codes. QR code (a), Data Matrix codes (b), and DotCode code (c). All represent the sentence "Indoor Positioning."

easily read the content of the code with a smartphone, but also with a camera. The main uses of QR codes are (nonexhaustive list) as follows:

- the connection to a website
- connecting to a WiFi access point
- triggering a phone call action or sending an SMS (short message system)
- sending an email
- the exchange of data like addresses or virtual business cards
- the exchange of data, usually short, of any type (and geographical, in particular, to localize a meeting point on a map, for example)
- update his agenda, etc.

Specific standardization has emerged and QR codes are extremely inexpensive to achieve. Many software programs simply allow you to create your own codes. In addition, it is also possible to generate codes containing a large number of data: these codes are then more complex to read and the conditions of contrast and luminosity become important. Such an example is given in Figure 5.12a.

Two additional features are quite interesting:

1. There are some error correction strategies that allow the recovery of the content of a code even if part of it is corrupted. Thus, it is possible to include a logo in your code (see Figure 5.12b, for such an example).
2. Because of the ability of some readers (cameras mainly) to adapt the reading to rather bad conditions of luminosity and contrast, it is possible to directly mark some parts with typically Dot Codes or Matrix Codes (see Figure 5.13).

The various parameters considered for this technology in the scope of positioning are given in Chapter 4. There is a great similarity with bar codes, however. The main idea is simply to distribute QR codes characterizing a location throughout an area. By decoding it, you know the location of the QR code, hence your own location (Table 5.6).

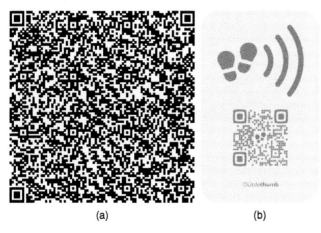

(a)   (b)

Figure 5.12 A larger QR code (a) and a credit card format positioning passive tag including both NFC and QR codes (b). Source: Courtesy of LittleThumb.

Figure 5.13 Direct marking of parts for industrial logistic purposes. Source: Courtesy of Leuze electronic.

Table 5.6 Summary of the main parameters for QR codes.

Infrastructure complexity	Infrastructure maturity	Infrastructure cost	Terminal complexity	Terminal maturity	terminal cost	Smartphone	Calibration complexity
None	Existing	Low	Low	Software development	Low	Existing	None

Positioning type	Accuracy	Reliability	Range	Sensitivity to environment	Positioning mode	In/out transition	Calibration needed
Absolute	Decimeter	Very high	Proximity	Low	User action needed	Easy	None

Infrastructure complexity, maturity, and cost: The price of a QR code, intrinsically very low, must be increased in the case of a positioning system because of the need to "fortify" it somehow. Indeed, it must be resistant to shocks and resistant over time. This remains, however, a reduced cost. It is the installation and the integration with the cartography that represents the biggest investment. Note also that it will be relatively easy to vandalize the system, simply by rendering the codes unreadable. A solution would be able to put the codes out of reach, but that would then require a specific reader. This could be imagined in the case of glasses with cameras, but these devices have not found their market for the moment. The advantage of such an approach would be more perceptible because the use of the codes could then be simultaneous for several users, as well as potentially automated. Indeed, several works relate to image processing, which, with good performance, makes it possible to detect a QR code within an image, even a complex image.

Terminal complexity, maturity, and cost: All the modern mobile terminals include a camera and processing capabilities. This is enough to deal with any QR code.

Calibration complexity: There is no calibration in the strict sense of the term. However, it is necessary to correctly position the codes in the cartography: it is the sine qua noncondition for the positioning provided by the system to be acceptable. This would require sophisticated mapping tools (see Chapter 13 for some details and discussions on this fundamental point).

Positioning type: It only depends on the coordinate system used to define the location of the QR code. It can actually be whatever one wants. The easier and most efficient is probably to keep a world coordinate system, like the one used by global navigation satellite system (GNSS).

Accuracy: It is very good as long as the codes are located at the right places in the cartography and that they have not been moved.

Reliability: The same remark as for "accuracy" applies. In addition, even a vandalized code does not jeopardize the positioning system. It simply renders one of the codes unusable. The user will go to another to locate himself.

Sensitivity to environment: As previously stated, in crowded places, the reading of a code should be more difficult. The principle in this case is probably to deploy a sufficient number of codes in order not to generate jamming.

In/out transition: This is probably a great advantage of such systems. The transition is absolutely transparent. Nevertheless, as the outdoors is well covered by GNSS, a real system should probably be limited to the interior and its surroundings.

5.6 Discussion of Other Technologies

The boundary is not so clear between the technologies concerning their range: it is easy to make the confusion between range and coverage, and it will sometimes be the case in the book indeed. Thus, a real deployment may of course use a technology that has the ability to cover a range R_1 in a situation where a reduced range R_2 is actually implemented. The other way round is more complex but can also occur: for instance, one can easily imagine a whole system of NFC tags that are scattered through a very large area, even outdoors, in order to provide a large geographical area. Thus, the parameter "range" is usually used in this book in order to characterize the unit performances of a technology, but it is necessary to take a step back in order to design a complete system of positioning.

Bibliography

1 Zhou, C. and Liu, X. (2016). The study of applying the AGV navigation system based on two dimensional bar code. In: *2016 International Conference on Industrial Informatics – Computing Technology, Intelligent Technology, Industrial Information Integration (ICIICII)*, Wuhan, 206–209. IEEE.

2 Li, Z. and Huang, J. (2018). Study on the use of Q-R codes as landmarks for indoor positioning: preliminary results. In: *2018 IEEE/ION Position, Location and Navigation Symposium (PLANS)*, Monterey, CA, 1270–1276. IEEE.

3 Razak, S.F.A., Liew, C.L., Lee, C.P., and Lim, K.M. (2015). Interactive android-based indoor parking lot vehicle locator using QR-code. In: *2015 IEEE Student Conference on Research and Development (SCOReD)*, Kuala Lumpur, 261–265. IEEE.

4 Lei, F. (2011). Design of QR code-based Mall shopping guide system. In: *International Conference on Information Science and Technology*, Nanjing, 450–453. IEEE.

5 Tang, S., Tok, B., and Hanneghan, M. (2015). Passive indoor positioning system (PIPS) using near field communication (NFC) technology. In: *2015 International Conference on Developments of E-Systems Engineering (DeSE)*, Dubai, 150–155. IEEE.

6 Kim, K., Jeong, S., Kim, W. et al. (2017). Design of small mobile robot remotely controlled by an android operating system via bluetooth and NFC communication. In: *2017 14th International Conference on Ubiquitous Robots and Ambient Intelligence (URAI)*, Jeju, 913–915. IEEE.

7 Edwan, E., Bourimi, M., Joram, N. et al. (2014). NFC/INS integrated navigation system: the promising combination for pedestrians' indoor navigation. In: *2014 International Symposium on Fundamentals of Electrical Engineering (ISFEE)*, Bucharest, 1–5. IEEE.

8 Bonzani, N., Kang, E., Yu, C., and Yun, M. (2015). Smart guide: mid-scale NFC navigation system. In: *2015 IEEE MIT Undergraduate Research Technology Conference (URTC)*, Cambridge, MA, 1–4. IEEE.

9 Ozdenizci, B., Ok, K., Coskun, V., and Aydin, M.N. (2011). Development of an indoor navigation system using NFC technology. In: *2011 Fourth International Conference on Information and Computing*, Phuket Island, 11–14. IEEE.

10 Nandwani, A., Edwards, R., and Coulton, P. (2012). Contactless check-ins using implied locations: a NFC solution simplifying business to consumer interaction in location based services. In: *2012 IEEE International Conference on Electronics Design, Systems and Applications (ICEDSA)*, Kuala Lumpur, 39–44. IEEE.

11 Cai-mei, H., Zhi-kun, H., Yue-feng, Y. et al. (2014). Design of reverse search car system for large parking lot based on NFC technology. In: *The 26th Chinese Control and Decision Conference (2014 CCDC)*, Changsha, 5054–5056. IEEE.

12 Kim, M.S., Lee, D.H., and Kim, K.N.J. (2013). A study on the NFC-based mobile parking management system. In: *2013 International Conference on Information Science and Applications (ICISA)*, Suwon, 1–5. IEEE.

13 Huang, J.C., Lin, Y., Yu, J.K. et al. (2015). A wearable NFC wristband to locate dementia patients through a participatory sensing system. In: *2015 International Conference on Healthcare Informatics*, Dallas, TX, 208–212. IEEE.

14 Hiramoto, M., Ogawa, T., and Haseyama, M. (2004). A novel image recognition method based on feature-extraction vector scheme. In: *2004 International Conference on Image Processing, 2004. ICIP '04*, Singapore, vol. 5, 3049–3052. IEEE.

15 Huang, Y., Jiang, H., and Yang, J. (2008). Research on genetic algorithm based on tabu search for landmark image recognition. In: *2008 7th World Congress on Intelligent Control and Automation*, Chongqing, 9270–9275. IEEE.

16 Greenspan, H., Porat, M., and Zeevi, Y.Y. (1992). Projection-based approach to image analysis: pattern recognition and representation in the position-orientation space. *IEEE Transactions on Pattern Analysis and Machine Intelligence* 14 (11): 1105–1110.

17 Lee, J.A. and Yow, K.C. (2007). Image recognition for mobile applications. In: *2007 IEEE International Conference on Image Processing*, San Antonio, TX, VI-177–VI-180. IEEE.

18 Gao, H., Chen, X., and Ren, Z. (2002). Algorithm design for a position tracking sensor based on pattern recognition. In: *IEEE 2002 28th Annual*

Conference of the Industrial Electronics Society. IECON 02, Sevilla, vol. 3, 2173–2178. IEEE.

19 Yamada, K., Takeuchi, T., Goto, T., and Hirano, S. (2016). Image recognition for automatic traveling wheelchair. In: *2016 IEEE 5th Global Conference on Consumer Electronics*, Kyoto, 1–2. IEEE.

20 Tsai, C. and Hsu, K. (2016). An application of using Bluetooth indoor positioning, image recognition and augmented reality. In: *2016 IEEE 13th International Conference on e-Business Engineering (ICEBE)*, Macau, 276–281. IEEE.

6

Room-Restricted Technologies: Challenges and Reliability

Abstract

A few technologies are limited to confined areas because the physics they use in order to carry out the measurements are unable to "pass through" walls, whatever the material is. This is notably the case with optical-based technologies, as well as sound-based ones. Thus, it is not surprising to find image, lidar, or ultrasound technologies in the list for the present chapter. Note that it does not mean that the range is short as a laser can reach several hundreds of meters of range indoors.

Keywords *Room range; image; lidar; sound; walls*

The classification described in Chapter 4 led to the following table concerning these technologies (Table 6.1).

6.1 Image Markers

There are many image processing technologies that can be used to identify the position of an object in its environment. This can be realized in an absolute way if one is able to detect in the image "markers" whose positions are known in an absolute reference frame, or relative to reference points of the image. Using markers often simplifies the efficiency of the systems, as in the case when dealing with position estimation and camera calibration. Specific markers have also been used for robot positioning. Various approaches have been investigated, such as using two cameras or two images including the same markers taken from different locations. Using a single image with passive markers, which is the basic idea of this paragraph, has also been investigated. In such a case, the difficulty lies in the fact that there is an ambiguity concerning the projection plane. Let us consider that we know the exact location of the markers that are visible in the image, either in a relative or absolute coordinate system. What is then sought is to define the real position of any point of the current

Indoor Positioning: Technologies and Performance, First Edition. Nel Samama.

Table 6.1 Main "room" technologies.

Technology	Positioning type	Accuracy	Reliability	Range	Sensitivity to environment	Calibration needed	Positioning mode	Technique	Signal processing	Position calculation
Image markers	Absolute	<1 m	Medium	Room	Very high	Once	Almost continuous	Image(s)	A combination of	Matrices calculus
Infrared	Symbolic	A few meters	High	Room	Very high	None	Almost continuous	Physical	Detection	Zone determination
Laser	Absolute	<1 cm	Very high	Room	Very high	None	Almost continuous	Phase(s)	Propagation modeling	∩ Spheres
Lidar	Absolute	<1 cm	Very high	Room	Very high	Once	Almost continuous	Time(s)	Correlation	∩ Planes + distance(s)
Sonar	Relative	A few centimeters	Medium	Room	Very high	None	Continuous	Time(s)	Detection	∩ Planes + distance(s)
Ultrasound	Absolute	A few decimeters	Low	Room	Very high	None	Continuous	Time(s)	Propagation modeling	∩ Spheres

image. Unfortunately, as we are dealing with a single image, there is no real way to find out the location of any point in the image because the projection possibilities are countless without knowing the projection plane in which the object to be positioned is physically located. In fact, the geometric mathematical transformations that are involved are based on the so-called homographic transformations. They make it possible to pass from an image plane (the photograph itself) to a projected plane (the physical reality), the one in which the object to be positioned is located.

Thus, the basic idea is to find a transformation between the 3D coordinates of a physical point and its coordinates in the 2D plane of the image obtained from a camera. We know that a scaling factor must be considered corresponding to the fact that many real-world points will be superimposed on the same pixel in the image. Thus, one must specify, in one way or another, the real geographical plane one wants to work with. Thus, the principle consists in taking photos on which will be visible georeferenced markers, which have to allow us to find the coordinates of any point in the image.

Some constraints of use are, however, to be noted. As just said, the "georeferencing," with a single image, has to be made in a given plane, i.e. for a given altitude above the ground. As a matter of fact, in order to transform an N dimension space into another N dimension space, one needs $N + 2$ reference points. In our case, dealing with 2D spaces for simplicity of the mathematical description leads to four reference points being considered. We realize then a kind of correspondence between the image and the geographical planes (through the two so-called "projective" transformations, i.e. from an $(N + 1)$ dimension space to an N dimension space). Therefore, reference points must not be simply characterized by a couple (longitude and latitude), but have to take into account the height of the geographical plane in which one wishes to make the projection. In our case, it is specifically the height of the mobile terminal. Considering the case of pedestrians moving in an indoor environment, we can consider that the terminal, typically the smartphone, will be held in hand and will have, more or less, a determined altitude (between 1 and 1.5 m depending on the age and height of the person: we could retain an average value of 1.3 m, for example, to minimize the associated errors). Thus, the markers could be positioned at this altitude. Thus, the projection plane to be considered is defined in the image by four pixels, as described in Figure 6.1.

The main steps required in order to achieve such a goal are, respectively, as follows:

1. Calculating the conversion matrix (from image plane to geographical plane) for the considered marks.
2. Calculating the 2D position of the mobile terminal in the geographical plane based on the matrix chosen in step 1.

The real question is now: how to transform image pixel coordinates into geographical coordinates? Indeed, a possible approach consists in using an

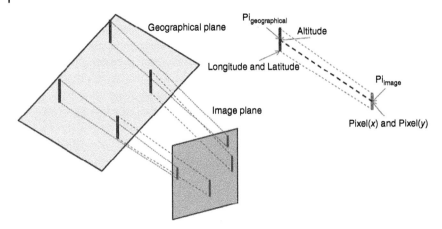

Figure 6.1 Diagram of the transformation we have to carry out, from image to geographical plane.

"intermediate" reference frame in which both coordinates (i.e. image and geographical) will be calculated. This intermediate frame is indeed a 3D reference frame as both planes (the image and the geographical ones) are projections of the real 3D world. Nevertheless, both planes are obtained through projections having different parameters. In order to match the two planes, we are going to transform both into the intermediate frame (through two different transformation matrices). Then, this intermediate reference frame allows us to translate the coordinates from one system to the other, through simple matrix calculations.

Let us call this intermediate reference frame Int_{RF} and the image and geographical reference frames Im_{RF} and Geo_{RF}, respectively. In order to define the image (resp. geographical) transformation matrix, we have to choose four points with known coordinates, say P_1 through P_4. A very strong constraint arises when one wants to use only one image. The matching between the image and the geographical coordinates can only be carried out in 2D (i.e. from the image plane to a real geographical plane). This means that the four markers must be located in the same plane.

The mathematics associated with the transformation from a plane to the intermediate reference frame are characterized by the resolution of the following system, where the unknown variables are the three transformation parameters λ, μ, and τ:

$$\begin{pmatrix} u_1 & u_2 & u_3 \\ v_1 & v_2 & v_3 \\ 1 & 1 & 1 \end{pmatrix} \begin{pmatrix} \lambda \\ \mu \\ \tau \end{pmatrix} = \begin{pmatrix} u_4 \\ v_4 \\ 1 \end{pmatrix} \tag{6.1}$$

where (u_i, v_i) are the coordinates of point P_i. The last row of "1" translates the fact that the intermediate reference frame is a 3D one when the considered points are in a plane. One can also note that Eq. (6.1) can also be written by introducing a matrix M leading to Eq. (6.2), as follows:

$$M = \begin{pmatrix} \lambda u_1 & \mu u_2 & \tau u_3 \\ \lambda v_1 & \mu v_2 & \tau v_3 \\ \lambda & \mu & \tau \end{pmatrix} \text{ and } \begin{pmatrix} u_4 \\ v_4 \\ 1 \end{pmatrix} = M \begin{pmatrix} 1 \\ 1 \\ 1 \end{pmatrix} \qquad (6.2)$$

Then, one can see that if considering the matrix M (see Eq. (6.2)), it is possible to "match" points P_1, P_2, and P_3 of the image with points $(1,0,0)$, $(0,1,0)$, and $(0,0,1)$, respectively, in Int_{RF}. This is thus our new basis. Note that P_4 now corresponds to $(1,1,1)$, as shown in Eq. (6.2).

The resolution of (6.1) or (6.2) is not very complex and gives

$$\lambda = \frac{[(v_2 - v_3)(v_3 u_4 - u_3 v_4) - (v_4 - v_3)(v_3 u_2 - u_3 v_2)]}{[(v_2 - v_3)(v_3 u_1 - u_3 v_1) - (v_1 - v_3)(v_3 u_2 - u_3 v_2)]}$$

$$\mu = \frac{1}{(v_2 - v_3)}[(v_4 - v_3) - (v_1 - v_3)\lambda]$$

$$\tau = 1 - \mu - \lambda$$

Now, one can obtain the intermediate reference frame coordinates of any point, given its plane coordinates, through the following formula (6.3):

$$\begin{pmatrix} x_i \\ y_i \\ z_i \end{pmatrix} = M^{-1} \begin{pmatrix} u_i \\ v_i \\ 1 \end{pmatrix} \qquad (6.3)$$

where (u_i, v_i) are the coordinates of point P_i in the plane considered and (x_i, y_i, z_i) are the coordinates of point P_i in the 3D intermediate reference frame Int_{RF}.

Furthermore, it is possible to carry out the same process concerning the same reference points for both the geographical plane and the image plane. Thus, we define two matrices, M_{pixel} and $M_{geograpical}$, by calculating the corresponding parameters λ_p, μ_p, and τ_p, and λ_g, μ_g, and τ_g, respectively. The systems to be solved are respectively given by Eqs. (6.4) and (6.5).

$$\begin{pmatrix} up_1 & up_2 & up_3 \\ vp_1 & vp_2 & vp_3 \\ 1 & 1 & 1 \end{pmatrix} \begin{pmatrix} \lambda_p \\ \mu_p \\ \tau_p \end{pmatrix} = \begin{pmatrix} up_4 \\ vp_4 \\ 1 \end{pmatrix} \qquad (6.4)$$

$$\begin{pmatrix} ug_1 & ug_2 & ug_3 \\ vg_1 & vg_2 & vg_3 \\ 1 & 1 & 1 \end{pmatrix} \begin{pmatrix} \lambda_g \\ \mu_g \\ \tau_g \end{pmatrix} = \begin{pmatrix} ug_4 \\ vg_4 \\ 1 \end{pmatrix} \qquad (6.5)$$

where (up_i, vp_i) (respectively (ug_i, vg_i)) are the coordinates of point P_i in the image (respectively geographical) plane.

Once all the λ_p, μ_p, and τ_p, and λ_g, μ_g, and τ_g parameters are calculated, we obtain the two matrices:

$$M_{pixel} = \begin{pmatrix} \lambda_p up_1 & \mu_p up_2 & \tau_p up_3 \\ \lambda_p vp_1 & \mu_p vp_2 & \tau_p vp_3 \\ \lambda_p & \mu_p & \tau_p \end{pmatrix} \text{ and } M_{geographical} = \begin{pmatrix} \lambda_p ug_1 & \mu_p ug_2 & \tau_p ug_3 \\ \lambda_p vg_1 & \mu_p vg_2 & \tau_p vg_3 \\ \lambda_p & \mu_p & \tau_p \end{pmatrix}$$

$$(6.6)$$

As we are interested in obtaining geographical coordinates of the point that is visible in an image from its pixel coordinates, we are interested in the transformation from pixel coordinates (u_{pi}, v_{pi}) to geographical coordinates (u_{gi}, v_{gi}). We know that

$$\begin{pmatrix} x_i \\ y_i \\ z_i \end{pmatrix} = M_{pixel}^{-1} \begin{pmatrix} u_{pi} \\ v_{pi} \\ 1 \end{pmatrix} \text{ and } \begin{pmatrix} x_i \\ y_i \\ z_i \end{pmatrix} = M_{geographical}^{-1} \begin{pmatrix} u_{gi} \\ v_{gi} \\ 1 \end{pmatrix}$$

$$(6.7)$$

where (x_i, y_i, z_i) being the intermediate coordinates of point P_i in the 3D intermediate reference frame Int_{RF}. Thus, the geographical coordinates can be obtained by applying formula (6.8):

$$\begin{pmatrix} u_{gi} \\ v_{gi} \\ 1 \end{pmatrix} = M_{geographical} M_{pixel}^{-1} \begin{pmatrix} u_{pi} \\ v_{pi} \\ 1 \end{pmatrix}$$

$$(6.8)$$

The only problem in this expression concerns the fact that the calculation (right part of the equation) does not always lead practically to a "1" as the third coordinate. This is due to the fact that we have to deal with the scaling factor w_{gi} briefly mentioned previously corresponding to the camera parameters, including the lens characteristics. Thus, instead of q. (6.8), we obtain equality (6.9).

$$\begin{pmatrix} u_{gi} \\ v_{gi} \\ w_{gi} \end{pmatrix} = M_{geographical} M_{pixel}^{-1} \begin{pmatrix} u_{pi} \\ v_{pi} \\ 1 \end{pmatrix}$$

$$(6.9)$$

The result sought is then

$$u_{gifinal} = u_{gi}/w_{gi} \text{ and } v_{gifinal} = v_{gi}/w_{gi}$$

$$(6.10)$$

Although it seems complex, it is not. These calculations can be carried out easily on any terminal. The most complicated aspect, and this remains valid for all techniques using images, lies in the quality required for the image. The latter must not be blurred and the markers must all be visible. One possibility would be the deployment of many markers in order to overcome the masking, but it will then be necessary to carry out more complex calculations and potentially lead to contradictions between calculations that will then have to be

Table 6.2 Summary of the main parameters for image markers.

Infrastructure complexity	Infrastructure maturity	Infrastructure cost	Terminal complexity	Terminal maturity	Terminal cost	Smart-phone	Calibration complexity
Low	None	Zero	Low	Software development	Zero	Existing	Medium

Positioning type	Accuracy	Reliability	Range	Sensitivity to environment	Positioning mode	In/out transition	Calibration needed
Absolute	<1 m	Medium	Proximity	Very high	Almost continuous	Easy	Once

dealt with. Finally, a fundamental point concerns the way markers are identified: indeed, this identification stage is fundamental as the calculation is based on an "a priori" knowledge of their positions. An identification error would almost certainly lead to an error in estimating the position of the mobile terminal.

Let us now come back to our parameter table. The various parameters considered for this technology are given in Chapter 4. However, a few could be discussed. Let us explain those that are quite questionable (in the positive sense of the word) (Table 6.2).

Infrastructure complexity, maturity, and cost: They are all at quite a low level as nothing is required except to place some markers in the environment. The only complexity lies in the fact that each marker should be identifiable. This could be achieved quite easily through the use of bar codes or QR codes. In such a way, let us imagine a combination of a purely QR code-based system, as described in Chapter 5, and an image marker one. In case only one or two codes are visible in the image, then the location of the user is set as being the position coded within the QR code (or a sort of combination of the locations of the two QR codes if two are visible). In case four QR codes are visible, then the approach described in this section is applied and a more accurate location can be provided to the user.

Terminal complexity and cost: It is almost zero as cameras are available in all smartphones and are improving rapidly in terms of quality (optical part) and performance (image processing).

Terminal maturity: This is the critical part today. Software developments are still required in order to overcome the classical difficulties encountered in camera-based approaches. They are mainly due to masking and blurring. Algorithms that will allow specific "forms" or "shapes" to be detected in a complex image are under development. Performances obtained with QR codes, for instance, are impressive. For example, the QR code located in the bottom left of the image shown in Figure 6.2 can easily be

Figure 6.2 A QR code included in an image. Source: Photo by Ray Rui on Unsplash.

detected and analyzed. One could imagine such a system implemented on connected glasses.

Calibration complexity: The calibration of a camera-based system is often a problem as its optical parameters are not necessarily known (and thus a lens calibration is required). Nevertheless, there are also approaches that could help in achieving a sort of "self-calibration." Let us consider that the markers are "normalized" in size and form: this can be used in order to calibrate the image parameters.

Positioning type: With image markers, the positioning is intended to be provided in an absolute way.

Accuracy: It can be very good, down to a few centimeters in good conditions, but is typically in the range of a few decimeters to 1 m.

Reliability: It is linked to the ability to have the markers in the image and to be able to detect them. Thus, environmental conditions are of uppermost importance and many real-life situations are difficult to be solved (presence of people around, impossibility to have four markers in the image, smog, etc.).

Range: One should be able to "see" the markers. Of course, the range will depend on the size of the markers, but it will be limited to line-of-sight (LOS) positioning, hence the "proximity" level considered.

Sensitivity to environment: Very high, due to potential obstacles.

Positioning mode: In case markers are distributed in the whole area, there is no reason the positioning could not be continuous, in a similar manner as is achieved in the case of simultaneous localization and mapping (SLAM) approaches (see Chapter 8).

6.2 Infrared Sensors

Infrared radiation, like radio waves, is an electromagnetic wave that is propagated at the speed of light. Time measurements are quite complex to carry out indoors, unless a precise time-distributed clock is deployed, which is still too expensive. Although different techniques could have been implemented, the one most largely used is just the detection approach: it consists simply in determining whether a receiver is present in the detection range of the device (see Figure 6.3).

The type of positioning is slightly different from the previous ones. Symbolic means that the positioning is no longer given in terms of spatial coordinates, in an absolute or relative manner, but rather in terms of room numbers or names. One would then know in which office, corridor, or conference room someone is, rather than the absolute positioning.

The various parameters considered for this technology are given in Chapter 4. However, a few could be discussed. Let us explain those that are quite questionable (Table 6.3).

Infrastructure complexity and cost: There is clearly the need for a specific infrastructure. Despite this, the costs of infrared components are quite low, leading to rather low cost for the global system. As usual, such deployments are easier in building under construction rather than in rehabilitation situations.

Infrastructure maturity: Infrared positioning is currently not a considered solution indoors and thus optimized components are not really available for that purpose.

Figure 6.3 The "active badge" system: beacon (a) and transmitters (b). Source: Courtesy of the Computer Laboratory of the University of Cambridge, United Kingdom.

(a) (b)

Table 6.3 Summary of the main parameters for infrared sensors.

Infrastructure complexity	Infrastructure maturity	Infrastructure cost	Terminal complexity	Terminal maturity	Terminal cost	Smart-phone	Calibration complexity
Medium	Development	Medium	Medium	Integration	Medium	Near future	None

Positioning type	Accuracy	Reliability	Range	Sensitivity to environment	Positioning mode	In/out transition	Calibration needed
Symbolic	A few meters	High	Room	Very high	Almost continuous	Easy	None

Terminal complexity, maturity, and cost: Although current smartphone cameras can detect near-infrared signals, an efficient system would certainly require a dedicated sensor.

Smartphone: Nevertheless, it would not be so difficult to include in smartphones. In addition, LiFi approaches (see Chapter 8) are bound to address the same problem: although current sensors have the ability to cope with the signals, dedicated sensors would allow a significant improvement of the systems.

Positioning type and accuracy: The positioning is clearly provided under the symbolic way, leading to an accuracy that is difficult to quantify (we consider "a few meters" in order to characterize the size of a typical room).

Reliability: It is a characteristic of the symbolic approaches that they are highly reliable. Another recurrent question is the relationship between the accuracy and the reliability of any positioning system.

Sensitivity to environment: It highly depends on the deployment. Usually, the "base station" is located on the ceiling in order to minimize the masking-related difficulties. Nevertheless, the orientation, position, and posture of the receiver, either on a smartphone or on any other device, remain a challenge for the detection. On the other hand, the symbolic approach is a factor of simplification concerning the propagation difficulty matters.

Positioning mode: Thus, the infrared-based positioning, as described in this paragraph, can be considered as a continuous system, although the positioning itself is discrete (symbolic).

6.3 Laser

One of the most impressive optical components is certainly the laser. The very clear transmission can be used for many different applications, ranging from telecommunications to medicine. One positioning-related application of the laser is telemetry. The basic principle of telemetry is to achieve distance measurement by measuring the flight time of a pulse. In the case of laser telemetry, this pulse is an optical one, generated by a laser, thus is at a very precise frequency. The principle is identical to that of radar with an accuracy that lies easily in the "few millimeters" range. Modern laser systems also apply "carrier phase" measurements, often using different carrier frequencies in order to achieve approximate and fine measurements. The two difficulties associated with laser telemetry for positioning are the pointing and the sensitivity to the environment (physical occultation), as for all optical systems.

In some specific environments, knowing the location of obstacles can enable complex laser-based systems to carry out self-positioning. Let us imagine a system composed of three laser beams as described in Figure 6.4 in a closed environment indicated by the polygon. The laser telemetry system allows d_1, d_2, and d_3 to be obtained with an accuracy that can reach a few millimeters, even when the rays are not perpendicular to the reflected surfaces.[1] Knowing the shape of the confined zone, it is possible to carry out computations in order to define the location of the laser system and its orientation. Of course, this example is only 2D, but a similar approach can be taken with additional measurements in order to achieve 3D positioning and orientation.

The main difficulty is of course that potential obstacles can certainly lead to wrong distance measurements. This is also the case with open doors in indoor environments or with windows (although not always true in this latter case). A solution could be to have the system pointing at the "sky," i.e. up to the ceiling, for example, like the radio satellite-based systems. In such a case, more measurements are required as the "sky" is a perfect plane. This approach could present some interest in static environments where the positions of objects and the structures are well defined.

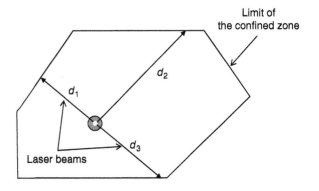

Figure 6.4 Possible laser positioning system and environment.

Limit of the confined zone

d_2

d_1

d_3

Laser beams

---

1 Note that this is a major difference from ultrasound telemetry systems.

Table 6.4 Summary of the main parameters for laser.

Infrastructure complexity	Infrastructure maturity	Infrastructure cost	Terminal complexity	Terminal maturity	Terminal cost	Smart-phone	Calibration complexity
None	None	Zero	Medium	Existing	Medium	Future	None

Positioning type	Accuracy	Reliability	Range	Sensitivity to environment	Positioning mode	In/out transition	Calibration needed
Absolute	<1 cm	Very high	Room	Very high	Almost continuous	Difficult	None

The various parameters considered for this technology are given in Chapter 4. However, a few could be discussed. Let us explain those that are quite questionable (in the positive sense of the word) (Table 6.4).

Infrastructure complexity, maturity, and cost: There is indeed no specific infrastructure.

Terminal complexity, maturity, and cost: A laser is not compulsorily an expensive component, but when it comes to telemetry, measurements are more complex, as is processing. However, this does not entail high costs. It is possible to find good-quality laser rangefinders for a $100. Of course, a positioning system like the one described above adds a little more complexity. The difficulty here rests more in the fact that today, the material would be dedicated to a use, like the building, for example, and not intended for use by the general public (but that can change quickly).

Smartphone: It is not so much the integration in the smartphone that is likely to pose a problem, as the way to use such a system. This is the reason why we imagine the latter in a specific use.

Calibration complexity: No calibration of the "sensor" is required, but it is, as is often the case indoors, necessary to have a map that often requires in-depth work for its realization.

Accuracy: The accuracy is at an excellent level because of the high accuracy of the elementary distance measurements. As a matter of fact, once the measurement errors are reduced to a minimum, the positioning is then quite easy. See Chapter 12 for details concerning the relationship between the accuracy of the measurements and the accuracy of the positioning.

Reliability: It depends mainly on the quality of the modeling (map) of the environment and the presence of masking.

Range: Clearly limited to LOS environments.

Sensitivity to environment: This is the most difficult problem to solve in non-professional environments.

In/out transition: This technology for a complete positioning, without mixing it with another one such as rotary coding wheels (as in a theodolite system described in Chapter 8), is not intended to work outdoors.

6.4   Lidar

The Lidar (light detection and ranging) is the "optical" version of the radar whose principles will be discussed in Chapter 7. It can be considered as a rotating laser in three dimensions. Figure 6.5 gives a schematic representation of a three-dimensional lidar.

Being able to perform successive measurements with a high repetition rate (several hundred hertz, even kilohertz) can reproduce a "depth map" of its environment. In two dimensions, this is translated, for example, in the form given in Figure 6.6 where each point is obtained by measuring the distance between the lidar and the first object encountered by the beam.

The spatial resolution being related to the wavelength of the beam, the three-dimensional representations that it is possible to obtain are remarkable, as shown in Figure 6.7, which represents a reconstruction of a few buildings associated with a geographical information system (GIS). The details are of a really high resolution.

The various parameters considered for this technology are given in Chapter 4. However, a few could be discussed. Let us explain those that are quite questionable (in the positive sense of the word) (Table 6.5).

Infrastructure complexity, maturity, and cost: As in the case of the laser, there is no prior deployment to carry out in the case of lidar. Only masking can reduce performance, sometimes drastically.

Terminal complexity, maturity, and cost: Such equipment is quite expensive today and is not likely to be included as it is in nonspecific terminals. Figure 6.8 shows a lidar from Trimble.

Figure 6.5 Typical lidar architecture.

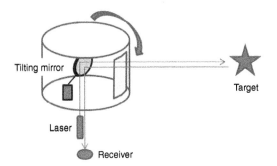

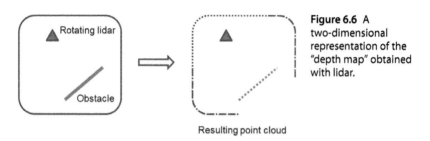

Figure 6.6 A two-dimensional representation of the "depth map" obtained with lidar.

Resulting point cloud

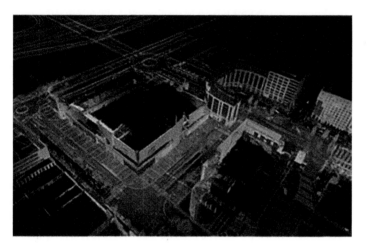

Figure 6.7 A three-dimensional representation of the "depth map" obtained for a city district. Source: © Trimble Inc.

Table 6.5 Summary of the main parameters for Lidar.

Infrastructure complexity	Infrastructure maturity	Infrastructure cost	Terminal complexity	Terminal maturity	Terminal cost	Smart-phone	Calibration complexity
None	None	Zero	High	Existing	Very high	Almost impossible	Medium

Positioning type	Accuracy	Reliability	Range	Sensitivity to environment	Positioning mode	In/out transition	Calibration needed
Absolute	<1 cm	Very high	Room	Very high	Almost continuous	Difficult	Once

Figure 6.8 TX8 lidar from Trimble.
Source: © Trimble Inc.

Calibration complexity: This technology is a professional one, thus leading to the requirement of a minimum skill in order to use the equipment that actually requires an initial calibration.

Accuracy: The positioning accuracy is one of the best thanks to the combined use of high-quality laser rangefinders and high-precision elevation and azimuth encoding wheels. This accuracy depends, as is often the case for systems involving angle measurements, on the distance measured, but subcentimeter accuracy is achievable.

Reliability: It is excellent as long as one is carrying out the right measurement (i.e. no masking or obstacle).

Range and sensitivity to environment: It can go up to a few hundreds of meters (and even more in good conditions of light) but cannot cope with nonline of sight measuring, hence the "room" assignment in this book.

Positioning mode: It can be continuous, also in motion with current equipment that can be motorized in order to "follow" a target that is generally in the form of a 360° prism. Note that in this case, from a known position of the station (as described in Figure 6.7), the station will "aim" at the target and follow it in its movement to extract the instantaneous position at each moment. The current possible refresh rates are in the range of 10–20 Hz (10–20 positions per second).

In/out transition: As it is the station that follows the movement of the target, the positioning between inside and outside is possible if there is a continuity in the aiming without having to move the station, otherwise a new calibration of the latter is necessary. In fact, it would be possible to maintain a moving position of the station, in particular by using a centimetric global navigation satellite system (GNSS) receiver, but then the absolute orientation for the encoding wheels would be lost (a system with two centimetric GNSS receivers, for example, would have to be considered to maintain this orientation, but this becomes a bit complicated and it is not certain that this is actually useful for current uses).

6.5 Sonar

Sonar (sound navigation and ranging) is the "sound" version of radar whose principles will be discussed in Chapter 7. By moving the sonar, whether it is a proper movement of the device or of its carrier, it is possible to make a "depth map." In two dimensions, this translates, as in the case of lidar, into a representation similar to that in Figure 6.6. The major difference with lidar is the beam width and associated measurement noises, which are much less fine for sonar. However, propagation can be carried out in more disturbed environments, for example, where the optical opacity does not allow good measurements with lidar. Finally, the cost of acoustic transmitters and receivers is also much lower. A typical operating mode of such a system is shown in Figure 6.9. The signal issued from the transmitter is reflected by any surface and comes back to the receiver. The measurement of the time of flight allows the distance to be determined.

The spatial resolution is then no longer due solely to the wavelength of the signal because it would then be too low. The transmitter is then moved to implement interferometric processes to reconstruct good resolution 3D representations. An example is given in Figure 6.10 in the context of underwater exploration (the marine environment is particularly well adapted to sonar, which has a decisive advantage over lidar or radar in terms of propagation).

The various parameters considered for this technology are given in Chapter 4. However, a few could be discussed. Let us explain those that are quite questionable (in the positive sense of the word) in an indoor environment (Table 6.6).

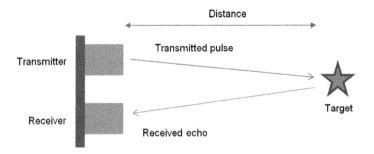

Figure 6.9 Image reconstruction from a sonar.

Figure 6.10 Image reconstruction from a sonar. Source: © MedSurvey.

Table 6.6 Summary of the main parameters for sonar.

Infrastructure complexity	Infrastructure maturity	Infrastructure cost	Terminal complexity	Terminal maturity	Terminal cost	Smart-phone	Calibration complexity
None	None	Zero	Low	Integration	Low	Difficult	None

Positioning type	Accuracy	Reliability	Range	Sensitivity to environment	Positioning mode	In/out transition	Calibration Needed
Relative	A few centimeters	Medium	Room	Very high	Continuous	Easy	None

Infrastructure complexity, maturity and cost: Once again, the system can be considered as "passive," i.e. not requiring the participation of the infrastructure.

Terminal complexity, maturity, and cost: As already mentioned, the unit cost of sound sensors remains low, but a full sonar requires the addition of signal shaping and processing units to produce a quality "picture."

Smartphone: The addition of sensors on the smartphone does not seem impossible, but it is more in use that the question would arise.

Calibration complexity: No calibration is required as all the electronics are colocated (hence the time synchronization system between the transmitter and the receiver).

Positioning type: This is clearly relative with respect to the sonar itself. Of course, as usual, it is often possible to add another system in order to achieve absolute positioning. When outdoors, it is a satellite-based system. When indoors, the easiest way is to set the sonar fixed at a known location and rotate it around two axes, as in the case of the lidar seen in the previous paragraph. Thus, we obtain a reconstruction of its environment in an absolute reference frame.

Accuracy: It can be at an excellent level, down to the centimeter, because of the nature of the waves used.

Reliability: Not so good because of the environmental conditions. First of all, masking is of major concern as the equivalent wavelength is not small enough in order to reduce diffraction problems. Then, some parameters are difficult to estimate. For example, the real propagation velocity of the wave can change in windy environments.

Range: Similar to optical systems, only LOS measurements are possible. In some implementations, this could be seen as an advantage as one knows nonline-of-sight measurements are not possible (removing the need for complex and sometimes not really efficient techniques which try to cope with this problem).

Sensitivity to environment: Thus, the sensitivity is quite high.

6.6 Ultrasound Sensors

There is a great similarity between the propagation of radio waves in free space and the propagation of sound in the air. The frequency of a typical radio wave is about 10^9 Hz (1 GHz) with a velocity of 3.10^8 m s^{-1}. The frequency of a typical sound wave is around 1 kHz and travels at around 300 m s^{-1}. The corresponding wavelengths,[2] which are a characteristic of all propagation signals, are respectively 30 and 30 cm: identical![3] Thus, even if the physical forms of both waves are fundamentally different,[4] both propagations are very similar. Nevertheless, there is a fundamental difference: the velocity of the waves, slowing down from 300 000 000 to 300 m s^{-1}, i.e. a factor of 10^6. When dealing with the time needed to define distances, it is much easier to achieve accurate measurements when the time constraint is loose: this is the case with sound waves. Thus, ultrasound positioning systems based on time measurements can be quite accurate. To illustrate the comparison in terms of required clock accuracy, it gives us, say 1 ms for an ultrasound system, which is equivalent to 1 ns for an electromagnetic wave. Furthermore, sound exhibits reflections, but once again, for a reflected ray that would travel 5 m more, it corresponds to a 16.7 ms delay,

---

2 The wavelength is defined by λ = velocity of the wave/frequency.
3 This is not completely true as the velocities are not the exact values, but this is quite nice to consider it for the purpose of the discussion.
4 A radio wave is an electromagnetic wave made of photons, whereas a sound wave is a mechanical vibration of molecules of air (or also an air pressure wave).

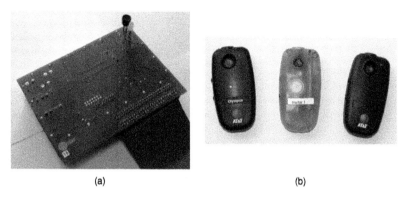

(a)   (b)

Figure 6.11 The Bat system: beacon (a) and transmitters (b). Source: Courtesy of the Computer Laboratory of the University of Cambridge, United Kingdom.

which is easily discriminated from the direct path. The reader can now understand the reason for the increased accuracy of ultrasound solutions with quite cheap electronics. An example of such a system was developed by the Computer Laboratory of the University of Cambridge, United Kingdom (illustrated Figure 6.11).

The various parameters considered for this technology are given in Chapter 4. However, a few could be discussed. Let us explain those that are quite questionable (in the positive sense of the word) Table 6.7.

Infrastructure complexity, maturity, and cost: Here, the idea is no longer to operate in a radar approach, but by analyzing the reception times of signals synchronized to the transmission. Unlike GNSS (see Chapters 8 and 12), the transmitter is a mobile terminal that the user owns. The infrastructure is then a set of receivers detecting the signal coming from the terminal. The analysis of the arrival time differences on the various receivers, supposedly synchronized because they are all physically connected, makes it possible to calculate

Table 6.7 Summary of the main parameters for ultrasound sensors.

Infrastructure complexity	Infrastructure maturity	Infrastructure cost	Terminal complexity	Terminal maturity	Terminal cost	Smart-phone	Calibration complexity
High	Existing	Medium	Medium	Integration	Low	Easy	None

Positioning type	Accuracy	Reliability	Range	Sensitivity to environment	Positioning mode	In/out transition	Calibration needed
Absolute	A few decimeters	Low	Room	Very high	Continuous	Easy	None

the position of the transmitter. The infrastructure is then well present and dedicated to positioning. The latter does not present any technical difficulty and the associated unit costs remain very low. However, any modification of a building generates costs to bring maps or electrical circuits into conformity, which are often the most important part of the system to be deployed. It is probably more economical to include the system in the construction.

Terminal complexity, maturity, and cost: An ultrasonic transmitter is inexpensive and its integration into an existing terminal is not difficult. However, the difficulty lies elsewhere: when using the positioning system, it will be necessary to have a clear propagation between the transmitter and at least three to four receivers. This is why they are often positioned high, on the ceiling, for example. In particular, the human body is an impassable obstacle and only direct LOS receptors will have the ability to detect. This requires a large number of receivers. Moreover, the geometric distribution of the receivers will not be very good for position calculation (see the passage on GNSS "DOP" (dilution of precision) in Chapters 8 and 12).

Smartphone: This should not be a problem, but what would be the type of use and implementation?

Calibration complexity: No calibration is required except the need to know the locations of the distributed receivers.

Positioning type: It is identical to the way in which receiver positions are given, i.e. relative in case the locations of the receivers coordinates are provided in a relative way, absolute if they are provided in an absolute way.

Accuracy: As for all sound-based positioning systems, the time-of-flight measurement is potentially quite accurate (a few centimeters). Then, positioning is based on multiple measurements and calculations in real environments, hence the "a few decimeters" level assigned in Table 6.7.

Reliability: It highly depends on the real conditions. In case of many reflections on close objects, then it will be poor. Otherwise, in well-mastered environments, mainly static ones, it could be quite good indeed.

Range: It still does not pass through walls.

Sensitivity to environment: Already discussed and very sensitive.

In/out transition: As it is based on the deployment of an indoor infrastructure, it is not likely to be used outdoors. Another means should then be used.

Bibliography

1 Pribula, O. and Fischer, J. (2011). Real time precise position measurement based on low-cost CMOS image sensor: DSP implementation and sub-pixel measurement precision verification. In: *2011 18th International Conference on Systems, Signals and Image Processing*, Sarajevo, 1–4.

2 Borstell, H., Pathan, S., Cao, L. et al. (2013). Vehicle positioning system based on passive planar image markers. In: *International Conference on Indoor Positioning and Indoor Navigation*, Montbeliard-Belfort, 1–9.

3 Mochizuki, Y., Imiya, A., and Torii, A. (2007). Circle-marker detection method for omnidirectional images and its application to robot positioning. In: *2007 IEEE 11th International Conference on Computer Vision*, Rio de Janeiro, 1–8.

4 Bousaid, A., Theodoridis, T., and Nefti-Meziani, S. (2016). Introducing a novel marker-based geometry model in monocular vision. In: *2016 13th Workshop on Positioning, Navigation and Communications (WPNC)*, Bremen, 1–6.

5 Teshima, T., Saito, H., Ozawa, S. et al. (2006). Vehicle lateral position estimation method based on matching of top-view Images. In: *18th International Conference on Pattern Recognition (ICPR'06)*, Hong Kong, 626–629.

6 Moreno, M.V., Zamora, M.A., Santa, J., and Skarmeta, A.F. (2012). An indoor localization mechanism based on RFID and IR data in ambient intelligent environments. In: *2012 Sixth International Conference on Innovative Mobile and Internet Services in Ubiquitous Computing*, Palermo, 805–810.

7 Xu, Z., Huang, S., and Ding, J. (2016). A new positioning Method for indoor laser navigation on under-determined condition. In: *2016 Sixth International Conference on Instrumentation & Measurement, Computer, Communication and Control (IMCCC)*, Harbin, 703–706.

8 Tilch, S. and Mautz, R. (2010). Current investigations at the ETH Zurich in optical indoor positioning. In: *2010 7th Workshop on Positioning, Navigation and Communication*, Dresden, 174–178.

9 Yao, Y., Lou, M., Yu, P., and Zhang, L. (2016). Integration of indoor and outdoor positioning in a three-dimension scene based on LIDAR and GPS signal. In: *2016 2nd IEEE International Conference on Computer and Communications (ICCC)*, Chengdu, 1772–1776.

10 Martínez-Rey, M., Santiso, E., Espinosa, F. et al. (2016). Smart laser scanner for event-based state estimation applied to indoor positioning. In: *2016 International Conference on Indoor Positioning and Indoor Navigation (IPIN)*, Alcala de Henares, 1–7.

11 Kokert, J., Höflinger, F., and Reindl, L.M. (2012). Indoor localization system based on galvanometer-laser-scanning for numerous mobile tags (GaLocate). In: *2012 International Conference on Indoor Positioning and Indoor Navigation (IPIN)*, Sydney, NSW, 1–7.

12 Tamas, L., Lazea, G., Popa, M. et al. (2009). Laser based localization techniques for indoor mobile robots. In: *2009 Advanced Technologies for Enhanced Quality of Life*, Iasi, 169–170.

13 Islam, S., Ionescu, B., Gadea, C., and Ionescu, D. (2016). Indoor positional tracking using dual-axis rotating laser sweeps. In: *2016 IEEE International*

Instrumentation and Measurement Technology Conference Proceedings, Taipei, 1–6.

14 Kim, B., Choi, B., Kim, E., and Yang, K. (2012). Indoor localization using laser scanner and vision marker for intelligent robot. In: *2012 12th International Conference on Control, Automation and Systems*, JeJu Island, 1010–1012.

15 Li, K., Wang, C., Huang, S. et al. (2016). Self-positioning for UAV indoor navigation based on 3D laser scanner, UWB and INS. In: *2016 IEEE International Conference on Information and Automation (ICIA)*, Ningbo, 498–503.

16 Chen, Y., Liu, J., Jaakkola, A. et al. (2014). Knowledge-based indoor positioning based on LiDAR aided multiple sensors system for UGVs. In: *2014 IEEE/ION Position, Location and Navigation Symposium – PLANS 2014*, Monterey, CA, 109–114.

17 Shamseldin, T., Manerikar, A., Elbahnasawy, M., and Habib, A. (2018). SLAM-based Pseudo-GNSS/INS localization system for indoor LiDAR mobile mapping systems. In: *2018 IEEE/ION Position, Location and Navigation Symposium (PLANS)*, Monterey, CA, 197–208.

18 Liu, S., Atia, M.M., Karamat, T. et al. (2014). A dual-rate multi-filter algorithm for LiDAR-aided indoor navigation systems. In: *2014 IEEE/ION Position, Location and Navigation Symposium – PLANS 2014*, Monterey, CA, 1014–1019.

19 Li, R., Liu, J., Zhang, L., and Hang, Y. (2014). LIDAR/MEMS IMU integrated navigation (SLAM) method for a small UAV in indoor environments. In: *2014 DGON Inertial Sensors and Systems (ISS)*, Karlsruhe, 1–15.

20 Yoshisada, H., Yamada, Y., Hiromori, A. et al. (2018). Indoor map generation from multiple LIDAR point clouds. In: *2018 IEEE International Conference on Smart Computing (SMARTCOMP)*, Taormina, 73–80.

21 Li, J.H., Kang, H.J., Park, G.H. et al. (2017). Sonar image processing based underwater localization method and its experimental studies. In: *OCEANS 2017 – Anchorage*, Anchorage, AK, 1–5.

22 Lee, Y., Choi, J., and Choi, H. (2015). Experimental results of real-time sonar-based underwater localization using landmarks. In: *OCEANS 2015 – MTS/IEEE Washington*, Washington, DC, 1–4.

23 Yeol, J.W. (2005). An improved position estimation algorithm for localization of mobile robots by sonars. In: *2005 Student Conference on Engineering Sciences and Technology*, Karachi, 1–5.

24 Guarato, F., Laudan, V., and Windmill, J.F.C. (2017). Ultrasonic sonar system for target localization with one emitter and four receivers: Ultrasonic 3D localization. In: *2017 IEEE SENSORS*, Glasgow, 1–3.

25 Huang, L., He, B., and Zhang, T. (2010). An autonomous navigation algorithm for underwater vehicles based on inertial measurement units and sonar. In: *2010 2nd International Asia Conference on Informatics in Control, Automation and Robotics (CAR 2010)*, Wuhan, 311–314.

26 Urdiales, C., Bandera, A., Ron, R., and Sandoval, F. (1999). Real time position estimation for mobile robots by means of sonar sensors. In: *Proceedings 1999 IEEE International Conference on Robotics and Automation (Cat. No.99CH36288C)*, Detroit, MI, USA, vol. 2, 1650–1655.

27 Cheng, X. and Wang, Y. (2017). Multi-target localization analysis based on nonparametric spectral estimation method for MIMO sonar. In: *2017 IEEE International Conference on Signal Processing, Communications and Computing (ICSPCC)*, Xiamen, 1–5.

28 Sosa-Sesma, S. and Perez-Navarro, A. (2016). Fusion system based on WiFi and ultrasounds for in-home positioning systems: The UTOPIA experiment. In: *2016 International Conference on Indoor Positioning and Indoor Navigation (IPIN)*, Alcala de Henares, 1–8.

29 Png, L.C., Chen, L., Liu, S., and Peh, W.K. (2014). An Arduino-based indoor positioning system (IPS) using visible light communication and ultrasound. In: *2014 IEEE International Conference on Consumer Electronics – Taiwan*, Taipei, 217–218.

30 Holm, S. (2012). Ultrasound positioning based on time-of-flight and signal strength. In: *2012 International Conference on Indoor Positioning and Indoor Navigation (IPIN)*, Sydney, NSW, 1–6.

31 Kitanov, A., Tubin, V., and Petrovic, I. (2009). Extending functionality of RF Ultrasound positioning system with dead-reckoning to accurately determine mobile robot's orientation. In: *2009 IEEE Control Applications, (CCA) & Intelligent Control, (ISIC)*, St. Petersburg, 1152–1157.

32 Wehn, H.W. and Belanger, P.R. (1997). Ultrasound-based robot position estimation. *IEEE Transactions on Robotics and Automation* 13 (5): 682–692.

33 Medina, C., Segura, J.C., and Holm, S. (2012). Feasibility of ultrasound positioning based on signal strength. In: *2012 International Conference on Indoor Positioning and Indoor Navigation (IPIN)*, Sydney, NSW, 1–9.

34 De Angelis, A., Moschitta, A., and Comuniello, A. (2017). TDoA based positioning using ultrasound signals and wireless nodes. In: *2017 IEEE International Instrumentation and Measurement Technology Conference (I2MTC)*, Turin, 1–6.

35 Lindo, A., García, E., Ureña, J. et al. (2015). Multiband waveform design for an ultrasonic indoor positioning system. *IEEE Sensors Journal* 15 (12): 7190–7199.

7

"Set of Rooms" Technologies

Abstract

The limit for only one room is quite clear because of the physical limitations of the physics used. For coverage of "a few rooms," it is a little bit less obvious. Where is the boundary between a few rooms and a whole building? What are the real performances of the technologies when passing through walls? Are techniques to detect the presence of obstacles available? And what are the real impacts of such obstacles? All these questions are important ones and are the main difficulties associated with these techniques. Basically, from "a single room" to "a few rooms," one changes the physical quantities we are playing with. From optical and sound-based physics, we move to radio-based ones. Although optical and radio waves are identical in principle (Maxwell's equations), their propagation properties are quite different.

Keywords *Radio based positioning; RFID; UWB; Radar*

The classification described in Chapter 4 led to the following table concerning these technologies (Table 7.1).

7.1 Radar

Besides the fact that the word radar is a palindrome, this system allows a great diversity of uses. The first is the detection and measurement of a distance (this is the use we will make of it), but allows many other things: the identification of a target, the measurement of its angle of arrival, and the determination of its relative speed in relation to the radar (Doppler), which is very useful for us. The technique is described in detail in Chapter 3.

Let us now come back to our parameter table. The various parameters considered for this technology are given in Chapter 4. However, a few could be discussed. Let us explain those that are questionable (Table 7.2).

Indoor Positioning: Technologies and Performance, First Edition. Nel Samama.
© 2019 The Institute of Electrical and Electronics Engineers, Inc. Published 2019 by John Wiley & Sons, Inc.

Table 7.1 Main technologies for "a few rooms."

Technology	Positioning type	Accuracy	Reliability	Range	Sensitivity to environment	Calibration needed	Positioning mode	Technique	Signal processing	Position calculation
Radar	Absolute	A few centimeters	Medium	A few rooms	High	None	Continuous	Phase(s)	A combination of	∩ Plans + distance(s)
RFID	Absolute	Decimeters	High	A few rooms	Low	None	Discrete	Physical	Detection	Spot location
UWB	Absolute	A few centimeters	Medium	A few rooms	High	None	Continuous	Time(s)	A combination of	∩ Spheres

Table 7.2 Summary of the main parameters for Radar.

Infrastructure complexity	Infrastructure maturity	Infrastructure cost	Terminal complexity	Terminal maturity	Terminal cost	Smartphone	Calibration complexity
Medium	Research	Medium	High	Research	Medium	Difficult	None

Positioning type	Accuracy	Reliability	Range	Sensitivity to environment	Positioning mode	In/out transition	Calibration needed
Absolute	A few centimeters	Medium	A few rooms	High	Continuous	Easy	None

Infrastructure complexity, maturity, and cost: Radars like the ones used in airport, or even for road speed control, are quite expensive with respect to a positioning system intended to be used for mass market deployment. Thus, a positioning system based on the radar principle (such as the ones described in this section) is indeed a little bit different. They are based on an array of typically four antennas defining two directions (usually orthogonal ones) and thus two plans (usually a vertical one and a horizontal one) in which the two angles defining a three-dimensional direction will be obtained (see Figure 7.1 for detail). In order to obtain a position, one needs an additional measurement.

There are typically two approaches now: either you duplicate the system described above and you carry out a second three-dimensional direction of arrival or you implement a complementary system in order to measure a range

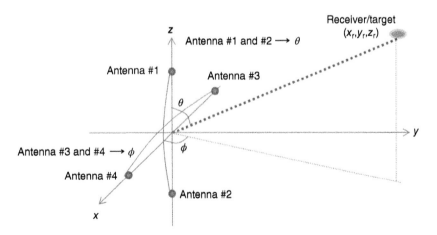

Figure 7.1 Three-dimensional direction of arrival measurement principle.

(the distance between the radar antennas described in Figure 7.1) and the mobile terminal. In both cases, one needs to know the location of the radar system(s) and to carry out some mathematical calculations.

Various techniques exist to carry out the measurements. We can cite two main ones:

- In case the terminal is "active" in the sense that it sends a signal, then the two antennas of the same direction allow a differential measurement of the received phases, which gives precisely this direction of incidence.
- In case the terminal is passive, then the radar must transmit a signal (from the central antenna shown in Figure 7.1). It will be the signal reflected on the terminal that will allow, according to a process identical to that described above, the direction of arrival to be measured.

It is thus necessary to carry out the two differential phases in order to know the direction in three dimensions.

Then, in the case of a single system, a distance measurement will be necessary. The latter is classically obtained by using a signal of variable frequency at transmission (always from the central antenna of the system). By measuring the frequency difference between transmission and reception, the propagation time is known and the distance from the target can be derived. The advantages of such systems are numerous: the absence of the need to synchronize the radar (also in the case of the use of two radars, for example), the fact that the signals come from and are marked by the same clocks, and in the case of an integrated distance measurement, the fact of needing only one radar. The latter systems described are at the research stage and are reasonably complex.

Terminal complexity and cost: The principle would be that the terminal would be completely standard. However, in the case of widely distributed systems, whether personal or industrial, it is sometimes necessary to use an active terminal. For example, the latter can amplify the received signal in order to extend the range, but can also change the frequency or delay the retransmission of the signal in order to reduce both the direct reflection effects on the target and the disturbances associated with multiple paths. In this case, the cost and maturity of the terminal are no longer zero.

Terminal maturity: When passive targets are used, there is no subject. In case the terminal is active, then although not expensive, the design has to be carried out. It is mainly at a research level today.

Calibration complexity and need: A limited calibration is required because when accurate phase differences have to be measured, even very small mismatching between antennas or connections could lead to significant error. Such a calibration is not difficult, but needed.

Smartphone: It is not currently planned to have such a system integrated into smartphones, except in the case of certain approaches using ultrawide band (UWB) signals (see Section 7.3 below).

Positioning type: It is absolute in the sense that once the location of the radar is given, the location of the terminal can be obtained (more or less accurately) in the same coordinate frame. Basically, it is not designed to provide relative positioning, although it could be imagined (through Doppler measurements, for instance).

Accuracy: As usual for radio systems, in good propagation conditions (no multipath, line of sight (LOS), etc.), the localization accuracy can reach the centimeter level. In real conditions, it is a little bit less accurate but the potential is at a high level.

Reliability: It depends mainly on the propagation conditions. A few research works are reporting nonline-of-sight (NLOS) detection with quite good performance, but it is mainly imagined in LOS conditions in order to take advantage of the fact one uses only one radar.

Range: With the restriction mentioned above concerning LOS and NLOS conditions, one can reach a few rooms depending on the transmitted signal power level and the sensitivity of the receiver. Nevertheless, regulations sometimes exist in certain frequency bands reducing the real possible range. The maximum range can be obtained with sophisticated signals including coding, for example.

Sensitivity to environment: Quite high, as with all radio-based approaches.

7.2 RFID

Combining a radio system for its ease and low cost of implementation and a symbolic positioning, one can reach the radio frequency identification (RFID) concept. An RFID is an electronic label that can be associated with any object. The principle is to use a "reader" that can supply power directly to the label, allowing it not to be active in the sense of electric power (see Figure 7.2). Thus, it is possible to realize the so-called "RFID tags" at very low cost and

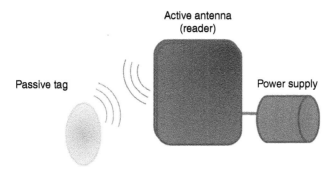

Figure 7.2 A typical RFID system architecture.

of various sizes and shapes. The tag can be identified because once powered by the "reader," it can transmit data, such as an identification number or a location where it is placed.

An RFID-based positioning system consists of positioning some tags at various locations. As a tag is a very low-cost element, it is possible to place a large number of tags all over a large area. The mobile terminal consists of a reader that can interrogate the tags when passing nearby. Current readers can act as far as 1 m (sometimes farther, but this is quite unusual), thus leading to an equivalent accuracy of about 1 m. The positioning method consists in getting the identifier back from the tag and reading a location database to make the link between the identifier and a location. Of course, one can imagine that the tag indicates the location directly to the reader. One can also think that the mobile terminal could be the tag rather than the reader: it is less expensive for the terminal, but the new system then requires the location to be transmitted to the tag (this means that the tags have to be programmable).

One of the advantages of RFID tags lies in their ability to be produced in all kinds of sizes and shapes. It is also possible to include these tags in clothing, for example, opening the way to the imagination of innovative approaches.

Let us now come back to our parameter table. The various parameters considered for this technology are given in Chapter 4. However, a few could be discussed. Let us explain those that are quite questionable (in the positive sense of the word) (Table 7.3).

Infrastructure complexity, maturity, and cost: The complexity and cost of the infrastructure, which is essential here, depends on how the system is deployed. In general, tags, passive or active, are deployed in the environment. In this case, the cost remains measured for the "material" aspects. The cost of installation, deployment, and maintenance may not be negligible. In the opposite case, where the readers are the fixed infrastructure, the costs are then much higher because a power supply, a network link, etc., must be provided. Note, however, that in such a case, the mobile terminal is only a

Table 7.3 Summary of the main parameters for RFID.

Infrastructure complexity	Infrastructure maturity	Infrastructure cost	Terminal complexity	Terminal maturity	Terminal cost	Smart-phone	Calibration complexity
Low	Existing	Low	Low	Software development	Low	Low	None

Positioning type	Accuracy	Reliability	Range	Sensitivity to environment	Positioning mode	In/out transition	Calibration needed
Absolute	decimeters	High	Proximity	Low	Discrete	Easy	None

tag. All the elements of such a system, whatever the approach chosen, are industrially mature.

Terminal complexity, maturity, and cost: In general, the user's terminal is the smartphone (for consumer applications) and constitutes the reader. The advantage of this solution lies in the simplicity of installation of the infrastructure but also in the fact that it is then up to the user to provide the necessary energy for the proper functioning of the system. In this case, the tags distributed can be either passive (reduced range) or active for longer distance operation.

Calibration complexity and need: The only calibration required here is to know the position of the infrastructure elements.

Positioning type: As usual, it depends on the way the locations of the infrastructure elements are provided. In order to facilitate the in/out transition, it is common to use global navigation satellite system (GNSS)-compatible coordinates.

Accuracy: As the range is limited, the accuracy can be seen as inversely proportional to the range. Thus, one could consider the accuracy for passive tags is better than that for active tags because the uncertainty will be higher in the latter case. This reasoning is undoubtedly too simple, but it nevertheless gives a specific vision to systems whose localization coverage is not continuous in space.

Reliability: The corollary to precision is reliability. Given the requirement to be in an area close to an element of the infrastructure, the reliability of positioning increases as accuracy improves. The limit of this happens when a piece of infrastructure breaks down or, still worse, is moved. These two cases are not difficult to deal with. For the first, the system does not provide a position: it is annoying, but not catastrophic within the framework of a general public use of comfort, for example. In the second case, it is absolutely necessary to provide an additional system to detect the movement of such an element and to deactivate it in this situation. The effect of providing a bad position when the system is supposed to be very reliable is not acceptable.

Range: We have chosen here the level "proximity," but it is then the functioning of the unitary element of the system. A broad deployment is of course what is conventionally targeted. Such a deployment is quite possible over a very large area, both indoors and outdoors.

Sensitivity to environment: It is more than moderate because if it is true that magnetic obstacles or materials are likely to cause interference and reduce performance, the need to be close reduces the range of these disturbances and "involves" the user who should be able to find parades by himself.

Positioning mode: This is probably the most disturbing element in the world as it is now. Global positioning system (GPS) has completely shaped society's perception of positioning. The latter must be continuous in time and space, as we have already recalled in the first chapters of the book. RFID is nevertheless providing a discontinuous positioning, except in case of a very large deployment or when using very long-range components (which is not the way we mean it in this section). However, it seems clear that the need is absolutely not that, at least in many cases (probably the majority) of mainstream applications. One could even go a step further and say that in fact, what matters is probably being able to get an indication of one's position "on demand," but instantly, which GPS is not able to provide.

7.3 UWB

Within the wireless personal area networks (WPANs), the UWB approach has a specific status, both because it uses a time-based approach, compared to the classical frequency approach, and because its objective is to provide either low data rates (IEEE 802.15.4a standard planned for June 2007) or very high data rates (IEEE 802.15.3a with typically 480 MB s^{-1} in order to achieve a wireless USB link[1]). The interesting feature for positioning is that UWB uses very short pulses. Thus, time is an embedded feature of UWB, unlike other typical radio systems (WiFi, Bluetooth, GSM, UMTS, GPS, etc.). Moreover, the first application of a UWB system was in radar, because of two factors that will also be of great help for positioning: firstly, the capacity of these wide band signals to get past obstacles[2] and secondly because of the sharp time discrimination that is possible to achieve.[3] The UWB approach of positioning is thus based on time measurements from four transmitters ("T_1" to "T_4" in Figure 7.3) and is bound to be very accurate because of the very short pulse duration, typically less than 1 ns.

The basic principle of this is to carry out time measurements: one knows that the intrinsic measurements can be very accurate, but we are left with the problem of global synchronization,[4] which is achieved through a fifth UWB module (referred to as "B" in Figure 7.3). After having made the time measurements, it is possible to calculate either propagation time delays or difference of propagation time delays from a UWB-equipped mobile terminal and the various UWB system modules. Then, a trilateration method is employed, similar to that for GNSS systems, in order to determine the indoor location. Related to accurate

---

1 Called WUSB! Note that another standard, IEEE 802.15.3c, intends to use the 57–64 GHz band.
2 Because of the wide frequency band, there is always only a part of the signal that is disturbed by the obstacle, thus allowing the rest of the signal to pass. It obviously gives better results than classical narrow band signals.
3 Note that the radar has a specificity that does not exist with other positioning systems considered here: the transmitter and the receiver electronics are driven by the same clock!
4 This is similar to the GNSS constellations for which the ground segment is in charge of this global synchronization. The delays between the perfect time and the specific times of each satellite are sent to the receivers through the navigation message.

Figure 7.3 A typical UWB indoor positioning configuration.

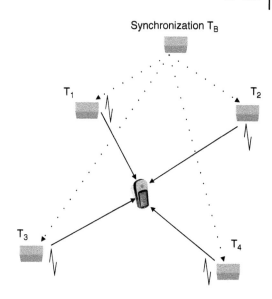

Synchronization T_B

T_1 T_2

T_3 T_4

time measurements, the resulting accuracy of the positioning is quite good: results as accurate as a few centimeters have been reported by different teams.

UWB telecommunication systems are facing some difficulties arising from the jungle of wireless standards. The IEEE decided not to issue a standard yet and to leave the two possibilities (single-band and multiband) open. The bandwidth has also been a subject of discussion: the GPS community is largely the cause of the bandwidth being moved from 1–10 GHz to 3–10 GHz. Because of the very low level of GPS signals, GPS people were afraid of this new disturbing signal coming into the GPS band and put some pressure on the standardization organization in order to move the UWB upward. Unfortunately, others followed and the final band now starts at 3 GHz, with some other potential fears, like, for instance, the WiFi "a" that lies in the 5 GHz band. To try to avoid probable new difficulties, the multiband approach has been proposed: as the official definition of a UWB signal is a signal whose bandwidth is greater than 500 MHz, why not consider channels of, say, 528 MHz and use these channels to fill the entire UWB band? In such a way, 11 bands are available, and this method allows the removal of some bands when needed (for example, to avoid collision with WiFi "a" or in some countries where other bands might not be available).

Unfortunately, this approach put some constraints on the form of the pulse that is bound to reduce the positioning accuracy to a few decimeters rather than a few centimeters for subnanosecond pulses. Furthermore, as with wireless local area network (WLAN) and Bluetooth positioning systems that do not correspond to the telecommunication network, the UWB positioning capabilities are excellent but are not achieved with the power levels allowed in telecommunication systems. If this were the case, the range would be largely reduced, down to a few meters, compared to a few tens of meters using power levels of a few hundred milliwatts.

Let us now come back to our parameter table. The various parameters considered for this technology are given in Chapter 4. However, a few could be discussed. Let us explain those that are quite questionable (in the positive sense of the word) (Table 7.4).

Infrastructure complexity, maturity, and cost: The great difficulty in establishing standards, not in the sense of the norm but in the sense of production and deployment of a technology, makes things complicated and relatively expensive. In fact, it remains very paradoxical compared to current technology, which has an incredible competitive advantage: it uses manufacturing techniques that are those of nanotechnologies. That is, they are very well adapted to very large series, particularly well controlled and capable of producing at very low costs. It is then the volumes and especially the uses that do the rest. In the case of UWB, these are not yet stabilized. However, the potential remains important for the localization field, even if the announced performances are not always there in real environments.

The fundamental aspect of the UWB approach is based on flight time measurement and is thus associated with all the problems related to the synchronization of the various elements of the solution (a detailed discussion on this point is available in Chapter 11 on GNSS systems). In this case, many approaches have been proposed over the past 20 years. Without detailing them all, it is important to understand the philosophy underlying current synchronization techniques, and more broadly flight time measurement. As it is very complex to synchronize all the system components, the idea is based on a clever time difference measurement.

Two-way ranging has been invented to provide the optimum in-flight time measurement (see Figure 7.4).

In such a case, the transmitter T measures the so-called round-trip time t_{rt} including, as shown in Figure 7.4, twice the time of flight and the reply time of

Table 7.4 Summary of the main parameters for UWB.

Infrastructure complexity	Infrastructure maturity	Infrastructure cost	Terminal complexity	Terminal maturity	Terminal cost	Smart phone	Calibration complexity
Low	Development	High	Medium	Integration	Medium	Near future	None

Positioning type	Accuracy	Reliability	Range	Sensitivity to environment	Positioning mode	In/out transition	Calibration needed
Absolute	A few centimeters	Medium	A few rooms	High	Continuous	Easy	None

Figure 7.4 A typical two-way ranging approach.

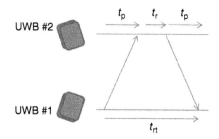

the receiver R. Then, considering the reply time t_r is a known value previously defined by the system, it is possible to obtain the average propagation time (t_p in the Figure 7.4) using Eq. (7.1):

$$t_p = \frac{t_{rt} - t_r}{2} \tag{7.1}$$

This approach eliminates the nonsynchronization of the two devices' clocks. Nevertheless, there is still clock drift in each device. As low-cost oscillators are usually used in such systems, the error induced by the drifts can reach several decimeters quite easily. Thus, a symmetric double-sided two-way ranging protocol has been designed, as described in Figure 7.5.

In such a case, there are indeed two two-way ranging approaches that are measured successively with two round-trip times t_{rt1} and t_{rt2} and two associated reply times t_{r1} and t_{r2}, as described in Figure 7.5. It can be shown that choosing reply times that are larger than the propagation times can lead to a reduction of the impact of the clock drifts, typically by a factor of 2.

Terminal complexity and cost: In this context, the terminal is not very complex but remains dedicated equipment. The cost of the latter, given the techniques used to produce it, is no different from other conventional radio technologies (WiFi, Bluetooth, etc.).

Calibration complexity and need: Indeed, no specific calibration is required, this being achieved automatically by the system protocol mentioned above.

Smartphone and terminal maturity: Even if the vast majority of current smartphones do not include this UWB technology, some manufacturers wishing

Figure 7.5 A typical symmetrical double-sided two-way ranging approach.

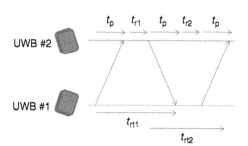

to promote the technology do offer it. The difficulty of integrating new standards into current smartphones is mainly due to two aspects: technically, there is not much free space in the latest smartphones and it is therefore necessary to demonstrate beforehand the interest in terms of uses and therefore new functionalities. It is a difficult and energetic exercise.

Positioning type: The positioning is provided in an absolute manner.

Accuracy: Regarding precision, a discussion is often necessary. This is also true for most technologies (and not only for radio approaches): this is what theory can predict, under almost optimal conditions, and there is the reality on the ground, often cruel. In the case of UWB, many theoretical aspects are going in the right direction: very wide frequency band, hence potential excellent temporal resolution, initial use of the technology to realize radar, ability to cross obstacles (because of the very wide band), performance in rejection of multiple paths because of the impulse form of signals, etc. In reality, the limitations of the authorized power levels, clock drifts, pulse deformation, etc., reduce, as always, the expected theoretical performances. However, many works have made it possible to approach these performances.

Reliability: The propagation of radio signals cannot be predicted with certainty, even if relatively traditional dual transmission techniques today make it possible to evaluate the propagation channel used in near-real time. Moreover, the calculation of a position requires several measurements that are carried out under sometimes very different conditions, which has an impact on reliability.

Range: Here again not totally obvious. In principle, the signals have the ability to penetrate walls (hence the level "a few rooms" retained). In reality, the power levels currently used do not really allow such penetration, except in structures transparent to electromagnetic waves in these frequency bands.

Sensitivity to environment: Linked to propagation aspects, as with all radio systems (at least), the sensitivity to obstacles, people around, walls, etc., is quite high.

Bibliography

1 Bahl, P. and Padmanabhan, V.N. (2000). RADAR: an in-building RF-based user location and tracking system. In: *Proceeding IEEE INFOCOM 2000*, 775–784. IEEE.

2 Fontana, R.J. (2004). Recent system applications of short-pulse ultra-wideband (UWB) technology. *IEEE Transactions on Microwave Theory and Techniques* 52 (9): 2087–2104.

3 Fontana, R.J. and Gunderson, S.J. (2002). Ultra-wideBand precision asset location system. In: *IEEE Conference on UWB Systems and Technologies*. IEEE.

4 Frazer, E. (2003). Indoor positioning using ultrawideband techniques – analysis and experimental results. In: *11th IAIN World Congress*, Berlin, Germany. German Institute of Navigation (DGON).

5 Gezici, S., Tian, Z., Giannakis, G.B. et al. (2005). Localization via ultra-wideband radios – a look at positioning aspects of future sensor networks. *IEEE Signal Processing Magazine* 22: 70–84.

6 Ni, L.M., Liu, Y., Lau, Y.C., and Patil, A.P. (2003). LANDMARC: indoor location sensing using active RFID. In: *Proceedings of the First IEEE International Conference on Pervasive Computing and Communications, 2003. (PerCom 2003).*, Fort Worth, TX, 407–415. IEEE.

7 Zhang, C., Kuhn, M.J., Merkl, B.C. et al. (2010). Real-time noncoherent UWB positioning radar with millimeter range accuracy: theory and experiment. *IEEE Transactions on Microwave Theory and Techniques* 58 (1): 9–20.

8 DiGiampaolo, E. and Martinelli, F. (2014). Mobile robot localization using the phase of passive UHF RFID signals. *IEEE Transactions on Industrial Electronics* 61 (1): 365–376.

9 Wang, G., Gu, C., Inoue, T., and Li, C. (2014). A hybrid FMCW-interferometry radar for indoor precise positioning and versatile life activity monitoring. *IEEE Transactions on Microwave Theory and Techniques* 62 (11): 2812–2822.

10 Zhang, C., Kuhn, M., Merkl, B. et al. (2006). Development of an UWB indoor 3D positioning radar with millimeter accuracy. In: *2006 IEEE MTT-S International Microwave Symposium Digest*, San Francisco, CA, 106–109. IEEE.

11 Errington, A.F.C., Daku, B.L.F., and Prugger, A.F. (2010). Initial position estimation using RFID tags: a least-squares approach. *IEEE Transactions on Instrumentation and Measurement* 59 (11): 2863–2869.

12 Waldmann, B., Weigel, R., and Gulden, P. (2008). Method for high precision local positioning radar using an ultra wideband technique. In: *2008 IEEE MTT-S International Microwave Symposium Digest*, Atlanta, GA, USA, 117–120. IEEE.

13 Silva, B., Pang, Z., Åkerberg, J. et al. (2014). Experimental study of UWB-based high precision localization for industrial applications. In: *2014 IEEE International Conference on Ultra-WideBand (ICUWB)*, Paris, 280–285. IEEE.

14 Chattopadhyay, A. and Harish, A.R. (2008). Analysis of low range indoor location tracking techniques using passive UHF RFID tags. In: *2008 IEEE Radio and Wireless Symposium*, Orlando, FL, 351–354. IEEE.

15 Wang, G., Gu, C., Inoue, T., and Li, C. (2013). Hybrid FMCW-interferometry radar system in the 5.8 GHz ISM band for indoor precise position and motion detection. In: *2013 IEEE MTT-S International Microwave Symposium Digest (MTT)*, Seattle, WA, 1–4. IEEE.

16 Wehrli, S., Gierlich, R., Huttner, J. et al. (2010). Integrated active pulsed reflector for an indoor local positioning system. *IEEE Transactions on Microwave Theory and Techniques* 58 (2): 267–276.

17 Gierlich, R., Huttner, J., Ziroff, A. et al. (2011). A reconfigurable MIMO system for high-precision FMCW local positioning. *IEEE Transactions on Microwave Theory and Techniques* 59 (12): 3228–3238.

18 Luo, R.C., Chuang, C., and Huang, S. (2007). RFID-based indoor antenna localization system using passive tag and variable RF-attenuation. In: *IECON 2007 – 33rd Annual Conference of the IEEE Industrial Electronics Society*, Taipei, 2254–2259. IEEE.

19 Ebelt, R., Hamidian, A., Shmakov, D. et al. (2014). Cooperative indoor localization using 24-GHz CMOS radar transceivers. *IEEE Transactions on Microwave Theory and Techniques* 62 (9): 2193–2203.

20 Waldmann, B., Weigel, R., Gulden, P., and Vossiek, M. (2008). Pulsed frequency modulation techniques for high-precision ultra wideband ranging and positioning. In: *2008 IEEE International Conference on Ultra-Wideband*, Hannover, 133–136. IEEE.

21 Dardari, D., Conti, A., Ferner, U. et al. (2009). Ranging with ultrawide bandwidth signals in multipath environments. *Proceedings of the IEEE* 97 (2): 404–426.

22 Marano, S., Gifford, W.M., Wymeersch, H., and Win, M.Z. (2010). NLOS identification and mitigation for localization based on UWB experimental data. *IEEE Journal on Selected Areas in Communications* 28 (7): 1026–1035.

23 Jourdan, D.B., Dardari, D., and Win, M.Z. (2008). Position error bound for UWB localization in dense cluttered environments. *IEEE Transactions on Aerospace and Electronic Systems* 44 (2): 613–628.

24 Guvenc, I., Chong, C., and Watanabe, F. (2007). NLOS identification and mitigation for UWB localization systems. In: *2007 IEEE Wireless Communications and Networking Conference*, Kowloon, 1571–1576. IEEE.

25 Zetik, R., Sachs, J., and Thoma, R.S. (2007). UWB short-range radar sensing – the architecture of a baseband, pseudo-noise UWB radar sensor. *IEEE Instrumentation & Measurement Magazine* 10 (2): 39–45.

26 Wang, C. and Chen, C. (2014). RFID-based and Kinect-based indoor positioning system. In: *2014 4th International Conference on Wireless Communications, Vehicular Technology, Information Theory and Aerospace & Electronic Systems (VITAE)*, Aalborg, 1–4. IEEE.

27 Gharat, V., Colin, E., Baudoin, G., and Richard, D. (2017). Indoor performance analysis of LF-RFID based positioning system: comparison with UHF-RFID and UWB. In: *2017 International Conference on Indoor Positioning and Indoor Navigation (IPIN)*, Sapporo, 1–8. IEEE.

8

Building Range Technologies

Abstract

As said in the previous chapter, the boundary between a few rooms and "building" is not so clear. Indeed, it depends mainly on the deployment considered and the performances one wants to obtain in terms of accuracy and reliability. The differentiation made in this book is based on two aspects: the theoretical positioning accuracy achievable (typically better for the technologies described in Chapter 7) and the methodologies that allow, for the optical-based technologies of this chapter, a whole building to be covered with a single device. Note that a specific chapter (Chapter 9) is devoted to indoor GNSS and thus is not dealt with here (bold in Table 8.1).

Keywords *Building range; SLAM; Inertial; Light based approaches; WLAN*

The classification described in Chapter 4 led to the following table concerning these technologies (Table 8.1).

8.1 Accelerometer

As already described in Chapter 3, acceleration is the rate of change of velocity. Although it is possible to use accelerometers in order to obtain the relative distance traveled, this is sometimes not a very efficient way because errors and biases are going to accumulate. When possible, in an automobile system, for instance, it is preferable to use direct displacement measurement through the use of odometers. In a car, this is simply a sensor that can count the rotation of the wheels and convert the value into a linear distance. For indoor purposes, the same idea applies for rolling objects. For pedestrians, it is different because it is not convenient to use wheels in any way. Then, accelerometers can be used to count the number of footsteps and then convert them into a distance.

Indoor Positioning: Technologies and Performance, First Edition. Nel Samama.
© 2019 The Institute of Electrical and Electronics Engineers, Inc. Published 2019 by John Wiley & Sons, Inc.

Table 8.1 Main "building" technologies.

Technology	Positioning type	Accuracy	Reliability	Range	Sensitivity to environment	Calibration needed	Positioning mode	Technique	Signal processing	Position calculation
Accelerometer	Relative	$f(t)$	Medium	Block	No impact	Often	Continuous	Physical	Detection	Math functions $(\int, \int\int, \int\int\int, \ldots)$
BLE	Absolute	A few meters	Medium	Building	High	Several times	Almost continuous	Physical	Pattern matching	Math functions $(\int, \int\int, \int\int\int, \ldots)$
Gyrometer	Relative	$f(t)$	Medium	Building	no impact	Often	Continuous	Physical	Detection	Math functions $(\int, \int\int, \int\int\int, \ldots)$
Image-relative displacement	Relative	<1 m	Medium	Building	High	Several times	Almost continuous	Image(s)	A combination of	Math functions $(\int, \int\int, \int\int\int, \ldots)$
Image SLAM	Relative	<1 m	Medium	Building	High	Several times	Almost continuous	Image(s)	A combination of	Math functions $(\int, \int\int, \int\int\int, \ldots)$
Indoor GNSS	Absolute	A few decimeters	Medium	Building	High	None	Continuous	Phase(s)	Correlation	∩ Hyperbolae
LiFi	Symbolic	A few meters	Low	Building	Very high	None	Almost continuous	Physical	Detection	Spot location
Light opp	Relative	100 m	Low	Building	Very high	Often	Almost continuous	Physical	Classification	Zone determination
Sound	Relative	>100 m	Medium	Building	High	Often	Continuous	Physical	Detection	∩ Circles
Theodolites	Absolute	A few centimeters	Very high	Building	Very high	Once	Continuous	Angle(s)	A combination of	∩ Plans + distance(s)
WiFi	Absolute	A few meters	Medium	Building	High	Several times	Continuous	Physical	Pattern matching	Math functions $(\int, \int\int, \int\int\int, \ldots)$
WLAN symbolic	Symbolic	Decameter	Very high	Building	Low	None	Continuous	Physical	Propagation modeling	Zone determination

SLAM, simultaneous localization and mapping.
Only the bold lines are treated in this chapter.

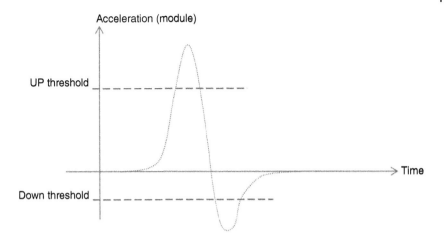

Figure 8.1 A typical acceleration response of a pedestrian's footstep.

Figure 8.1 gives the typical form of the acceleration when a pedestrian is walking. This form greatly depends on the location of the accelerometer on the body: the foot is the best place, the hand probably the worst, and the hip is somehow a good compromise. Note that we are using the module of the acceleration, i.e. the square root of the sum of the squares of the accelerations on each axis. In current smartphones, an accelerometer is indeed composed of three individual accelerometers, each one associated with one axis of the terminal. The axes are associated with the screen of the smartphone and measurements are then completely relative, i.e. relative to the smartphone but also relative to its posture. By detecting that the module is successively crossing the up threshold and the down threshold (see Figure 8.1) within a given interval of time (not too long in order to correspond to a typical step), it is possible to consider that a new footstep has been achieved. A step counter has thus been designed. In order to define the distance traveled, one has to know the length of the step. Different approaches are possible: either you use the acceleration values and double integrate them (over a short time interval, thus leading potentially to reduced errors, note that this is not so sure) or you can assign to a step a value that would depend on your size, for example (this is the usual way).

Let us now come back to our parameter table. The various parameters considered for this technology are given in Chapter 4 (Table 8.2).

Infrastructure complexity, maturity, and cost: There is no need for any infrastructure, and this is the major advantage of an inertial system. Another way to say it would be to consider that the infrastructure is inherent to the earth, thus available everywhere and at any time.

Terminal complexity, maturity, and cost: Inertial personal devices for pedestrians are complex to implement, in comparison with those for automobiles,

Table 8.2 Summary of the main parameters for accelerometers.

Infrastructure complexity	Infrastructure maturity	Infrastructure cost	Terminal complexity	Terminal maturity	Terminal cost	Smart-phone	Calibration complexity
None	None	Zero	Medium	Hardware development	Medium	Existing	Medium

Positioning type	Accuracy	Reliability	Range	Sensitivity to environment	Positioning mode	In/out transition	Calibration needed
Relative	$f(t)$	Medium	Block	No impact	Continuous	Already exist	Often

mainly because of the much larger range of physical measurements required and the nonconstant attitude of the mobile terminal.[1] For instance, let us imagine a mobile phone equipped with a global navigation satellite system (GNSS) receiver and also with an inertial system in order to allow dead reckoning when GNSS signals are no longer available (indoors, for instance). As a mobile terminal, the phone is subject to many small but violent movements such as rotation, hand shaking, or even falls. All these movements must be analyzed by the inertial system and should not lead to the accumulation of errors. Unfortunately, as the basic principle is integration, all the errors are added up: as these kinds of motions are frequent, the resulting error can be significant. Moreover, the other types of movements that are bound to be difficult for pedestrian mobile terminals are also very slow and hesitant. For example, when one is moving from one leg to the other very gently, the resulting signals are almost identical whether there is a real displacement or not. In this second case, the errors are not the major concern, but the interpretation of the motion is. Thus, although some remarkable implementations have been achieved, the use of inertial systems for pedestrian mobile terminals has mainly been deployed for step counting.

Calibration complexity and need: Calibration is absolutely compulsory with accelerometers. Its complexity greatly depends on the performance one is expecting for. In the case of a step counter, the two required calibrations are, respectively, the one concerning the thresholds (see Figure 8.1) and one that concerns the length of a step. The first one can probably be achieved once and for all, even for all types of accelerometers, as rather a large error margin (in the values of the thresholds) can be acceptable. For the second one, an empirical link between morphological characteristics of a person and her or his step length can also leads to really good results. When one copes with distance evaluation from a double integration process, calibration

---

1 Within the car, the horizontal plane should roughly remain unchanged.

is much more complex and should happen regularly. The main problem to be addressed is the drift of the distance estimated with respect to the accumulation of noise (which is inherent to the integration). A current operation mode consists in estimating the drift by analyzing the difference between a measured and a reference one. The difficulty lies in the fact that the drift is not always constant over time and that it is not so easy to have reference distances around. Thus, real implementations are always constraints in the deployment.

Positioning type: It is "highly" relative. As explained in the introduction, calculations are inherently relative to an initial state. In the case of current smartphones, this relativity is enhanced by the fact that the accelerometers are fixed within the terminal. In such a case, measurements are relative to the posture of the terminal, as well as with its initial state. This is a sort of "double relativity."

Accuracy: Quite good when counting steps and depending on time when estimating distances by the integration process. Note that the user (and usage) requirement is also not comparable. Accuracy is searched for in the second case, whereas a small percentage of error is quite acceptable in the first one.

Reliability: It is not so good because of the useless, but noisy, movements the accelerometer is undergoing and that are not the movements it wants to estimate. Nevertheless, the measurements themselves are relatively reliable. This is the reason that works are being carried out mainly in order to reduce the influence of the noise.

Range: It is usually accepted to say that the best smartphone-based accelerometers are capable of limiting the position drift to 1 m min^{-1} (integration process). This leads to a range at the "building" level considering that a person spends a few tens of minutes in the place. It should be re-estimated depending on the real usage.

Sensitivity to environment: There is no impact, unless one is close to a black hole (which is not so often in fact).

Positioning mode: It is completely continuous. In fact, the time continuity is also compulsory. If the measure is stopped, a new process starts with an initial reference being the last state of the previous measurement.

In/out transition: As the environment has no impact, the transition is seamless.

8.2 Bluetooth and Bluetooth Low Energy

Within the WPAN (wireless personal area network) domain, Bluetooth was originally designed in order to get rid of cables in the near vicinity of a terminal (namely a computer). Thus, the ranges afforded by Bluetooth systems were rather small, typically 10 m. As a matter of fact, three "classes" exist, depending on the power level of the signal transmitted, with corresponding ranges of a few meters, a few tens of meters, and finally 100 m. The main advantage of Bluetooth compared to radio frequency identification (RFID), for instance, is the longer range and the radio effect that allows propagation through walls. However, if the same cell identification method is used, then the resulting accuracy is also weakened. New techniques have been imagined when a few base stations are available at a given point. Then, the power level-based method can be applied in order to achieve indoor systems with an accuracy of a few meters. The major drawbacks are the need for a large number of Bluetooth modules to cover any given place and the relative complexity to fill the database required to implement a "nearest neighbor" algorithm. The database needs to be modified whenever an indoor element, such as a desk or a cupboard,[2] is moved. As this "initiation step" is really time-consuming, a lot of effort has been made in order to find a method so that the database filled in automatically with permanent continuous measurements, as people are connecting to the Bluetooth network (or WiFi as long as the problem is absolutely identical). In order to overcome the time needed for the database and the uncertainty introduced by moving objects in the environment, current techniques are implementing a sort of "real-time monitoring" of the propagation conditions. In fact, the idea consists in adding either "sniffers" or additional receivers and to use them as calibration components. The propagation (or the database) is updated depending on the power values received at these specific locations. It needs to have a fine understanding of the main propagation parameters in order to define adequate sniffer locations.

With the advent of Bluetooth 4.0 (also named Bluetooth Low Energy or BLE), things were largely simplified. Indeed, previous versions presented a difficulty in implementing the above-mentioned approaches. The time required to establish the communication between two Bluetooth modules was rather long and one had to wait in order to achieve a power measurement with an acceptable stability. BLE changed this aspect drastically.[3]

Concerning the possible application and deployment scheme, a lot of proposals have already been made by many authors. A typical example is taken from the Osaka Guide concept, which is a Bluetooth-based positioning system for museum visits. Figure 8.2 shows the way it is proposed to work with a Cell-Id triangulation approach (AP stands for Access Point).

---

2 Note also that even normal daily operations are likely to change the environment, such as opening windows or closing doors. The fact of taking these actions into account leads to including error margins on the power levels and, as a direct impact, dramatically decreases the resulting positioning accuracy.

3 Note another "practical" point: indoor positioning applications using WiFi networks (see the last sections of this chapter for details) were not possible with Apple products as the corresponding measurements were not available. Although WiFi presented some advantages at that time, Bluetooth was preferred for this reason also.

Figure 8.2 A typical indoor Bluetooth-based deployment.

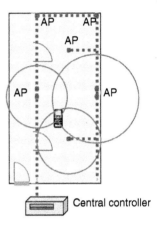

Central controller

As already discussed, the required network is larger than the telecommunication network, in that the number of access points must be greatly increased. One could argue that telecommunication needs are bound to increase, leading to a natural increase in the number of access points and hence fill the gap described above, but we also know that if increased rate capabilities are really required, manufacturers, helped by technicians, will be able to find solutions with a reduced set of access points in order to stay within acceptable costs. Then, the compromise is, as usual, difficult to be optimized.

Let us now come back to our parameter table. The various parameters considered for this technology are given in Chapter 4. However, a few could be discussed. Let us explain those that are quite questionable (in the positive sense of the word) (Table 8.3).

Infrastructure complexity, maturity, and cost: The infrastructure consists in deploying some transmitters throughout a building (note that it can also be

Table 8.3 Summary of the main parameters for Bluetooth.

Infrastructure complexity	Infrastructure maturity	Infrastructure cost	Terminal complexity	Terminal maturity	Terminal cost	Smart-phone	Calibration complexity
Low	Existing	Low	None	Software development	Zero	Existing	Medium

Positioning type	Accuracy	Reliability	Range	Sensitivity to environment	Positioning mode	In/out transition	Calibration needed
Absolute	A few meters	Medium	Building	High	Almost continuous	Easy	Several times

used outdoors, although it presents a limited interest). The current approach is based on autonomous modules with a battery as power supply (their lifetime in real operational conditions is estimated to be several years). The main advantages are the ease of placement of the modules and the reduced cost of installation: one only needs to "glue" them. In order to achieve a few meters of accuracy (down to 1 m reported in certain cases), the number of modules should be rather high, say typically about one every 10 linear meters. Current cost (2018) of such a module is in the range of $20 and the infrastructure, even if a large number of modules should be deployed, is thus not so expensive.

Terminal complexity, maturity, and cost: Almost all the current smartphones include Bluetooth 4.0 and the terminal is always the smartphone. In some cases (visioguides in museums, for instance), specific devices have been designed, but the smartphone is the ideal one.

Calibration complexity and need: Calibration can be seen in different manners for BLE. There is often the need to know the location of the transmitting module, but the accuracy of it depends on the positioning accuracy being looked for. In the case of an adaptation of the Cell-Id concept, when you only want to know roughly where you are, the corresponding calibration is light. It consists only in positioning the module on a map and drawing a rough radio coverage area around it (this can be done in many different ways but is not a challenge). When a better accuracy is wanted, then the typical fingerprint approach is usually implemented, requiring a more complex and time-consuming calibration, as explained in the introduction of this paragraph;

Positioning type: It is linked to the way the locations of the transmitting modules have been defined, in an absolute or relative reference frame. One could easily imagine an "inverted" approach where the infrastructure consists in receiving the signals transmitted by the smartphones, but this is not the current mode.

Accuracy: Often not as good as said, it can reach a few meters. In order to go below this value, the problem of reliability arises. It greatly depends on the density of deployed modules: to a certain extent, one can consider the accuracy is better when the number of modules increases, but this has a limit where too many modules lead to conflicts in the determination of the location when considering different sets of modules (mainly due to measurement errors). Usually, the compromise is obtained "experimentally" by deploying a network of modules and analyzing the positioning performance of the system. In case there is a specific need in some specific locations, new modules are deployed. Prediction tools exist, but the real complexity of indoor environments requires a fine description of the environment (materials of walls, size of windows, etc.), which is too high with respect to the corresponding gain (in time and real help in the deployment).

Reliability: BLE is an interesting technology for a discussion about reliability. When a Cell-Id approach is implemented, the reliability can be excellent indeed, as long as one takes sufficient margins concerning the covered area of the radio modules. The problem is that usually the comparison with global positioning system (GPS) leads to a competition in terms of accuracy. Often accuracies of 1–2 m are reported, leading to a very high dependency of the performance to the environmental conditions, and thus in turn drastically decreasing the reliability of the positioning.

Range: Current modules can be detected over several floor levels and, on the same floor, to a few tens of meters. In free space (but this is of no interest to us), several hundreds of meters can be reached.

Sensitivity to environment: The same discussion as for reliability applies. It is at an acceptable level for the Cell-Id approach and increases rapidly when one wants to reach 1 m of accuracy.

In/out transition: Quite easy to achieve, but the interest is reduced to very specific cases when GPS is indeed influenced by many reflected paths leading to really poor performances. This can be the case in old city centers with narrow streets or in modern city centers or business centers with high buildings around.

8.3   Gyrometer

As stated in Chapter 3, the output of a gyroscope is the rate of change of the angle of its axis. The principle of the measurements is based on the gyroscopic effect (the one that helps you to keep equilibrium when riding a bicycle). This is used in order to measure the relative movement of the terminal or the variation of the direction of the mobile terminal. However, the drift of gyroscopes needs to be compensated for regularly and that is the main difficulty in using this sensor efficiently.

Let us come back to our parameter table. The various parameters considered for this technology are given in Chapter 4. However, a few could be discussed. Let us explain those that are quite questionable (in the positive sense of the word; Table 8.4).

Infrastructure complexity, maturity, and cost: This is the main advantage of inertial sensors and the reason they are seen as ideal candidates for fusion with radio or optical systems. There is no infrastructure required and measurements are completely local and attached to the terminal. The main drawback is then that the measurements cannot lead to an absolute positioning but are only relative to an initial position. In turn, this makes it necessary to have an external aid (for this initial location) if one wants an absolute value.

Terminal complexity, maturity, and cost: Many different levels of performance are achievable, depending on the complexity and physical realization of the

Table 8.4 Summary of the main parameters for gyrometer.

Infrastructure complexity	Infrastructure maturity	Infrastructure cost	Terminal complexity	Terminal maturity	Terminal cost	Smart-phone	Calibration complexity
None	None	Zero	Medium	Hardware development	Medium	Existing	Medium

Positioning type	Accuracy	Reliability	Range	Sensitivity to environment	Positioning mode	In/out transition	Calibration needed
Relative	$f(t)$	Medium	Building	No impact	Continuous	Already exist	Often

gyrometer. Thus, prices can range from a few dollars to thousands of dollars. The current ones included in smartphones are based on MEMS (microelectromechanical systems), very low cost electronic integrated systems. Thus, integration is easy as such systems are indeed integrated chips. The ones included in current smartphones, unfortunately, are not of sufficient quality in order to fully solve the continuity problem of the positioning.[4]

Calibration complexity and need: This is the major difficulty linked to the inertial systems in general. In the case of a gyro, the calibration is of medium complexity but requires updating quite often as the various drifts involved are not constant in time. In addition, such a calibration requires reference to a known value of the rate of angle variation, which is not so easy to obtain. The current approaches are based on conditions that make this variation zero. It can be the case when a foot is on the ground and where the assumption is made that it cannot move.

Smartphone: The majority of current smartphones are already equipped as gyros usually come with the standard "inertial sensors set."

Positioning type: As already stated, it cannot provide the user with an absolute positioning. For example, in case you shut down your terminal and turn it on a little bit later, the system is not able to give you a position (except in the case where you know you have not changed the position).

Accuracy: It is indeed time dependent and also related to the quality of the calibration and its updating. The use of only gyros is never implemented as the indication of the distance traveled is not available. Thus, gyros are not considered as a positioning means, but a useful complementary technology.

Reliability: It depends both on the quality of the sensor and that of the calibration. It also depends on the way it is integrated within the positioning system. With current smartphones, one cannot consider it is a good measurement

---

4 Note that it is a chance for the author as this book would not have had any sense or interest.

sensor. Nevertheless, it can help to choose between classes of movement of the terminal.

Range: Another great advantage of inertial systems is their independence with respect to infrastructure. Of course, depending on the technology used, some environmental conditions can have a disturbing effect, but once correctly used, they are independent of the environment. Thus, the range is indeed unlimited. The reason we decided to use the "building" level here concerns in fact the accuracy and reliability dependency of the measurements with time. If one considers staying within a building for a few tens of minutes, the drifts will be high enough to endanger the efficiency of the positioning.

Sensitivity to environment: As already said, there is no impact.

Positioning mode: It works in a continuous mode.

In/out transition: There is no impact on whether indoors or outdoors.

8.4 Image-Relative Displacement

The relative displacement positioning then consists in "following" in successive images the same characteristic form (or forms). For instance, imagine you have identified a door in the first image and let us consider you also know about the orientation and the relative displacement (from image 1 to the next one) of your camera. Then, you will be able to start to "reconstruct" your trajectory (considering you have the information of the relative positioning and orientation between the camera and you). Well, this does not seem to be so simple and in fact, this is not! The reason is that one needs to know a few practical aspects (the ones cited above) and to accept a non-negligible calculation burden. In addition, as usual for optical-based technologies, masking is probably the most challenging problem, although in case a large number of successive images are taken into consideration, then the presence of obstacles could be eliminated (at the cost of an even increased computation complexity).

Let us now come back to our parameter table. The various parameters considered for this technology are given in Chapter 4. However, a few could be discussed. Let us explain those that are quite questionable (in the positive sense of the word; Table 8.5).

Infrastructure complexity, maturity, and cost: No infrastructure is required. This is an analogous advantage as with inertial systems. As the measurements and the calculations are carried out at the terminal level in a "passive" way (i.e. without the need for specific signals to be sent), there is no need for any infrastructure. This is clearly a great advantage for a positioning system.

Terminal complexity, maturity, and cost: Things are a little bit more complex when it comes to the terminal. Of course, cameras are available on almost all mobile terminals, when needed. The cost of such components is now at

Table 8.5 Summary of the main parameters for image-relative displacement.

Infrastructure complexity	Infrastructure maturity	Infrastructure cost	Terminal complexity	Terminal maturity	Terminal cost	Smart-phone	Calibration complexity
None	None	Zero	Low	Software development	Zero	Existing	Medium

Positioning type	Accuracy	Reliability	Range	Sensitivity to environment	Positioning mode	In/out transition	Calibration needed
Relative	<1 m	Medium	Building	High	Almost continuous	Moderate	Several times

a minimum value and this is not even a subject to deal with. In terms of complexity, the comparison with GPS makes sense: the hardware is a piece of highly integrated electronics of very high level, but as the cost is very low, one usually considers it is standard. Finally, the complexity of the terminal makes the difference: software developments are required in order to allow current terminals to achieve such image technologies. Some current devices are able to carry out such a task, but they are the top ones with optimized algorithms.

Calibration complexity and need: There is no need for calibration in the sense meant until now, i.e. a specific measurement phase dedicated to the calibration. Nevertheless, there is usually the need to estimate some camera parameters, such as the focal length of the lens or the zoom value. Of course, these parameters are quite important in case one wants to evaluate a real distance from an image measurement.

Smartphone: Already available on smartphones, cameras are not a problem. This is the way this technology could be used with smartphones, although their computing capabilities might be limiting.

Positioning type: Unless some recognizable signs are visible in the image and associated with absolute coordinates in a standardized reference frame, the positioning with images is relative to the environment.

Accuracy: It could reach very good values thanks to the repetition of images and the high number of corresponding forms from image to image.

Reliability: It is at a rather high level indeed once a positioning is provided, but masking is subject to reducing the efficiency of the positioning, hence the "medium" level used here.

Range: The "building" level is clearly not appropriate to the camera because light cannot pass through walls. Nevertheless, current approaches allow the positioning system to work continuously from room to room throughout a building, hence the level considered.

Sensitivity to environment: Very high, as already stated.

Positioning mode: It is almost continuous as the transitions from room to room could be a problem because the number of remarkable forms could be reduced and potentially not sufficient in order to provide the required continuity.

In/out transition: It requires a significant increase in the computational burden as fundamental forms are of different natures.

8.5 Image SLAM

The principle is identical to the "image-relative displacement" technology described above. However, this time, at the same time as the positioning is done, we also try to create the cartography of the places. The successive images must then be close to each other in order to allow sufficient overlap. This should make the construction of the cartography more reliable.

Let us now come back to our parameter table. The various parameters considered for this technology are given in Chapter 4.

Table 8.6 is identical to Table 8.5 and the same comments apply. Note that the comments mainly concern the positioning part, and here, we do not discuss the mapping process.

8.6 LiFi

LiFi (for Light Fidelity) is the equivalent of WiFi for light waves. The principles are still the same, but it is now a question of using visible light (we have already seen in Chapter 6 approaches based on the use of infrared, for example). In particular, the rapid development of LED-based lighting (light-emitting diodes) allows signals to be transmitted at the same time as light (the latter will in fact

Table 8.6 Summary of the main parameters for image SLAM.

Infrastructure complexity	Infrastructure maturity	Infrastructure cost	Terminal complexity	Terminal maturity	Terminal cost	Smart-phone	Calibration complexity
None	None	Zero	Low	Software development	Zero	Existing	Medium

Positioning type	Accuracy	Reliability	Range	Sensitivity to environment	Positioning mode	In/out transition	Calibration needed
Relative	<1 m	Medium	Building	High	Almost continuous	Moderate	Several times

"modulate" the frequency of the light signals). Thus, we have the possibility to use the lights as wireless signal transmitters.[5] The field of telecommunications is of course the first targeted, but why not also offer a "positioning" approach? This is what underlies the LiFi. As always, it will be necessary to have a "map" of the light fixtures installed, each being identified with a specific address. Let us note then that it will be necessary that these lights are in operation, i.e. switched on, so that the transmission is effective. Note that this is absolutely identical in the case of WiFi (the only difference being that in this last case, this is not "visible").

The basic principle then consists in demodulating the received light in order to extract the transmitted information. This can be done in various ways, more or less simple today to implement on a smartphone. The first approach is simply to use the camera that is capable of detecting the modulation. However, performance remains limited. A second approach consists in using a dedicated terminal (or an adapter installed on the smartphone, if necessary).

One of the first applications developed with LiFi was a guidance system in buildings. The range is sufficient (typically between low-power Bluetooth and WiFi) to allow relatively efficient Cell-Id approaches.

Let us now come back to our parameter table. The various parameters considered for this technology are given in Chapter 4 (Table 8.7).

Infrastructure complexity, maturity, and cost: Although some commercial products exist, the technology is a new one, and applications, services, and business are not completely stabilized. Thus, the need to change the light bulb is a constraint. We know that the LED technology is meant to be deployed at a large scale and one can consider it will be "naturally" available

Table 8.7 Summary of the main parameters for LiFi.

Infrastructure complexity	Infrastructure maturity	Infrastructure cost	Terminal complexity	Terminal maturity	Terminal cost	Smart-phone	Calibration complexity
Medium	Development	Medium	Medium	Integration	Medium	Near future	None

Positioning type	Accuracy	Reliability	Range	Sensitivity to environment	Positioning mode	In/out transition	Calibration needed
Symbolic	A few meters	Low	Room	Very high	Almost continuous	Easy	None

---

5 Note that currently, the communication is unidirectional, in a descending way. This makes quite a difference with WiFi.

almost everywhere in the next few years. Nevertheless, this is a preliminary requirement (and such a preliminary is often a drawback for a new technology). Thus, the infrastructure is not so complex, but its maturity is not yet at the level of other wireless technologies, such as WiFi, for example. Concerning the cost, the interesting point is that changing light bulbs to LEDs is bound to provide a rapid and direct investment return because the power consumption will be reduced immediately, thus providing the LiFi technology with a real market advantage. For those having already jumped into the LED technology, in certain cases, the cost will be reduced when LEDs are compatible with the required modulators, but the compatibility is not general today.

Terminal complexity, maturity, and cost: The main difference is the use of a standard camera for the signal detection or the need for a specific device. It can be predicted that, depending on the real usefulness of the technology, integration within current terminal should not be a problem.

Calibration complexity and need: In case a Cell-Id-like approach is implemented, there is no real need for calibration except the compulsory need to associate the identifier of each LED with a position on a map (the "map" problem is dealt with in Chapter 13). In the case where flight times are measured in order to calculate a more accurate position, the need for positioning the LEDs is then essential.

Smartphone: Technically speaking, there is no problem using the camera of the smartphone to achieve LiFi positioning. The difficulty lies indeed in the way one is using the device. The propagation constraints are higher than those for radio wireless system as nonline of sight is unsolvable. Considering there are no walls (or other obstacles) between the bulb and the terminal, the user can also be considered as an obstacle (note that it is similar for radio signals, but not at the same level). These comments are also highly related to the positioning technique envisaged. In case of a Cell-Id approach, the problem is largely reduced as a nondetected bulb will probably be replaced by another one. The accuracy will be potentially reduced, but the positioning will still be available with probably an acceptable reliability. In case one wants to implement a time-of-flight approach, it is much more of a problem and the final performance of the system will rely greatly on an increased complexity of the infrastructure.

Positioning type: It can be absolute or relative, depending on the way the location of the bulbs is considered. Note also that this technology is sometimes used outdoors where it is relatively simple to obtain GNSS-compatible coordinates for an absolute positioning in continuity with satellite navigation systems. Nevertheless, the most common positioning approach is based on Cell-Id, which provides the user with a symbolic value, i.e. an area of highly probable presence.

Accuracy: As usual, theory and practice do not always correspond. Propagation matters are mainly responsible for this.

Reliability: It can be excellent if accuracy is not the only parameter to optimize.

Range: It is currently commonly accepted to consider a range lying between that of Bluetooth (when using a low power) and WiFi, i.e. typically greater than 10 m, to a few tens of meters at a maximum.

Sensitivity to environment: As with all optical-based technologies, the sensitivity to the environment is very high. A solution can be to increase the level of complexity of the infrastructure, but this has a corresponding cost. In all cases, the environment and the way the positioning system should be used have a dramatic impact.

Positioning mode: There is no problem for a perfect continuity in the positioning. Of course, it means there are LED bulbs everywhere.

In/out transition: No problem, as long as the lights are "on." Although indoors, this is often the case, which is not the same outdoors. The question is then: Are we ready to accept leaving the lights on during the day for LiFi (whatever the application is envisaged)? Not sure the answer is "yes," although once again this is the case for radio networks that are "on" 24 hours a day.

8.7 Light Opportunity

As the problem of indoor positioning is a real challenge, many original solutions have been investigated. Among others, the measurement of the level of the surrounding light level is an unusual way of applying fingerprinting. This technique relies on the observation, under certain conditions, of the variation of a given parameter (here the light levels). Following a calibration phase, where measurements are carried out throughout the whole place, a database is established. Then, an instantaneous light-level measurement allows pattern matching recognition in the database and possibly location determination. This general approach can easily be extended to a lot of physical parameters, such as the temperature or radio power levels, either locally generated (wireless local area network [WLAN]) or regionally generated (TV signals, for example). The efficiency of this approach highly depends on the physical quantity. In the case of light, a few external parameters are bound to have a significant role: close to the windows, outside luminosity is a major one. Even in the case of an approach, only based on the measurement of the variation of luminosity, the calibrations are not so easy to carry out. For example, day and night conditions are extremely different and must be taken into account.

Let us now come back to our parameter table. The various parameters considered for this technology are given in Chapter 4 (Table 8.8).

Infrastructure complexity, maturity, and cost: By definition of the technology, the infrastructure is already present. Nevertheless, only a few initiatives in

Table 8.8 Summary of the main parameters for light opportunity.

Infrastructure complexity	Infrastructure maturity	Infrastructure cost	Terminal complexity	Terminal maturity	Terminal cost	Smart-phone	Calibration complexity
None	Research	Zero	Low	Software development	Low	Near future	Light

Positioning type	Accuracy	Reliability	Range	Sensitivity to environment	Positioning mode	In/out transition	Calibration needed
Relative	100 m	Low	Room	Very high	Almost continuous	Moderate	Often

this field have been reported as the overall approach is not so obvious for a reliable positioning.

Terminal complexity and cost: Luminosity sensors are very cheap and can thus be available everywhere they are needed.

Terminal maturity: The only problem could be the calibration of the sensor, or more precisely the calibration of your specific sensor with respect to the one(s) used for calibrating the system. This is where efforts are still required.

Calibration complexity and need: It consists in moving around along the building and measuring the luminosity parameter. The problem, as explained above, is that because of the important influence of external conditions, this calibration should take into account many configurations. Two options are then possible: either a complex initial phase in order to allow for further classification of conditions and the adaptation of the algorithm to the considered class, or more simple calibrations carried out quite often.

Smartphone: The current smartphones all embark luminosity sensors in order to adapt the screen luminosity (for power saving purposes). In addition, the range of measurements is quite sufficient for positioning purposes.

Positioning type: It is intrinsically relative when considering the differential measurement method but can be absolute thanks to the overlying algorithm providing the user with the positioning.

Accuracy: Reported values are in the range of a few meters, but this seems to have been obtained in very well mastered environments with very carefully carried out calibrations. Values of several tens of meters to hundreds of meters are more probable (leading to a real interrogation concerning the usefulness of such an approach indoors).

Reliability: The same as for the "accuracy" parameter applies.

Range: The "building" level considered here seems to be the right one.

Sensitivity to environment: Very high because of the high interference of external environmental conditions on the measurements.

Positioning mode: It can be continuous indoors.

In/out transition: Quite complex as outdoor calibration has no real sense indeed. Thus, this kind of approach has to be restricted to indoor environments.

8.8 Sound

Everything that has been said about ultrasound remains valid here, but the considered approach is not envisaged by measurements of distances. In fact, at a level similar to that of "ambient light" or LiFi (see above), the sounds are analyzed and, depending on the results of the analyses, an area with a higher probability of presence is defined. This can be relatively efficient in significant environments. In particular, imagine an exhibition building with a specific message for each room. Using a smartphone speaker would detect the room in question. This could also be considered with background music. Note that this is a kind of "geo-located Shazam.[6]"

Let us now come back to our parameter table. The various parameters considered for this technology are given in Chapter 4 (Table 8.9).

Infrastructure complexity, maturity, and cost: In the present case, we consider that the infrastructure exists and therefore that there are no additions to be made.

Terminal complexity, maturity, and cost: As previously mentioned, software development is required on current terminals, but without any difficulties, applications on current smartphones already offer such an approach.

Calibration complexity and need: As usual, a map including the correspondence between the audio signal and the location is required. One could also

Table 8.9 Summary of the main parameters for sound.

Infrastructure complexity	Infrastructure maturity	Infrastructure cost	Terminal complexity	Terminal maturity	Terminal cost	Smartphone	Calibration complexity
None	None	Zero	Low	Integration	Low	Existing	Light

Positioning type	Accuracy	Reliability	Range	Sensitivity to environment	Positioning mode	In/out transition	Calibration needed
Relative	>100 m	Medium	Building	High	Continuous	Difficult	Often

---

6 Shazam is a smartphone application allowing finding out the exact title and singer of any song by simply hearing the song, thanks to your speakers.

imagine an even simpler method, where the audio signal sends the location directly (similarly to a bar code or a quick response (QR) code for optical approaches).

Smartphone: It is already possible to develop the required software.

Positioning type: It is related to the way locations are associated with the map. It could be relative or absolute, but it is intrinsically relative to the audio signal transmitter.

Accuracy: If the "Cell-Id" approach is used, the accuracy can be considered as a room and thus depends on the local definition of the room. It can go from a small one to large conference rooms. It will also greatly depend on the audio message diffusion. An additional problem will potentially arise in both noisy and silent environments. In the first case, the loudspeaker could miss the message and not provide the location, while in the second case, the opposite could happen: detect a message that corresponds to a room in which the user is not.

Reliability: Because of the environmental noise, the considered level for this parameter is set to medium, although it could be rather high in case the different covered areas are not too close to each other.

Range: It can easily be imagined over a whole building and probably over several buildings indeed.

Sensitivity to environment: Mainly due to the ambient noise, it is relatively sensitive to the environment. In some specific and well-mastered areas, it should be more efficient.

Positioning mode: It can be continuous in the sense of Cell-Id, i.e. with a succession of areas of high probability of presence.

In/out transition: Due to outdoor noise, except in very specific cases (very loud audio, absence of external noise, etc.), this solution seems difficult to implement outside.

8.9 Theodolite

In its original definition, a theodolite is a precision measuring instrument used by surveyors to make angle measurements. This makes it possible to carry out "triangulation" processes, i.e. calculating the dimensions and angles of triangles (to determine areas or heights, for example). The instrument therefore consists of elements for measuring angles in the vertical and horizontal planes. Depending on the quality of the measurement elements, theodolites are more or less fine in their estimates and are generally very precise. In an improved modern version, the theodolite is coupled to a laser rangefinder allowing an additional measurement of distance, very useful if one wishes to "position" the object or the point which is the target.

Without going into too much detail, a principle of use is based on an initial "installation" of the theodolite in order to provide it with a reference direction. The angles in the horizontal plane will be provided by taking the value zero for that direction. With regard to the vertical angle, several cases are possible, but let us remember here that a level is available and so the horizontal is the reference of the zero (this is not always the case). Once the theodolite is installed on a fixed point (known or not), it is possible to "survey" points that will then be defined by a pair of horizontal and vertical angles, relative to the installation made. If you also have a range finder, it is also possible to know how far you are from the target point.

If one is now trying to position a sighting point, it is necessary, in addition to the installation described above, to know the position of the theodolite. The latter can be provided in an absolute reference frame if, for example, the theodolite is equipped with a GNSS receiver (in this case, the complete system including the theodolite and the GNSS is referred to as a "total station") or relative if the origin of the reference frame is considered to be the position of the theodolite. Note that all possible combinations of origin and repository choices are generally possible with the latest theodolite models. The target point then has as coordinates those of the theodolite to which must be added the projected values of the line joining the theodolite to the target point, as illustrated in Figure 8.3.

The theodolite coordinates are (x_t, y_t, z_t) in an absolute benchmark (x, y, z). The coordinates of the point referred to in the new theodolite proper locator (x', y', z') (defined by the point (x_t, y_t, z_t) and a direction that could be here x')

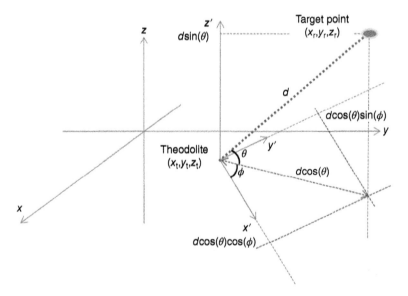

Figure 8.3 Theodolite positioning principle.

are then $(d\cos(\theta)\cos(\varphi), d\cos(\theta)\sin(\varphi), d\sin(\theta))$. In order to obtain, if necessary, the coordinates of the point in the initial reference frame (x, y, z), it would be advisable to apply transformation matrices: a translation (only one) and some rotations (potentially a maximum of three, only two in the case of Figure 8.3 because the z' axis is parallel to the z axis). Such matrices are further described in Chapter 9.

Let us now come back to our parameter table. The various parameters considered for this technology are given in Chapter 4 (Table 8.10).

Infrastructure complexity, maturity, and cost: As everything is measured from the theodolite, there is actually no infrastructure.

Terminal complexity, maturity, and cost: Current theodolites are absolutely not configured in order to propose indoor positioning. They remain professional equipment for precision measurement. Thus, their current configuration is very complex and expensive. They are not intended, for the moment, for general public use. However, we wanted to mention them because it is an approach that pushes to the limit the concepts used for indoor positioning. They can be seen as references.

Calibration complexity and need: The calibration is not so complex but requires experience in order to be achieved properly and efficiently. In addition, current theodolites include so many software capabilities that they are really professional tools.

Smartphone: It is not intended to be deployed on smartphones.

Positioning type: It is fundamentally a relative positioning with respect to the position of the theodolite but can easily be associated with an absolute location by applying simple matrix transformations.

Accuracy: Probably at the best level for distances up to a few hundreds of meters, in line of sight conditions.

Reliability: This parameter is also at the best possible level, provided it is used the right way.

Table 8.10 Summary of the main parameters for theodolite.

Infrastructure complexity	Infrastructure maturity	Infrastructure cost	Terminal complexity	Terminal maturity	Terminal cost	Smart-phone	Calibration complexity
None	None	Zero	Very high	Existing	Very high	Almost impossible	Medium

Positioning type	Accuracy	Reliability	Range	Sensitivity to environment	Positioning mode	In/out transition	Calibration needed
Absolute	A few centimeters	Very high	Building	Very high	Continuous	Difficult	Once

Range: The considered level here is "building" for indoor purposes, but in a finished building, its use is complex if you want some continuity of positions. Indeed, it will then be necessary to position successive bridges in visibility because it is impossible to make a measurement in optical nonvisibility. Thus, a change of floor could prove to be a delicate operation, for example.

Sensitivity to environment: Quite high due to the obligation to work in line of sight conditions.

Positioning mode: The latest materials have the capacity to be "mechanized," i.e. to be in fact motorized in order to follow the movement of a test pattern (typically). Thus, it is quite possible to achieve a continuous positioning of an object moving, for example (and this can even be done automatically).

In/out transition: Possible as long as the line of sight condition is respected. Thus, this limitation is quite high.

8.10 WiFi

The majority of indoor positioning work has been achieved with WiFi networks. The main difference compared with Bluetooth is the range of the transmitters, which is quite a bit larger in the case of WiFi. As the sensitivity of the devices is similar, the number of WiFi access points could probably be reduced compared to the same place using Bluetooth access points. Note that the transmitters could be either access points or other mobile terminals (this applies to both Bluetooth and WiFi). Although propagation time methods have been reported with WiFi, the received signal strength (RSS) approach is preferred (see Section 3.4.2 for details). Thus, the technique of positioning is similar to that with Bluetooth (see Figure 8.4).

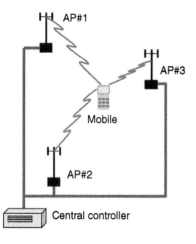

Figure 8.4 WiFi positioning system.

WiFi networks were primarily designed to allow a mobile terminal to connect to a wider network without wires, and WiFi deployment is going to be available everywhere indoors. Therefore, the range of any given access point is usually much larger than that of a Bluetooth access point,[7] designed for short-range connection. About 10 years ago, the deployment of a localization system was costly, and the number of access points was of prime concern. Thus, WiFi solutions were generally preferred, probably because dynamically reconfigurable networks had not yet been well developed.

Things have changed in recent years for different reasons. Indeed, it was not possible, on the first versions of the smartphone of a well-known company,[8] to have access to the levels of power. Thus, the solutions developed were not implementable on this flagship product. Thus, developers of indoor positioning solutions have turned to Bluetooth. Subsequently, as explained in Section 8.2, it was not until the advent of Bluetooth Low Energy that both the connection delay and consumption problems were resolved. By making the modules compatible for battery operation on BLE, WiFi was gradually abandoned. In the current situation, the two technologies coexist according to the needs and wishes of entrepreneurs and their customers.

Let us now come back to our parameter table. The various parameters considered for this technology are given in Chapter 4. It can be seen that it is quite similar to the Bluetooth Table 8.3 (the only difference is for the positioning mode with a very tiny step; Table 8.11).

Infrastructure complexity, maturity, and cost: WiFi networks are highly developed all over the world. This gives a definite advantage to this technology.

Table 8.11 Summary of the main parameters for WiFi.

Infrastructure complexity	Infrastructure maturity	Infrastructure cost	Terminal complexity	Terminal maturity	Terminal cost	Smart-phone	Calibration complexity
Low	Existing	Low	None	Software development	Zero	Existing	Medium

Positioning type	Accuracy	Reliability	Range	Sensitivity to environment	Positioning mode	In/out transition	Calibration needed
Absolute	A few meters	Medium	Building	High	Continuous	Easy	Several times

7 Although Bluetooth was designed for radio links between terminals, some developments have led to network access through the use of access points.

8 The one that in 2007, introduced a revolution concerning mobile phones and presented the first smartphone. Try to find the company. *Hint*: The logo is a fruit.

However, it turns out that site managers often prefer to have separate networks between telecommunications and positioning, thus reducing the initial advantage of network availability. There is no doubt about the maturity of these networks. The only question concerns the fact that the positioning is achieved at the terminal side, the infrastructure side, or at both sides.

Terminal complexity, maturity, and cost: Similarly, current terminals of all kinds make it easy to take the necessary measurements for a positioning system.

Calibration complexity and need: At the same complexity level as for Bluetooth, with greater ranges. Once again, a very important aspect is the way the map is provided to the users (see Chapter 13 for details).

Smartphone: Already available.

Positioning type: Absolute positioning is the standard way.

Accuracy: Excellent results, less than 1 m, are sometimes reported. It does not seem really reasonable, with this type of technique and with the fluctuations because of the environment and electronics, to retain these figures as significant ones for the technology. We preferred to characterize the accuracy by the level "a few meters," which still requires a significantly larger deployment than that required for telecommunication exchanges.

Reliability: This point is the Achilles' heel of these approaches. Very few systems include a real-time estimator of the accuracy or reliability of the positioning they offer. Thus, no verification can be conducted and poor results cannot be detected. Some filtering, averaging, or rating techniques are sometimes used, but this is still at a very high level of processing. In cases where basic measurements are of poor quality, these techniques are often flawed.

Range: We have all found out that WiFi access points are quite powerful and can be received throughout a whole building sometimes. Of course, it depends on the size of the building, but the considered level is clearly the right one.

Sensitivity to environment: At a high level because of the disturbances caused by people, walls, reflecting objects, and nonline of sight conditions.

Positioning mode: Can be continuous.

In/out transition: There is no difficulty in pursuing a positioning with WiFi outside, provided of course that you have also calibrated this environment. In general, there is a tendency to leave GNSS outdoors, but WiFi continuity over restricted and potentially poorly covered areas is quite possible.

8.11 Symbolic WiFi

The analysis of the various techniques of positioning indoors and in particular those relating to WLAN makes it possible to propose a synthesis of the principal characteristics of the systems proposed and to have a rather faithful vision of the evolutions. The following points emerged:

- The "nearest-neighbor search" approaches give good results in terms of accuracy.
- Propagation modeling appears to produce poorer results.
- The range of precision obtained is quite wide.
- The power measurement noise distribution is spread over approximately 10 db (a factor of 10).
- The orientation of the mobile in relation to the base is an important parameter.
- The number of detected transmitters has a significant influence.
- The use of complex search algorithms in databases does not bring any substantial improvement.
- The introduction of "privileged paths" leads to a very significant improvement in performance.[9]
- The increasing complexity of the infrastructure is likely to lead to both a simplification of the calibration phase (almost always essential) and an improvement in performance.

Based on this observation, comparative evaluations of the various techniques were conducted. The questions addressed were related to the importance of the necessary infrastructure, the possible accuracy, or the impact of the algorithms on the complete localization system. The positioning efficiency was evaluated for typical movements. The various studies have shown results consistent with what is described in the literature. However, the impression on the reliability of the positioning has never been very good, mainly because of the very high sensitivity of the power levels to the real environment. Moreover, the need for databases (and especially the measurements needed to fill them) seems to be a major drawback of the methods.

In addition, despite some rather good results, it seems that WiFi positioning is a delicate task in terms of absolute XY (and perhaps Z) coordinates. Approaches based on the use of power levels (RSSI [received signal strength indicator]) depend on the interior configuration and are not easily transposable to another building. They require an investment in terms of information collection during the calibration phase and then in terms of processing during position calculation, without taking into account the necessary redundancy with respect to the system required for telecommunication applications (number of stations). However, it is quite clear, for the observer of these power levels, whether he enters or leaves a room, for example. From this, a symbolic positioning (room, office, corridor, etc.) could be a better approach. This method requires only simple algorithms and is robust (compared to new types of environments or limited infrastructure). The idea is schematized in Figure 8.5.

---

9 Hence, the link with maps.

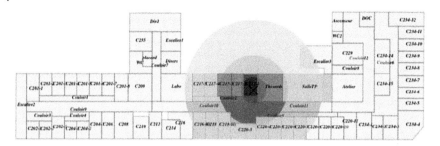

Figure 8.5 The symbolic WiFi positioning system.

This approach requires that the space considered be divided into symbolic surfaces, designated by rooms, offices, or corridors. The size and the very specific shape of certain parts led us to take into account their specificities, in particular in relation to their surfaces as well as to define a graph of the spatial neighborhood allowing us to specify the spatial organization of the parts (and in particular the neighbors). Figure 8.5 also shows the estimated coverage of a WiFi access point positioned in C217-1. This coverage is the result of a specific algorithm explained hereafter.

The basic principle of the symbolic approach is to characterize each access point by associated zones whose dimensions and shapes depend on the power level received by the mobile terminal.

The need to consider nonexclusive areas quickly became apparent: this reflects the fact that it is possible to have low reception power without being far from the access point. The relationship between power and distance is simply not univocal. The algorithm used divides the coverage area of an access point into three nonexclusive spaces. Note that the choice of three zones for each access point seems to be an acceptable compromise between complexity and positioning reliability (it would of course be possible to increase or decrease this number of zones). Each space is determined according to the surfaces of the rooms. For example, the propagation, and therefore the received power level, will not be the same in a large room and in a smaller one, the walls having a wave reflection effect. Thus, depending on where the access point is positioned, the size of the room in which it is located and the shapes of the surrounding rooms, the coverage will be different. The shapes of the three spaces in Figure 8.5 reflect this: the dark gray space is zone 1, the gray space zone 2, and finally the light gray space is zone 3. Once again, the three spaces are not exclusive and zone 3 therefore includes the entire coverage of the access point.

The implementation of the symbolic approach thus requires the definition, for each access point, of the associated spaces as well as the two thresholds (in this case of three spaces). Various studies have made it possible to propose an initial classification of indoor environments likely to be deployment areas. For each of them, rules for defining spaces 1, 2, and 3 have been proposed. Similarly, work has also been carried out on a three-dimensional approach, knowing that the access points on the lower or upper floors are also likely to be received, providing additional information.

Let us now come back to our parameter table. The various parameters considered for this technology are given in Chapter 4. As can be observed, the table is fundamentally different from the one associated with standard WiFi. In addition, please note that this symbolic approach could be deployed with many technologies, and not only with WiFi. Bluetooth is obvious, but many others are also possible: ultra wide band (UWB), LiFi, etc. (Table 8.12).

Infrastructure complexity, maturity, and cost: This is the first real difference with nonsymbolic WiFi. The basic idea is to use only the existing infrastructure, without any additions. Precision will suffer, but not reliability, which remains the main goal of the method. Thus, the cost associated with the infrastructure is reduced to a minimum.

Terminal complexity, maturity, and cost: All current terminals are compatible with both WiFi and Bluetooth. Of course, it is necessary to take into account the slightly different approach of the symbolic method, but this does not present any difficulty, whatever the operating system chosen.

Calibration complexity and need: This is an important new advantage of this symbolic approach. The creation and updating of the reception level database is no longer necessary because it has been replaced by a simple algorithm for determining the zones associated with each transmission point. The calculation of the position is no longer done by searching for the nearest neighbor in the database, but by geometric intersections of areas with a higher probability of presence.

Smartphone: It is not difficult to implement the symbolic approach on current smartphones.

Positioning type: It is clearly symbolic!

Table 8.12 Summary of the main parameters for symbolic WiFi.

Infrastructure complexity	Infrastructure maturity	Infrastructure cost	Terminal complexity	Terminal maturity	Terminal cost	Smart-phone	Calibration complexity
None	Existing	Zero	None	Software development	Zero	Existing	None

Positioning type	Accuracy	Reliability	Range	Sensitivity to environment	Positioning mode	In/out transition	Calibration needed
Symbolic	Decameter	Very high	Building	Low		Continuous	Easy	None

Accuracy: The fundamental point to understand for this type of approach is that precision is not the priority. Not that the latter is necessarily of poor quality (probably even "on the contrary" in many cases) but it is reliability that is sought. In order to do this, and given the nature of the RSS measurements carried out (with their large fluctuations in terms of equipment, conditions of use, and environments), it is interesting to take into account "fluctuation margins." These margins have a large impact on the estimates, whatever they may be, in order to provide a position. The symbolic approach takes these margins into account by relaxing the precision constraint. In general, this results in a much better consideration of reality and thus accuracies that are just as good as standard approaches, but with an incomparably better associated reliability.

Reliability: This reliability parameter is the starting point of the approach. How can this reliability be improved when poor-quality measurements are used, which are extremely dependent on a set of circumstances that it is impossible, in the envisaged cases of use, to control? The way it is improved is based on two aspects: the inclusion of margins in the measurements, and more specifically the fact that the zones associated with each access point are not exclusive. This last point is crucial: if considering exclusive zones from each other (i.e. zone 2 does not include zone 1 as is the case in the approach described here), then the performance, especially reliability, is no longer at a high level at all.

Range: The coverage is identical to that of traditional WiFi but requires that the position of the transmitters is known. Note that this is not necessary in the case of the constitution of databases in the case of a "fingerprint"-type approach that are achievable "blind."

Sensitivity to environment: Here is a new advantage, linked to the one on reliability. Taking into account the margins of error on the measurements considerably reduces the dependence to the environment. Again, this is paid for by less accuracy in theory. In practice, it is often the opposite, which happens because the theory is highly challenged by measurement errors that are much greater than expected. It is surprising from my point of view that these approaches have not been developed further, and we will have the opportunity to discuss this point in Chapter 12 at the end of the book.

Positioning mode: No problem to be continuous, as for WiFi.

In/out transition: The transition with outdoors is possible, but will not propose good performances because the approach is effective only when one takes into account the reflections in the buildings. One could consider that on the outside, the Cell-Id is a kind of symbolic positioning, which is possible, but is not quite the spirit of this paragraph.

Bibliography

1 Jeon, J., Kong, Y., Nam, Y., and Yim, K. (2015). An indoor positioning system using Bluetooth RSSI with an accelerometer and a barometer on a smartphone. In: *2015 10th International Conference on Broadband and Wireless Computing, Communication and Applications (BWCCA)*, Krakow, 528–531. IEEE.

2 Liu, J., Chen, R., Pei, L. et al. (2010). Accelerometer assisted robust wireless signal positioning based on a hidden Markov model. In: *IEEE/ION Position, Location and Navigation Symposium*, Indian Wells, CA, 488–497. IEEE.

3 Hsu, C. and Yu, C. (2009). An accelerometer based approach for indoor localization. In: *2009 Symposia and Workshops on Ubiquitous, Autonomic and Trusted Computing*, Brisbane, QLD, 223–227. IEEE.

4 Sheng-lun, Y., Ting-li, S., and Xue-bo, J. (2017). Improved smartphone-based indoor localization via drift estimation for accelerometer. In: *2017 IEEE International Conference on Unmanned Systems (ICUS)*, Beijing, 379–383. IEEE.

5 Faragher, R. and Harle, R. (2015). Location fingerprinting with Bluetooth low energy Beacons. *IEEE Journal on Selected Areas in Communications* 33 (11): 2418–2428.

6 Jianyong, Z., Haiyong, L., Zili, C., and Zhaohui, L. (2014). RSSI based Bluetooth low energy indoor positioning. In: *2014 International Conference on Indoor Positioning and Indoor Navigation (IPIN)*, Busan, 526–533. IEEE.

7 Fard, H.K., Chen, Y., and Son, K.K. (2015). Indoor positioning of mobile devices with agile iBeacon deployment. In: *2015 IEEE 28th Canadian Conference on Electrical and Computer Engineering (CCECE)*, Halifax, NS, 275–279. IEEE.

8 Ji, M., Kim, J., Jeon, J., and Cho, Y. (2015). Analysis of positioning accuracy corresponding to the number of BLE beacons in indoor positioning system. In: *2015 17th International Conference on Advanced Communication Technology (ICACT)*, Seoul, 92–95. IEEE.

9 Lohan, E.S., Talvitie, J., Figueiredo e Silva, P. et al. (2015). Received signal strength models for WLAN and BLE-based indoor positioning in multi-floor buildings. In: *2015 International Conference on Location and GNSS (ICL-GNSS)*, Gothenburg, 1–6. IEEE.

10 Basiri, A., Peltola, P., Figueiredo e Silva, P. et al. (2015). Indoor positioning technology assessment using analytic hierarchy process for pedestrian navigation services. In: *2015 International Conference on Location and GNSS (ICL-GNSS)*, Gothenburg, 1–6. IEEE.

11 Kyritsis, A.I., Kostopoulos, P., Deriaz, M., and Konstantas, D. (2016). A BLE-based probabilistic room-level localization method. In: *2016 International Conference on Localization and GNSS (ICL-GNSS)*, Barcelona, 1–6. IEEE.

12 Antevski, K., Redondi, A.E.C., and Pitic, R. (2016). A hybrid BLE and Wi-Fi localization system for the creation of study groups in smart libraries. In: *2016 9th IFIP Wireless and Mobile Networking Conference (WMNC)*, Colmar, 41–48. IEEE.

13 Ichimura, T. (2016). 3D-odometry using tactile wheels and gyros: localization simulation of a bike robot. In: *2016 16th International Conference on Control, Automation and Systems (ICCAS)*, Gyeongju, 1349–1355. IEEE.

14 Li, D., Eckenhoff, K., Wu, K. et al. (2017). Gyro-aided camera-odometer online calibration and localization. In: *2017 American Control Conference (ACC)*, Seattle, WA, 3579–3586. IEEE.

15 Wei, Y.L. and Lee, M.C. (2011). Mobile robot autonomous navigation using MEMS gyro north finding method in Global Urban System. In: *2011 IEEE International Conference on Mechatronics and Automation*, Beijing, 91–96. IEEE.

16 Marck, J.W., Mohamoud, A., vd Houwen, E., and van Heijster, R. (2013). Indoor radar SLAM A radar application for vision and GPS denied environments. In: *2013 European Radar Conference*, Nuremberg, 471–474. IEEE.

17 Kim, H.-D., Seo, S.-W., Jang, I.-h., and Sim, K.-B. (2007). SLAM of mobile robot in the indoor environment with Digital Magnetic Compass and Ultrasonic Sensors. In: *2007 International Conference on Control, Automation and Systems*, Seoul, 87–90. IEEE.

18 Yamada, T., Yairi, T., Bener, S.H., and Machida, K. (2009). A study on SLAM for indoor blimp with visual markers. In: *2009 ICCAS-SICE*, Fukuoka, 647–652. IEEE.

19 Albrecht, A. and Heide, N. (2018). Mapping and automatic post-processing of indoor environments by extending visual SLAM. In: *2018 International Conference on Audio, Language and Image Processing (ICALIP)*, Shanghai, 327–332. IEEE.

20 Chang, H., Lin, S., and Chen, Y. (2010). SLAM for indoor environment using stereo vision. In: *2010 Second WRI Global Congress on Intelligent Systems*, Wuhan, 266–269. IEEE.

21 Liu, J., Chen, Y., Jaakkola, A. et al. (2014). The uses of ambient light for ubiquitous positioning. In: *2014 IEEE/ION Position, Location and Navigation Symposium – PLANS 2014*, Monterey, CA, 102–108. IEEE.

22 Yoshino, M., Haruyama, S., and Nakagawa, M. (2008). High-accuracy positioning system using visible LED lights and image sensor. In: *2008 IEEE Radio and Wireless Symposium*, Orlando, FL, 439–442. IEEE.

23 Aguirre, D., Navarrete, R., Soto, I., and Gutierrez, S. (2017). Implementation of an emitting LED circuit in a visible light communications positioning system. In: *2017 First South American Colloquium on Visible Light Communications (SACVLC)*, Santiago, 1–4. IEEE.

24 Nakazawa, Y., Makino, H., Nishimori, K. et al. (2014). LED-tracking and ID-estimation for indoor positioning using visible light communication. In: *2014 International Conference on Indoor Positioning and Indoor Navigation (IPIN)*, Busan, 87–94. IEEE.

25 Yang, Z., Fang, J., Lu, T. et al. (2017). An efficient visible light positioning method using single LED luminaire. In: *2017 Conference on Lasers and Electro-Optics Pacific Rim (CLEO-PR)*, Singapore, 1–2. IEEE.

26 Kim, Y., Hwang, J., Lee, J., and Yoo, M. (2011). Position estimation algorithm based on tracking of received light intensity for indoor visible light communication systems. In: *2011 Third International Conference on Ubiquitous and Future Networks (ICUFN)*, Dalian, 131–134. IEEE.

27 Xu, W., Wang, J., Shen, H. et al. (2016). Indoor positioning for multiphotodiode device using visible-light communications. *IEEE Photonics Journal* 8 (1): 1, 7900511–11.

28 Zhuang, Y., Hua, L., Qi, L. et al. (2018). A survey of positioning systems using visible LED lights. *IEEE Communications Surveys & Tutorials* 20 (3): 1963–1988, third quarter.

29 Huang, J., Ishikawa, S., Ebana, M. et al. (2006). Robot position identification by actively localizing sound beacons. In: *2006 IEEE Instrumentation and Measurement Technology Conference Proceedings*, Sorrento, 1908–1912. IEEE.

30 Pei, L., Chen, L., Guinness, R. et al. (2013). Sound positioning using a small-scale linear microphone array. In: *International Conference on Indoor Positioning and Indoor Navigation*, Montbeliard-Belfort, 1–7. IEEE.

31 Yan, J., Bellusci, G., Tiberius, C., and Janssen, G. (2008). Analyzing non-linearity effect for indoor positioning using an acoustic ultra-wideband system. In: *2008 5th Workshop on Positioning, Navigation and Communication*, Hannover, 95–101. IEEE.

32 Park, G., Chanda, P.S., and Kang, T.I. (2006). Implementation of a real-time 3-D positioning sound synthesis algorithm for a handheld device. In: *2006 8th International Conference Advanced Communication Technology*, Phoenix Park, 1493–1496. IEEE.

33 Feng, C., Au, W.S.A., Valaee, S., and Tan, Z. (2012). Received-signal-strength-based indoor positioning using compressive sensing. *IEEE Transactions on Mobile Computing* 11 (12): 1983–1993.

34 Yang, C. and Shao, H. (2015). WiFi-based indoor positioning. *IEEE Communications Magazine* 53 (3): 150–157.

35 Lim, C., Wan, Y., Ng, B., and See, C.S. (2007). A real-time indoor WiFi localization system utilizing smart antennas. *IEEE Transactions on Consumer Electronics* 53 (2): 618–622.

36 Feng, C., Au, W.S.A., Valaee, S., and Tan, Z. (2010). Compressive sensing based positioning using RSS of WLAN access points. In: *2010 Proceedings IEEE INFOCOM*, San Diego, CA, 1–9. IEEE.

37 Figuera, C., Rojo-Alvarez, J.L., Mora-Jimenez, I. et al. (2011). Time-space sampling and mobile device calibration for WiFi indoor location systems. *IEEE Transactions on Mobile Computing* 10 (7): 913–926.

38 Bisio, I. et al. (2014). A trainingless WiFi fingerprint positioning approach over mobile devices. *IEEE Antennas and Wireless Propagation Letters* 13: 832–835.

39 Le Dortz, N., Gain, F., and Zetterberg, P. (2012). WiFi fingerprint indoor positioning system using probability distribution comparison. In: *2012 IEEE International Conference on Acoustics, Speech and Signal Processing (ICASSP)*, Kyoto, 2301–2304. IEEE.

9

Building Range Technologies: The Specific Case of Indoor GNSS

Abstract

Indoor GNSS is in fact everyone's quest. In the past 20 years, satellite navigation systems have gone from an emerging system to a daily use system that no one pays much attention to. This is true not only for the technical component but also for the implementation or use. However, technologies actually seeking to implement domestic GNSS remain relatively complex in the technicality they require. Finally, this is a theme that our research group at the Institut Mines-Telecom has particularly worked on and for which quite nice technical solutions have been proposed. All these reasons are at the origin of this specific chapter.

Keywords   *Indoor GNSS; pseudo-satellite; repeaters; repealites; Grin-Locs*

Th e classification described in Chapter 4 led to the following table concerning these technologies (Table 9.1).

9.1   Introduction

Although this chapter is devoted to infrastructure-based global navigation satellite system (GNSS) systems, other solutions have been studied. For example, high-sensitivity GNSS was designed to ensure continuity of service without additional infrastructure.

Thus, a solution could be to send the navigation message through telecommunication networks that are widely available indoors. Thus, knowing the message, the receiver is able to use the high sensitivity in order to acquire the GNSS signals and then is able to calculate a position as all the parameters needed (from the navigation message) are available. High sensitivity and assisted approaches are thus quite complementary.

The simple underlying idea is that the signals are always present inside, but even lower in the noise than outside. Thus, if a very sensitive receiver can be

Indoor Positioning: Technologies and Performance, First Edition. Nel Samama.
© 2019 The Institute of Electrical and Electronics Engineers, Inc. Published 2019 by John Wiley & Sons, Inc.

Table 9.1 Main "building" technologies.

Technology	Positioning type	Accuracy	Reliability	Range	Sensitivity to environment	Calibration needed	Positioning mode	Technique	Signal processing	Position calculation
Accelerometer	Relative	$f(t)$	Medium	Block	No impact	Often	Continuous	Physical	Detection	Math functions $(f, f, f, f, ...)$
BLE	Absolute	A few meters	Medium	Building	High	Several times	Almost continuous	Physical	Pattern matching	Math functions $(f, f, f, f, ...)$
Gyrometer	Relative	$f(t)$	Medium	Building	No impact	Often	Continuous	Physical	Detection	Math functions $(f, f, f, f, ...)$
Image-relative displacement	Relative	<1 m	Medium	Building	High	Several times	Almost continuous	Image(s)	A combination of	Math functions $(f, f, f, f, ...)$
Image SLAM	Relative	<1 m	Medium	Building	High	Several times	Almost continuous	Image(s)	A combination of	Math functions $(f, f, f, f, ...)$
Indoor GNSS	**Absolute**	**A few decimeters**	**Medium**	**Building**	**High**	**None**	**Continuous**	**Phase(s)**	**Correlation**	**∩ Hyperbolae**
LiFi	Symbolic	A few meters	Low	Building	Very high	None	Almost continuous	Physical	Detection	Spot location
Light opp	Relative	100 m	Low	Building	Very high	Often	Almost continuous	Physical	Classification	Zone determination
Sound	Relative	>100 m	Low	Building	High	Often	Continuous	Physical	Detection	∩ Circles
Theodolites	Absolute	A few centimeters	Very high	Building	Very high	Once	Continuous	Angle(s)	A combination of	∩ Plans + distance(s)
WiFi	Absolute	A few meters	Medium	Building	High	Several times	Continuous	Physical	Pattern matching	Math functions $(f, f, f, f, ...)$
WLAN Symbolic	Symbolic	Decameter	Very high	Building	Low	None	Continuous	Physical	Propagation modeling	Zone determination

SLAM, simultaneous localization and mapping; WLAN, wireless local area network.
Only the bold lines are treated in this chapter.

designed, it should be possible to install it indoors. A similar, but not identical, idea led to the design of the so-called Assisted-GNSS. The initial objective was to ensure indoor positioning by "helping" the receiver find signals in harsh environments. In such situations, one of the major problems of autonomous receivers is the impossibility of decoding the navigation message. Thus, one solution could be to send the navigation message via widely available telecommunication networks inside. Thus, knowing the message, the receiver is able to use high sensitivity to acquire GNSS signals and calculate a position, as all parameters are available. High sensitivity and assisted approaches are therefore fully complementary.

With this higher sensitivity, the receiver is now more easily jammed because of numerous reflected signals, which are now detectable. This is particularly true in urban canyons or indoors! Unfortunately, although these approaches have engendered real improvements for 90% of the situations, they are clearly not the right solutions for indoor positioning. This leaves the place for infrastructure-based approaches. In chronological order, the four technologies are the pseudolites, the repeaters, the repealites, and finally the Grin-Locs. They are described in the same order.

9.2 Concept of Local Transmitters

The first ideas for global positioning system (GPS)-type signal transmitters were born in the 1980s, when the limits of GPS were taken into account. Questions such as "how to use a receiver when fewer than three or four satellites are available?" or "how to improve the vertical precision of the system?" were raised.

The first response could be to increase the number of satellites by a factor of 2 or 3. However, the cost associated with the reduced performance improvement was deemed unsustainable. The idea of setting up GPS-type signal generators that can be deployed locally then appears: pseudolites were born.

A pseudo-satellite (contracted into pseudolite) is a generator that transmits GNSS signals. However, it is not a satellite. A pseudolite is easy to deploy in order to "augment" (that is the established term) constellations where the need is felt (reduced number of visible satellites or specific need for performance improvement). The addition of a pseudolite allows, for example, in the case of an open pit mine, the continuity of the positioning service to be ensured even for the actions carried out at the very bottom of the mine.

A similar approach was developed for local area augmentation systems (LAAS) where the problem was to provide good vertical accuracy for aircraft landing situations. This vertical precision is closely linked to a GNSS-specific parameter, the VDOP (VDOP stands for vertical dilution of precision. See

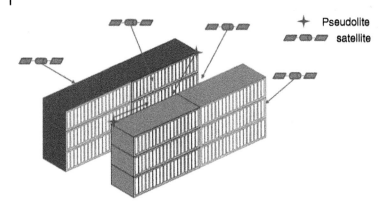

Figure 9.1 A typical urban canyon situation.

Ref. [40] for some details). Placing a pseudolite under the aircraft significantly improves this parameter.

Urban canyons are also complex environments for GNSS signals (see Figure 9.1). A receiver located between large buildings may have difficulty acquiring enough signals. A well-placed pseudolite can provide optimal performance.

In the examples cited so far, pseudolites "increase" GNSS, its coverage, and/or accuracy. It is possible to go one step further in the concept and implement a sufficient set of pseudolites in order to perform the location function independently of the satellites. This is the basic idea behind indoor positioning systems.

9.3 Pseudolites

The basic idea underlying the use of a local infrastructure is somehow to reconstruct a constellation associated with the local building. One could have called the local infrastructure a "terrestrial satellite constellation." As the problem of the indoor reception is mainly due to low signal levels, the fact of using generators can effectively solve the problem by providing higher level signals, i.e. powerful enough ones. The practical implementation of the pseudolite approach is to build complete signal generators. The shrewdness of pseudolites is to use similar signals to GNSS ones. In this way, standard receivers are already enabled to decode such signals without any hardware updates. Only software updates are required in order to tell the receiver to search for and to acquire these new "terrestrial satellites." For the GPS constellation, which is currently the only one that is fully operational, the C/A codes are reserved and it is not possible to use them. Thus, pseudolites have to choose other pseudo random noise. Fortunately, the 36 reserved codes,

Figure 9.2 Illustration of an indoor pseudolite configuration (P_i refers to pseudolite "i" and d_i to the distance between pseudolite "i" and indoor receiver).

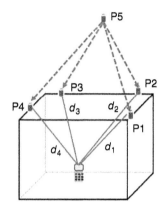

although exhibiting the best correlation features, are far from being the only possible ones. There are numerous Gold codes that can be used, with lower correlation quality, but as the power levels are higher for pseudolite generators and the distance from pseudolite to the receiver rather small compared to that between satellite and receiver, this is of no real difficulty. Each pseudolite is then considered as a signal generator, as illustrated in Figure 9.2.

The pseudolite concept is a very old one, imagined at the beginning of 1980s. It has been applied, in anticipation, to various types of missions, from augmentation to current GNSS constellations in difficult environments such as open cast mines or more recently urban canyons to a Mars exploration positioning system,[1] through local area augmentation system (LAAS) for the landing approach of planes. For an indoor system, there is the need for four pseudolites in order to allow three-dimensional positioning (see references for classical GPS equations).

There are two different signal generator approaches, depending on the accuracy required: meter or centimeter. As seen in Chapters 6 and 7, code positioning is less accurate than carrier phase positioning but requires less complex electronics. Figure 9.3 gives a diagrammatic representation of both techniques.

The code-based technique allows distances to be obtained without ambiguity, given the usual indoor ranges that are not likely to be greater than one code length, i.e. 300 km. The accuracy is then of a few meters as the running mode is identical to that of GPS, except for the various measurement errors occurring while the signal is crossing through the atmosphere. One has to remember that the main errors lie within this part of the propagation of the signals. Of course, to achieve such accuracy, like the standard constellation, the receiver needs to know about the location of the various pseudolites: this can be either local

---

1 This possibility was envisaged in order to provide the planet exploration with a precise deployable positioning system. The other possibility would have been to deploy a complete GPS-like constellation all around Mars.

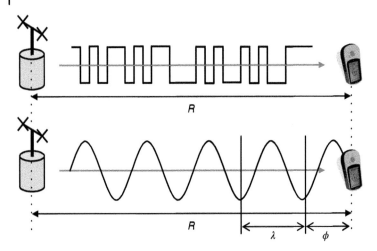

Figure 9.3 Code (top) and phase (bottom) techniques for pseudolite generators (R is the distance between the pseudolite and the receiver, λ the wavelength of the signal, and φ the measured carrier phase).

coordinates or global ones. The accuracy of the positioning of the pseudolite is clearly of importance, but with the code technique, this requirement is not too tight: a few tens of centimeters is enough.

The phase-based technique is more stringent for both the pseudolites and the receiver. As it deals with phase measurements, the ambiguities must be solved (see references for details concerning the problem of ambiguities), the resulting accuracy is a few centimeters. The counterpart is that the location of the pseudolites has to be known to within a few centimeters too.[2]

Nevertheless, although the theoretical concept is very clever, some difficulties exist in the practical deployment in the real world of pseudolites.

The first one is the synchronization required in order to allow the accuracies described above. One has to keep in mind that some satellites have onboard no fewer than four atomic clocks in order to supply time accuracy that can lead to a positioning accuracy of a few meters. Such an implementation is far too expensive for a locally deployed system for indoor location finding. The way to overcome this problem is through the use of an additional pseudolite that is used, as in the ultra wide band (UWB) system, for instance, as a base station for time reference.

The second is the so-called near-far effect. When indoors, the distances between the pseudolites and the receiver are a lot smaller than those between the satellites and a receiver (a few meters compared to more than 20 000 km).

---

2 Note that the determination of the indoor location of the pseudolites, when the precision required is of a few centimeters, is not such an easy case. This problem has not been investigated that much because it is not the major concern before the technical aspects are settled.

The important point is the relative difference between these distances, i.e. the fact that indoors, the ratio of one to other such distances can be rather high (as much as 20 or 30). The corresponding ratio for the satellite constellation is always less than $25\,600/20\,200 = 1.26$. The direct impact is that the power ratio of the received signals from pseudolites can be as far away as 20 or 30 to the square (considering a free space propagation model in d^2). Knowing that a ratio of 16 is equivalent to $24\,dB$ of power attenuation between two signals, one can anticipate the problem: as codes are pseudo random ones, there is a practical limitation in the discrepancy of power levels that can be processed by a receiver, mainly because low power signals are considered as noise. Therefore, high power pseudolites, i.e. those that are the nearest to the receiver, mean the other signals are considered as noise: they cannot be detected. This problem has an impact on the coverage area or the definition of the positioning of the infrastructure.

Thirdly, the phase-based pseudolites use a differential phase measurement technique, which is basically a relative positioning approach. Then, the determination of the starting point is of prime importance. As there is no real means, as long as the user is indoors, to obtain this first location with the required accuracy (a few centimeters), the pseudolite global system should provide this first point: this step can take quite a lot of time (10–20 minutes are typical values). On the other hand, one could state that this is only needed for high-accuracy indoor positioning, which is unlikely to be the typical pedestrian application, for instance. Thus, if a user requires centimeter accuracy, there is a time constraint in view to start the process: this can still be acceptable.[3]

A typical pseudolite-based positioning system is provided in Figure 9.4, and positioning results are provided in Figure 9.5. In this specific case, the

Figure 9.4 A typical indoor pseudolite configuration. Source: Courtesy of the School of Surveying and Spatial Information Systems, University of New South Wales.

---

3 This constraint existed for the first high-accuracy civil engineering systems and has been considered as acceptable.

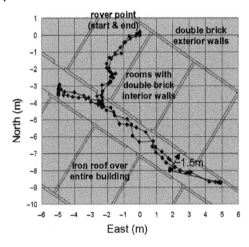

Figure 9.5 Typical results of an indoor pseudolite positioning system. Source: Courtesy of the School of Surveying and Spatial Information Systems, University of New South Wales.

pseudolites are installed outdoors, on the roof of the building. One has to remember that this implantation allows one to reduce the discrepancies between the various distances from the pseudolites to the receiver and thus reduce the near-far effect. The results obtained from different teams around the world, mainly implementing the phase measurement approach, are coherent and show that problems such as multipath are not so disturbing (although a complete study of indoor multipath for pseudolites still remains to be carried out).

9.4 Repeaters

Following the same philosophy as for pseudolites, i.e. the deployment of a terrestrial local constellation, the repeater's approach is a little bit different in the practical implementation. The basic idea comes from the fact that the need for an infrastructure is clearly a disadvantage, compared to high sensitivity GNSS (HS-GNSS) and assisted-GNSS (A-GNSS) solutions. However, it has been seen that there is probably a need for an infrastructure-based solution. Thus, to facilitate an eventual deployment, the infrastructure should be as simple as possible.[4]

The solution discussed here uses GPS repeaters that transmit signals from outdoors, where the receiving conditions are excellent, to indoors, where a current standard receiver is not able to compute a location. The two hardware configurations described below use repeaters that repeat all the incoming

---

4 A pseudolite or a repeater-based system is a specific infrastructure dedicated to positioning, as compared to WLAN or GSM/UMTS infrastructures for which positioning is only a by-product. The cost of the deployment is largely shared with telecommunication purposes (although not totally true for current WLAN indoor positioning systems as discussed in previous paragraphs).

Figure 9.6 A typical indoor "RnS" repeater configuration (S refers so satellites, R to repeaters).

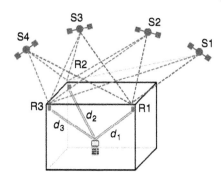

Figure 9.7 A typical indoor "R1S" repeater configuration (ρ refers to pseudo ranges).

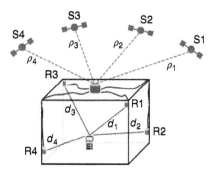

signals without processing (called "RnS," see Figure 9.6) or that repeat only the signal from one satellite (called "R1S," see Figure 9.7).

Obviously, the fact of using such a repeater leads to a difficulty, when trying to solve the navigation equations with a current standard GPS receiver. The signal propagation path is artificially "curved" through the repeater. Therefore, the actual measured propagation time does not give the real distance separating the satellite from the receiver, but the sum of the distance from the satellite to the repeater plus the distance from the repeater to the receiver (plus of course the various additional delays introduced by the repeater hardware). Therefore, the repeater approaches involve designing repeater system hardware and new receiver algorithms to solve the new set of navigation equations.

9.4.1 Clock Bias Approach

In the RnS approach, a repeater consists only of a receiving antenna, an amplifying chain, and a transmitting antenna. In this case, the signals from all the satellites are transmitted indoors. The computation of the navigation equations at the indoor receiver location (using any technique such as linearization, Kalman filtering, etc.) leads to a solution vector that gives the location of the receiving outdoor antenna together with the fourth component, which is the sum of all the common biases (see reference 50 for details on clock bias). Thus, it contains the clock bias but also the delays within the electronic structures of the repeater and finally also the free space propagation delay (a time equivalent to the distance d from the repeater to the receiver) from one repeater to the indoor receiver.

As we have no idea of ct_r (assumed to be the real clock bias of the receiver), it is impossible to go back to d. One solution is to carry out such a calculation four times by the way of a sequential process, using four repeaters that are located at different points. In such a case, it is possible to obtain two types of information for each calculation: the location of the repeater (the first three components) and the sum of the ct_r and the distance d_i between the repeater "I" and the indoor receiver. Therefore, after one "cycle," we have the following vector (taken from the four fourth components of the four solution vectors calculated for each repeater). Refer to Figure 9.6 and note that the following equations now deal with four repeaters, instead of only three given in Figure 9.6:

$$\begin{bmatrix} ct_1 \\ ct_2 \\ ct_3 \\ ct_4 \end{bmatrix} = \begin{bmatrix} ct_r + d_1 \\ ct_r + d_2 \\ ct_r + d_3 \\ ct_r + d_4 \end{bmatrix}. \tag{9.1}$$

where we would like to find d_1, d_{j2}, d_3, and d_4 to determine the actual location of the indoor receiver. This problem is identical to that of a classical outdoor location determination. As a matter of fact, the standard method or a "sphere expansion" method can be used. The latter is defined as follows: from the last set of equalities, it is possible to generate a new set by calculating the differences of the coordinates to eliminate ct_r, (which is not easy to determine). One obtains (assuming d_1 is the smallest distance from all the d):

$$\begin{bmatrix} d_2 - d_1 \\ d_3 - d_1 \\ d_4 - d_1 \end{bmatrix} = \begin{bmatrix} ct_2 - ct_1 \\ ct_3 - ct_1 \\ ct_4 - ct_1 \end{bmatrix}. \tag{9.2}$$

From that point, the method consists in choosing the d_1 so that the two largest spheres touch. The three spheres are defined by their radii ($d_{j_2} - d_{i_1}$), ($d_3 - d_1$), and ($d_4 - d_1$). Then, the expansion of the four spheres of radii d_1, d_2, d_3, and d_4 allows the actual intersection point to be determined, which is the position of the indoor receiver.

One needs to define the cycling time as well as the time, the start time, and the identification of the active repeater. Then, programs can be run to store the incoming raw data from the mobile receiver. In every case, the basic data required are

1. GPS time,
2. Computed clock bias (obtained from the current GPS location calculation process), and
3. Measured clock bias rate (obtained from the Doppler shift).

The mathematical computations are then quite easy to follow, based on the general theory previously described. The complete set of equations is given below.

$$\begin{bmatrix} ct_{cal}(t_{R_1}) = ct_{osc}(t_{R_1}) + \text{del}(R_1) + d_1 \\ ct_{cal}(t_{R_2}) = ct_{osc}(t_{R_2}) + \text{del}(R_2) + d_2 \\ ct_{cal}(t_{R_3}) = ct_{osc}(t_{R_3}) + \text{del}(R_3) + d_3 \\ ct_{cal}(t_{R_4}) = ct_{osc}(t_{R_4}) + \text{del}(R_4) + d_4 \end{bmatrix}$$

where

$t_{cal}(t_{R_i})$ is the computed clock bias at time t_{R_i}

$t_{osc}(t_{R_i})$ is the clock bias rate at time t_{R_i}

$\text{del}(R_i)$ is the induced delay of repeater i

d_i is the distance between the mobile receiver and repeater "i". (9.3)

The unknown variables are the various induced delays (which have to be calibrated in the laboratory), the distances d_i (which represent the x_r, y_r, and z_r coordinates of the mobile antenna), and the three clock bias rates. Therefore, the total number of unknowns is now 7 (x_r, y_r, and z_r, and the four clock bias rates). As a matter of fact, one needs to know about the short-term stability of the receiver's oscillator: if not known, it is still possible to follow it with a rather good accuracy through the Doppler shift. Thus, we can state that

$$ct_{osc}(t_{R_j}) = ct_{osc}(t_{R_i}) + \sum_{k=i+1}^{j} \Delta ct_{osc}(t_{R_k})$$

where

$\Delta ct_{osc}(t_{R_k})$ is the measured clock bias rate at time t_{R_k}. (9.4)

We now have to deal with a set of four equations with four unknown variables that are x_r, y_r, z_r, and $ct_{osc}(t_{R_i})$. The computation is then absolutely identical to that currently carried out by a GPS receiver: one can use the linearization method in order to compute a 3D positioning. As one may conclude, we have moved to a local frame of reference for convenience.

In a typical file, GPS time, clock bias, and clock bias rate are stored, and the difference between the evolution of the computed clock bias and the measured clock bias rate is calculated. This showed an interesting feature that caused a lot of trouble during the experiments: there are sometimes large skips, in one direction or another where one may consider that there has been a change of transmitting repeater. This is not necessarily the case: these two consecutive skips can be due to a change in the number of satellites used to compute the fix. One has to deal with this problem as one does not compute one's own navigation solution. The induced effect, if not taken into account, is to shift the point well outside the acceptable accuracy range (some computed indoor locations could be 4 or 5 m away from the real point). When removing these results, one obtains some quite good values, in the 1–2 m range.

Unfortunately, this problem of clock bias skips arose very often. The major reason is that the various filters within the mobile receiver were not removed. Thus, as one was expecting really raw data, the internal receiver algorithms were trying to smooth these data by either moving from one constellation to another (and generating the skips) or modifying the output data. In this latter case, the additional delay that we were seeking (corresponding to the distance d_i) was greatly modified. Another difficulty of this technique is linked to synchronization.

9.4.2 Pseudo Ranges Approach

These difficulties generated problems and another scheme was implemented, based on the direct use of the raw measurements. We know that it is not always possible to get the real raw data, i.e. to get rid of the various filters designed to "improve" the receiver solution. Therefore, it was decided to use the so-called "GNSS sensors," designed to deliver a lot of data from the receiver. It is also possible to modify some parameters such as loop smoothness, constellation configuration, and a lot of others.

As a matter of fact, mainly the Kalman filter parameters, the code loop and the frequency loop were tuned. As long as one just wanted to enable the continuity of the location service between outdoors to indoors, there is no interest in increasing the accuracy, as long as the goal of a few meters is still achieved (as will be shown below).

The principle of the new technique is simply to "follow" the evolution of the pseudo-distances when cycling from one repeater to the next (see Figure 9.8). As a matter of fact, one expects skips, of the value of the difference of distances from the transmitting indoor antenna of the first repeater to the indoor receiver, to the same distance from the second repeater to the receiver. In this case, after a complete cycle, we should get three differences that will characterize the current location of the receiver.

9.4.2.1 Theoretical Aspects

A typical curve showing the pseudo range evolution over a few cycles is given in Figure 9.9. To understand this curve, some assumptions have to be made:

- The delays induced by the repeaters, i.e. the electronic delays and the cable delays, have been previously calibrated or do not need to be calibrated;
- As long as the curve represents the difference from one time step to the next, values are equivalent to the drift of the pseudo-distances (or can also be seen as an acceleration);
- The remaining drift of the receiver's oscillator can be observed (the slight positive slope).

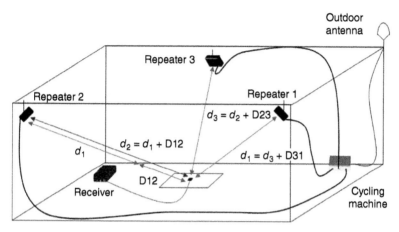

Figure 9.8 Indoor repeater-based positioning system.

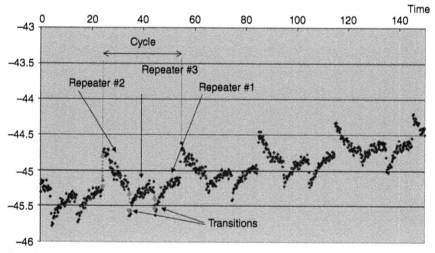

Pseudo-ranges differences

Figure 9.9 A typical pseudo range evolution.

It appears quite clearly that the receiver "sees" the change from one repeater to the next as acceleration. It is also possible to observe that after the transition in the pseudo-distance from one repeater to the next, the receiver "comes back" to the natural drift. The time needed for this return depends on the tuning parameters. These parameters are also very important to be able to follow the skips from one repeater to the next. Some tunings do not allow the observation and others do.

Signal switching from one repeater to the next results in an offset in the phase of the signal received. As in GPS positioning, which needs four satellites to achieve 3D positioning, four repeaters are necessary to have four phase offsets. It is thanks to these values of phase offsets (or phase jumps) that the calculation of the indoor location of the receiver is possible. To understand this, let us write the new indoor pseudo range expression (PR_i) between a satellite and an indoor GPS receiver. One should remember that the signal is received through a repeater R_i:

$$\rho_i(t) = D_{S \to A}(t) + d_i + c \cdot t(t) + \Delta \text{prop}(t) + \text{delay} \tag{9.5}$$

where $D_{S \to A}$ is the line-of-sight distance between the satellite and the outdoor antenna located on the roof, d_i is the distance between repeater R_i and the GPS receiver located indoors, t is the receiver clock bias, $\Delta \text{prop}(t)$ is the propagation error delay, and delay is the propagation time of the signal received in the cable connecting the outdoor antenna and repeater R_i. We suppose that these cables are identical for the four repeaters. In this case, this delay will be the same for all paths through R_1 to R_4.

Note that the above equation also depends on multipath error delay that is not considered here. Let us suppose that at time t, the signal switches from repeater R_i to repeater R_j. The measured pseudo range at time $t + dt$ through repeater R_j becomes.

$$\rho_i(t + dt) = D_{S \to A}(t + dt) + d_i + c \cdot t(t + dt) + \Delta \text{prop}(t + dt) + \text{delay} \tag{9.6}$$

Therefore, the pseudo range difference from time $t + dt$ to time t is given by

$$\Delta \rho(t + dt) = \rho_i(t + dt) - \rho_i(t)$$
$$= \Delta D_{S \to A}(t + dt) + \Delta d + c \cdot \Delta[t(t + dt)] + \Delta[\Delta \text{prop}(t + dt)] \tag{9.7}$$

where

$$\Delta D_{S \to A}(t + dt) = D_{S \to A}(t + dt) - D_{S \to A}(t)$$
$$\Delta d = d_j - d_i$$
$$\Delta[t(t + dt)] = t(t + dt) - t(t)$$
$$\Delta[\Delta \text{prop}(t + dt)] = \Delta \text{prop}(t + dt) - \Delta \text{prop}(t) \tag{9.8}$$

We are interested in Δd, which is the difference between the distances separating the receiver from the two repeaters R_i and R_j. Here, we cannot access its value because it is very small compared with $\Delta D_{S \to A}(t + dt)$ and $\Delta[t(t + dt)]$ values. As Δd appears only at transition time between successive repeaters, we proceed to a second time difference to detect Δd. In fact, at time t, the signal was still transmitted by R_i; therefore, the pseudo range difference between times t and $t - dt$ does not include Δd and it follows:

$$\Delta \rho(t) = \rho_i(t) - \rho_i(t - dt) = \Delta D_{S \to A}(t) + c \cdot \Delta[t(t)] + \Delta[\Delta \text{prop}(t)] \tag{9.9}$$

To extract Δd, we calculate the second variation of the pseudo range. The following equation is obtained:

$$\Delta\rho(t+dt) - \Delta\rho(t) = [\Delta D_{S\rightarrow A}(t+dt) - \Delta D_{S\rightarrow A}(t)]$$
$$+ c \cdot [\Delta[t(t+dt)] - \Delta[t(t)]]$$
$$+ [\Delta[\Delta\text{prop}(t+dt)] - \Delta[\Delta\text{prop}(t)]] + \Delta d \quad (9.10)$$

If the signal transition between repeater R_i and R_j is very fast (i.e dt is very small), double differences of $D_{S\rightarrow A}$, t, and Δprop (first, second, and third terms in the above equation, respectively) will be negligible and the double pseudo range difference will finally be reduced to $\Delta ji = \Delta d = d_j - d_i$.

At the end of each cycle, we have the following system:

$$\Delta 12 = d_1 - d_2$$
$$\Delta 23 = d_2 - d_3$$
$$\Delta 34 = d_3 - d_4$$
$$\Delta 41 = d_4 - d_1 \quad (9.11)$$

Note that the sum of first three equations gives the last one. This is the reason only three differences are retained from the four: for example, the first three. While expressing distances "d_i" by the coordinates of receiver (x_r, y_r, z_r) and those of repeaters R_i $(x_{R_i}, y_{R_i}, z_{R_i}, i = 1, \ldots, 4)$, it follows that:

$$\Delta 12 = d_1 - d_2$$
$$= \sqrt{(x_{R_1} - x_r)^2 + (y_{R_1} - y_r)^2 + (z_{R_1} - z_r)^2}$$
$$- \sqrt{(x_{R_2} - x_r)^2 + (y_{R_2} - y_r)^2 + (z_{R_2} - z_r)^2}$$
$$\Delta 23 = d_2 - d_3$$
$$= \sqrt{(x_{R_2} - x_r)^2 + (y_{R_2} - y_r)^2 + (z_{R_2} - z_r)^2}$$
$$- \sqrt{(x_{R_3} - x_r)^2 + (y_{R_3} - y_r)^2 + (z_{R_3} - z_r)^2}$$
$$\Delta 34 = d_3 - d_4$$
$$= \sqrt{(x_{R_3} - x_r)^2 + (y_{R_3} - y_r)^2 + (z_{R_3} - z_r)^2}$$
$$- \sqrt{(x_{R_4} - x_r)^2 + (y_{R_4} - y_r)^2 + (z_{R_4} - z_r)^2}$$
$$\Delta 41 = d_4 - d_1$$
$$= \sqrt{(x_{R_4} - x_r)^2 + (y_{R_4} - y_r)^2 + (z_{R_4} - z_r)^2}$$
$$- \sqrt{(x_{R_1} - x_r)^2 + (y_{R_1} - y_r)^2 + (z_{R_1} - z_r)^2} \quad (9.12)$$

which is a typical hyperbolic system of positioning equations. In this system, the unknowns are (x_r, y_r, z_r) – position of the receiver. The resolution is done either by linearization or using a hyperbolic solving algorithm (see reference

[50] for details). The problem of indoor repeater-based positioning thus consists in measuring the differences Δji. These differences correspond to the code phase jumps while switching from one repeater to the next. The main goal is then to detect and measure these code phase jumps at the time of signal transition between two successive repeaters.

The advantages of repeaters over pseudolites are the simplicity of the electronics and the absence of the "near-far effect" because of simultaneous transmissions from several transmitters (there is only one transmitting at a given time for repeaters). Another advantage is the "time difference" measurement implemented. The accuracy obtained is better than that allowed by the GPS correlation function. However, the shorter distances and the differential effect are not sufficient to explain this improvement. It is necessary to take into account the intimate operation of the electronic tracking loops whose performances in terms of detection of "jumps" are excellent. This is what is highlighted in the repeater method because it is these jumps that we are looking for and the loops are designed to extract these peaks.

The main disadvantage of repeaters lies in the impossibility (or more precisely in the great difficulty) to carry out measurements of the carrier phase of the signal. Indeed, these measurements allow a significant gain in accuracy and are in particular used outside for centimetric GNSS. We have thus devised an approach that makes it possible to carry out these measurements, without, however, returning entirely to pseudolites.

9.5 Repealites

By combining the advantages of pseudolites and repeaters, we obtain the "repealites[5]" with the objective of improving indoor performance. The advantages are the possibility to perform carrier phase measurements ("pseudolite" aspect) and to use only one transmitted signal for all repealites ("repeaters" aspect). The problem then is that repealites must transmit simultaneously, which leads to the near-far effect already described.

9.5.1 Proposed System Architecture

The main idea consists in applying the repeater system but replacing the sequential mode by a delayed mode. In order to avoid interferences, repealites are time delayed in relation to each other in such a way that the received

---

5 Part of this paragraph is based on a publication: Nel Samama (03 February 2012). Indoor positioning with GNSS-like local signal transmitters, global navigation satellite systems, Shuanggen Jin, IntechOpen.

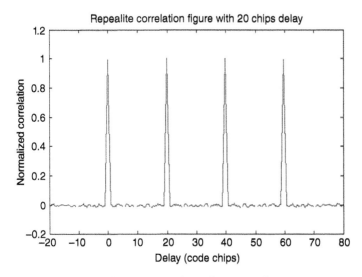

Figure 9.10 The resulting autocorrelation function at the receiver.

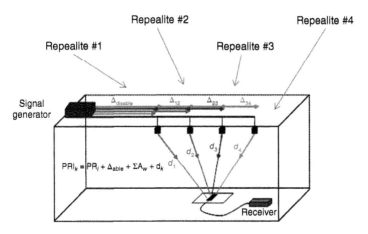

Figure 9.11 The repealite system.

signals at the receivers' antenna do not overlap. In such a case, we obtain a new autocorrelation function (provided in Figure 9.10): there is no longer only one peak for one code duration, but n peaks if n is the number of repealites.

Figure 9.11 proposes an implementation using a GNSS-like signal generator that involved a good synchronization of the transmitted signals. Note that as for repeaters, a single signal is sufficient.

The theory is also quite simple. Considering four delayed repealites, the terminal can measure four pseudo ranges. The equations to be solved are then the following ones (with the notations of Figure 9.11):

$$
\begin{cases}
PR_1 = d_1 + \Delta_{\text{cable}} \\
PR_2 = d_2 + \Delta_{\text{cable}} + \Delta_{12} \\
PR_3 = d_3 + \Delta_{\text{cable}} + \Delta_{12} + \Delta_{23} \\
PR_4 = d_4 + \Delta_{\text{cable}} + \Delta_{12} + \Delta_{23} + \Delta_{34}
\end{cases}
\tag{9.13}
$$

where one has

- PR_k the pseudo ranges measured,
- Δ_{cable} the cable delay (concerns the cable lying between the generator and the first repealite. Note that it should include both error and clock biases),
- Δ_{uw} the delays between repealites R_u and R_w
- d_k the distances between the receiver and repealite R_k.

There is the need to know[6] about the locations of the transmitters. In addition, the position is obtained in a local reference frame using indeed a classical GNSS algorithm. Note that (as long as the clock drift is unique for the four repealites) velocity can also be computed in the same referential, just like GNSS do outdoors.

9.5.2 Advantages

As repealites transmit continuously, it is possible to carry out carrier phase measurements of the signal. We know from GNSS experience that it can lead to a potential improvement in the positioning accuracy. This feature also makes it possible to cope with dynamic positioning quite easily with comparison to repeaters. In addition, the transition between indoors and outdoors is dramatically simplified as the same running mode is applied in the two environments.

Last, but absolutely not least (on the contrary!), the synchronization matters are highly simplified with respect to pseudolites. With repealites, there is no problem of synchronization between the transmitters as they are "driven" by a single generator. The drawback lies in the need for cables (or by the way of optical fibers[7]). The need for an initial calibration of the system could be required in order to compensate for variations of cable characteristics, but it remains at a reasonable level.

---

6 Some works are under consideration in order to propose methods for autopositioning the transmitters.
7 Optical fibers are also considered for the physical realization of the time delays between repealites.

9.5.3   Limitations

Multipath and the famous "near-far" effect are the two main remaining limitations. Multipath is dealt with through the use of either specific mitigation techniques (complex signal processing), averaging techniques over the time and the movement of the receiver, or in increasing the number of transmitters. This latter solution is very efficient when the transmitters are not fixed (case of the satellites, for instance), but not really useful when they are not (typically indoors). Thus, as it is not possible to be sure the receiver will be moving, there is the need to improve multipath signal processing techniques. The near-far effect is quite different as it depends highly on the codes that are used. A few research works have shown that quite efficient new codes could be found in order to significantly reduce, if not eliminate this problem completely.

9.6   Grin-Locs

All these transmitter-based systems nevertheless present a certain complexity of implementation in comparison to WiFi or Bluetooth approaches for which it is sufficient to "install" the transmitters. Of course, measurements are much more informative than power levels, but it would be good to further simplify the deployment. This is what we are aiming for with the latest technology in this chapter, the Grin-Locs.

Figure 9.12 represents the current version of such a system. In a typical configuration, every Grin-Loc transmits two synchronized signals on the same carrier frequency, each signal being characterized by a specific code. One can imagine a simple realization where each antenna transmits a GPS satellite

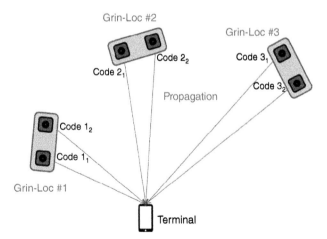

Figure 9.12 Standard Grin-Loc-based positioning system.

signal. In case one wants to get rid of the near-far effect, the two Grin-Locs could transmit on two frequencies.

A Grin-Loc is indeed an inverted radar and uses the simple technique of the interferometer. The main difference with the existing approaches is simply due to the fact that this is the receiver that carries out the measurements and thus the positioning. The other difference is that the transmitter transmits two coded signals. It allows a single receiver to carry out the differences of carrier phase measurements. Note that in the case of GNSS-like signals transmitted, a completely standard and current GNSS receiver is already enabled to work with Grin-Locs.

9.6.1 Double Antenna

The positioning is based on the differences of phase measurements from each Grin-Loc. This spatial diversity is then used to carry out geometrical intersections of lines (in case angle measurements are used) or quadrics. Indeed, there are two approaches for the calculations. The first one consists in the determination of an angle of arrival of the signal on the receiver antenna. The second is an analytical determination of the equation of the hyperboloid on which the receiver is supposed to lie. In this case, no hypotheses have to be made concerning the distance of the receiver to the Grin-Loc, but it leads to much more complex mathematics (as can be seen below). In all cases, it is thus compulsory to know about the location of the Grin-Locs. In addition, when quadrics are used, the orientation and posture of the Grin-Locs are also required. Figure 9.13 shows the typical configuration of the problem and the associated main parameters (used in the following equations). The coordinates of the two antennas of the Grin-Loc are (x_{a_1}, y_{a_1}) and (x_{a_2}, y_{a_2}), respectively, that of the receiver being (x_r, y_r).

The distance between the two antennas of a Grin-Loc should be less than or equal to one wavelength (λ) of the signal transmitted. This leads to nonambiguous measurements at the receiver. The nonambiguity means that a location corresponds to a difference of phase, and two different locations cannot be characterized by the same measurement. This is fundamental for simplifying the computation burden of the positioning.

The angle α is defined to the center of the Grin-Loc, i.e. in the middle of the two antennas. They are also the focal points of the hyperbola. In this latter case, the major difficulty is the very high sensitivity of the calculations to the measurement errors.

9.6.1.1 Angle Approach

Let us consider the two antennas of a given Grin-Loc are separated by l. Let now $\delta\varphi$ be the difference of phases between the signals from antenna 1 and antenna

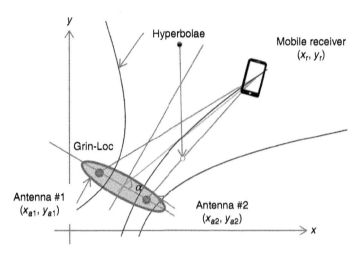

Figure 9.13 Standard geometry between a Grin-Loc and the receiver.

2 on the receiver. Then, it can be calculated that the angle α between the line of antennas and the receiver (as shown in Figure 9.13), is equal to

$$\alpha = \text{Arc}\cos\left(\frac{\delta\varphi}{\lambda}\right) \tag{9.14}$$

A simple formula, but unfortunately valid only if a hypothesis is made concerning the angles of arrival from the two antennas: they must be identical. This means the receiver should be far enough from the antennas. This hypothesis is usually considered as satisfied only if the distances between the antennas and the receiver are greater than typically 10 times the spacing between the transmitting antennas, i.e. in our present case 10λ (around 2 m when considering signals in the 1/1.5 GHz band). In order to calculate a position, it is then compulsory to carry out the intersection of several lines of sight and thus to use several Grin-Locs (as described in Figure 9.14).

9.6.1.2 Quadrics Approach

Considering the notations of Figure 9.13 with $\delta\varphi$ being the difference of carrier phases, one obtains

$$\delta\varphi = d_2 - d_1 = \sqrt{(x_{a_2} - x_r)^2 + (y_{a_2} - y_r)^2} - \sqrt{(x_{a_1} - x_r)^2 + (y_{a_1} - y_r)^2} \tag{9.15}$$

where d_2 and d_1 are the distances between the receiver and the antennas 1 and 2, respectively, of the Grin-Loc. Another form of (9.15) can be written as

$$d_2 = \delta\varphi + d_1 \tag{9.16}$$

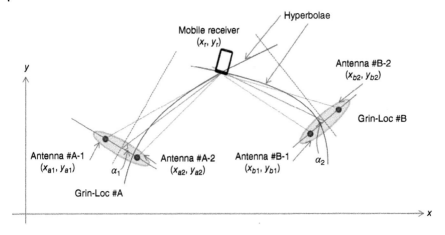

Figure 9.14 Positioning with two Grin-Locs.

which leads to

$$4\delta\varphi^2 d_1^{\;2} = (d_2^{\;2} - d_1^{\;2})^2 + \delta\varphi^4 - 2\delta\varphi^2(d_2^{\;2} - d_1^{\;2}) \tag{9.17}$$

where

$$
\begin{aligned}
d_2^{\;2} - d_1^{\;2} &= (x_{a_2} - x_r)^2 + (y_{a_2} - y_r)^2 \\
&\quad - [(x_{a_1} - x_r)^2 + (y_{a_1} - y_r)^2] \\
&= (x_{a_2}^{\;2} - x_{a_1}^{\;2}) + 2(x_{a_1} - x_{a_2})x_r + (y_{a_2}^{\;2} - y_{a_1}^{\;2}) \\
&\quad + 2(y_{a_1} - y_{a_2})y_r
\end{aligned}
\tag{9.18}
$$

By introducing the following intermediate variables,

$$d_2^{\;2} - d_1^{\;2} = 2\Delta X_{12}x_r + 2\Delta Y_{12}y_r + \Delta^2 X_{21} + \Delta^2 Y_{21} \tag{9.19}$$

$$
\begin{aligned}
(d_2^{\;2} - d_1^{\;2})^2 &= 4\Delta X_{12}^{\;2}x_r^{\;2} + 4\Delta Y_{12}^{\;2}y_r^{\;2} + (\Delta^2 X_{21} + \Delta^2 Y_{21})^2 \\
&\quad + 8\Delta X_{12}\Delta Y_{12}x_r y_r + 4\Delta X_{12}(\Delta^2 X_{21} + \Delta^2 Y_{21})x_r \\
&\quad + 4\Delta Y_{12}(\Delta^2 X_{21} + \Delta^2 Y_{21})y_r
\end{aligned}
\tag{9.20}
$$

with additionally

$$\Delta X_{12} = x_{a_1} - x_{a_2} \quad \text{and} \quad \Delta Y_{12} = y_{a_1} - y_{a_2} \tag{9.21}$$

$$\Delta^2 X_{21} = x_{a_2}^{\;2} - x_{a_1}^{\;2} \quad \text{and} \quad \Delta^2 Y_{21} = y_{a_2}^{\;2} - y_{a_1}^{\;2} \tag{9.22}$$

Then, Eq. (9.15) can be modified and finally be proposed under the following final form (9.23):

$$
\begin{aligned}
& x_r^{\;2}[\delta\varphi^2 - \Delta X_{12}^{\;2}] \\
&\quad + x_r[\delta\varphi^2 \Delta X_{12} - 2\delta\varphi^2 x_{a_1} - \Delta X_{12}(\Delta^2 X_{21} + \Delta^2 Y_{21})] \\
&\quad + y_r^{\;2}[\delta\varphi^2 - \Delta Y_{12}^{\;2}]
\end{aligned}
$$

$$+ y_r[\delta\varphi^2 \Delta Y_{12} - 2\delta\varphi^2 y_{a_1} - \Delta Y_{12}(\Delta^2 X_{21} + \Delta^2 Y_{21})]$$

$$- 2\Delta X_{12}\Delta Y_{12}x_r y_r$$

$$+ \delta\varphi^2(x_{a_1}^{\ 2} + y_{a_1}^{\ 2}) + \frac{\delta\varphi^2}{2}(\Delta^2 X_{21} + \Delta^2 Y_{21})$$

$$- \frac{1}{4}(\Delta^2 X_{21} + \Delta^2 Y_{21})^2 - \frac{\delta\varphi^4}{4} = 0. \tag{9.23}$$

As the generic form of a quadric is as follows (9.24):

$$A(x_r - x_{\text{ref}})^2 + B(y_r - y_{\text{ref}})^2 + C x_r y_r + D = 0 \tag{9.24}$$

where A, B, C, and D, and x_{ref} and y_{ref}, are coefficients of the quadric. Note that A to D are determined once the geometry of the problem is defined, i.e. the locations of the various Grin-Locs defined, and the difference of phases $\delta\varphi$ measured. Equation (9.24) is the typical shape of the quadric with a $x_r y_r$ crossed term (2D case; in 3D, the expression is a little bit more complex, but remains similar).

9.6.2 Resolution in Case of Several Double Antennas

A positioning will need the signal processing from several Grin-Locs, located at known positions. Both angle (in case the receiver is far enough from the transmitters) and quadric approaches are possible. Although two-dimensional and three-dimensional positioning are achievable, the equations below will only deal with the 2D case as the complexity of the 3D case is not useful for the understanding. Note also that Grin-Loc measurements are completely independent from each other (with respect to two different Grin-Locs), making the synchronization process between Grin-Locs totally useless. This is a fundamental improvement with regard to pseudolites as no cables or whatsoever are now required between Grin-Locs.

It leads to the need for two Grin-Locs or a 2D positioning and hence three for a 3D positioning. Figure 9.14 is an example of a typical geometry in a 2D configuration.

9.6.2.1 Positioning with the Angle Approach

Once the two angles mentioned in Figure 9.14 are measured, one has just to solve the system of equations composed of the following mathematical expressions:

$$\alpha 1_{21} = \text{Arc} \cos\left(\frac{\delta\varphi 1_{21}}{\lambda}\right)$$

$$\alpha 2_{21} = \text{Arc} \cos\left(\frac{\delta\varphi 2_{21}}{\lambda}\right) \tag{9.25}$$

It is possible to add a third measurement in order to define the positioning by intersection of three lines, and not only two. The real risk is to be unable to evaluate the real accuracy of the positioning with only two measurements as two lines always cross (unless parallel, but this case is highly improbable here). In such a case, an $\alpha 3_{21}$ angle would be measured, and the positioning would provide us with a possible triangle.

9.6.2.2 Positioning with the Quadric Approach

In the case of three Grin-Locs (the third one not being visible in Figure 9.14), the system of equations to be solved is given hereafter (9.26):

$$A_1(x_r - x_{ref1})^2 + B_1(y_r - y_{ref1})^2 + C_1 x_r y_r + D_1 = 0$$
$$A_2(x_r - x_{ref2})^2 + B_1(y_r - y_{ref2})^2 + C_2 x_r y_r + D_2 = 0$$
$$A_3(x_r - x_{ref3})^2 + B_1(y_r - y_{ref3})^2 + C_3 x_r y_r + D_3 = 0 \qquad (9.26)$$

where the A_i, B_i, C_i, D_i, and (x_{refi}, y_{refi}), $i = 1$–3, are the characteristics of the three quadrics. Of course, these parameters are determined once the complete geometry of the system, Grin-Locs and receiver, is given.

Let us now come back to our parameter table. The various parameters considered for this technology are given in Chapter 4. We have seen that several indoor GNSS technologies are available and present some differences. This is particularly true for the main parameters discussed in Table 9.2. However, the technology considered for the discussion is Grin-Loc, mainly because it seems to be the most promising one.

Infrastructure complexity, maturity, and cost: Although this technology seems interesting, it is still at the research level. Commercial pseudolites exist but indoor positioning using such transmitters has not yet been deployed on a significant scale. Thus, an important effort is still required for deployment. In particular, all the "logistical" associated problems on installation have not been dealt with yet. Nevertheless, the infrastructure is limited to transmitters with double antennas distributed throughout a building.

Table 9.2 Summary of the main parameters for indoor GNSS.

Infrastructure complexity	Infrastructure maturity	Infrastructure cost	Terminal complexity	Terminal maturity	Terminal cost	Smart-phone	Calibration complexity
Medium	Research	Medium	Low	Software development	Zero	Existing	None

Positioning type	Accuracy	Reliability	Range	Sensitivity to environment	Positioning mode	In/out transition	Calibration needed
Absolute	A few decimeters	Medium	Building	High	Continuous	Easy	None

Almost all difficulties associated with pseudolites or repeaters have been eliminated (such as synchronization between transmitters, synchronization with the receiver, quality of the measurements with respect to multipath, accuracy of individual measurement, etc.). Thus, the infrastructure complexity and cost are reduced. There remains the question of its maturity, which still asks for specific works.

Terminal complexity, maturity, and cost: Every GNSS receiver already carries out the required measurements. Indeed, the approach is rather easier than that currently used outdoors since ambiguity does not need to be estimated (this is the main difficulty for outdoors). In addition, the computations are either much simpler (case of angles) or of an equivalent (case of quadrics) complexity as outdoors. Furthermore, there are no cost-related issues as all receivers must carry out these measurements, from the cheapest to the most expensive.

Calibration complexity and need: There is no need for a calibration of the positioning system. Nevertheless, each Grin-Loc should be carefully manufactured as any mis-matching between the two paths from the signal generator to the antennas will create errors. Thus, at the factory, the Grin-Locs require a high level of engineering. Then, once installed in the considered environments, the classical constraint still exists: knowing where the antennas are deployed in the appropriate map (see Chapter 13).

Smartphone: Thanks to the Android operating system (since version 7.0), it is possible to have access to the carrier phase measurements. As it is intended in the case of outdoor GNSS, there will be a problem of ambiguity in the measurements once indoors. However, as we know that the Grin-Locs are providing nonambiguous measurements, it is easy to overcome this issue. Thus, Android-based smartphones are "Grin-Loc-based indoor positioning" ready. Unfortunately, operating systems are not the end of the story as chipsets should also make these values available (they are necessarily available in the chipsets in order for the GNSS receiver to work properly outdoors, but are not always available at the chip output). More and more chipsets are compatible, but a new problem arises: in order to reduce the power consumption of GNSS receivers embedded in smartphones, chipset manufacturers often implement a technique called "duty cycle" for the receivers. The induced effect consists in having the receiver working discontinuously, typically for only 10% or 20% of each second. This reduces the consumption but makes it impossible for the receiver to carry out "standard" (i.e. outdoors) carrier phase measurements. For Grin-Locs, this is of no importance as it can accommodate this mode, but for a standard receiver, it leads to no carrier phase availability (because of the impossibility to solve the ambiguity), thus leading to the unavailability of the carrier phase, even with the latest version of the operating system.

Positioning type: It is relative to the locations of the Grin-Locs, but the idea is clearly to provide the users with a complete continuity of service with outdoor GNSS. Thus, the principle is to give the Grin-Loc locations in GNSS-compatible coordinates. In such a case, a full continuity is achieved and absolute positioning is available all the time.

Accuracy: Centimeter accuracy is theoretically achievable, but "a few decimeters" appears to be the right level. Although individual differences of phases measurements are very good (down to a few millimeters in realistic conditions including multipath and people around, but not walls), even small errors can cause large positioning errors when equivalent angles are close to 90°, for example. The other difficulty is that individual Grin-Loc errors are combined for the calculation of a position of the receiver, increasing the potential inaccuracies. However, although not yet implemented, smoothing functions could be developed in order to reduce the impact of punctual measurement errors.

Reliability: Not so good because of the environments. In case one knows about it, then reliability could rise drastically. In the case of mass market applications, this is unfortunately never the case.[8]

Range: The "building" level considered is a little bit optimistic as wall crossing degrades the time-of-flight measurements. Nevertheless, at the cost of a sufficient power level transmitted, it is possible to carry out positioning through several walls and several floor levels. The accuracy and reliability decreases, but it is possible.

Sensitivity to environment: This is a radio-based technology, and for this reason, masking and nonline of sight conditions are of concern. Techniques are available to overcome these problems, but their efficiency is not at a sufficient level today.

Positioning mode: The continuity is indeed important in order to carry out good differences of phases. It is not compulsory as the measurements are nonambiguous and that there is not the necessity to continuously follow the carrier phases, but it is clearly better for allowing for error averaging.

In/out transition: Easy but not simple. The receiver should switch from the indoor mode with Grin-Locs to the outdoor one with satellites. This is not difficult to implement but a specific software should be developed.

Bibliography

1 Berkovich, G. (2014). Accurate and reliable real-time indoor positioning on commercial smartphones. In: *2014 International Conference on Indoor Positioning and Indoor Navigation (IPIN)*, 670–677. Busan: IEEE.

2 Chen, C.Y., Luo, T.H., Hwang, R.C., and Wang, S.T. (2013). A six-antenna station based indoor positioning system. In: *2013 2nd International*

---

8 This explains the differences observed between research results and field reality. The conditions are often well mastered for research purposes. Thus, such performances have to be seen as " optimal" ones, and absolutely not as " standard" ones.

Symposium on Instrumentation and Measurement, Sensor Network and Automation (IMSNA), 919–921. Toronto, ON: IEEE.

3 Chen, L.H., Wu, E.H.K., Jin, M.H., and Chen, G.H. (2014). Intelligent fusion of Wi-Fi and inertial sensor-based positioning systems for indoor pedestrian navigation. *IEEE Sensors Journal* 14 (11): 4034–4042.

4 Cullen, G., Curran, K., Santos, J. et al. (2014). To wireless fidelity and beyond—CAPTURE, extending indoor positioning systems. In: *2014 Ubiquitous Positioning Indoor Navigation and Location Based Service (UPINLBS)*, 248–254. Corpus Christ, TX: IEEE.

5 Crocoll, P., Caselitz, T., Hettich, B. et al. (2014). Laser-aided navigation with loop closure capabilities for Micro Aerial Vehicles in indoor and urban environments. In: *2014 IEEE/ION Position, Location and Navigation Symposium – PLANS 2014*, 373–384. Monterey, CA: IEEE.

6 Dovis, F., Chiasserini, C.F., Musumeci, L., and Borgiattino, C. (2014). Context-aware peer-to-peer and cooperative positioning. In: *International Conference on Localization and GNSS 2014 (ICL-GNSS 2014)*, 1–6. Helsinki: IEEE.

7 Gentner, C. and Jost, T. (2013). Indoor positioning using time difference of arrival between multipath components. In: *2013 International Conference on Indoor Positioning and Indoor Navigation (IPIN)*, 1–10. Montbeliard-Belfort: IEEE.

8 He, Z., Petovello, M., and Lachapelle, G. (2014). Indoor doppler error characterization for high sensitivity GNSS receivers. *IEEE Transactions on Aerospace and Electronic Systems* 50 (3): 2185–2198.

9 Hellmers, H., Norrdine, A., Blankenbach, J., and Eichhorn, A. (2013). An IMU/magnetometer-based indoor positioning system using Kalman filtering. In: *2013 International Conference on Indoor Positioning and Indoor Navigation (IPIN)*, 1–9. Montbeliard-Belfort: IEEE.

10 Herrera, J.C.A., Plöger, P.G., Hinkenjann, A. et al. (2014). Pedestrian indoor positioning using smartphone multi-sensing, radio beacons, user positions probability map and indoorOSM floor plan representation. In: *2014 International Conference on Indoor Positioning and Indoor Navigation (IPIN)*, 636–645. Busan: IEEE.

11 Hou, Y., Xue, Y., Chen, C., and Xiao, S. (2015). A RSS/AOA based indoor positioning system with a single LED lamp. In: *2015 International Conference on Wireless Communications & Signal Processing (WCSP)*, 1–4. Nanjing: IEEE.

12 Khider, M., Jost, T., Robertson, P., and Abdo-Sánchez, E. (2013). Global navigation satellite system pseudorange based multisensor positioning incorporating a multipath error model. *IET Radar, Sonar & Navigation* 7 (8): 881–894.

13 Lindgren, T. and Akos, D.M. (2008). A multistatic GNSS synthetic aperture radar for surface characterization. *IEEE Transactions on Geoscience and Remote Sensing* 46 (8): 2249–2253.

14 Lindo, A., García, E., Ureña, J. et al. (2015). Multiband waveform design for an ultrasonic indoor positioning system. *IEEE Sensors Journal* 15 (12): 7190–7199.

15 Martin-Gorostiza, E., Meca-Meca, F.J., Lázaro-Galilea, J.L. et al. (2014). Infrared local positioning system using phase differences. In: *2014 Ubiquitous Positioning Indoor Navigation and Location Based Service (UPINLBS)*, 238–247. Corpus Christ, TX: IEEE.

16 Murata, S., Yara, C., Kaneta, K. et al. (2014). Accurate indoor positioning system using near-ultrasonic sound from a smartphone. In: *2014 Eighth International Conference on Next Generation Mobile Apps, Services and Technologies*, 13–18. Oxford: IEEE.

17 Navarro, M., Closas, P., and Nájar, M. (2013). Assessment of direct positioning for IR-UWB in IEEE 802.15.4a channels. In: *2013 IEEE International Conference on Ultra-Wideband (ICUWB)*, 55–60. Sydney, NSW: IEEE.

18 del Peral-Rosado, J.A., Bavaro, M., Lopez-Salcedo, J.A. et al. (2015). Floor detection with indoor vertical positioning in LTE femtocell networks. In: *2015 IEEE Globecom Workshops (GC Wkshps)*, 1–6. San Diego, CA: IEEE.

19 Poudereux, P., García, E., Hernández, A. et al. (2013). Performance comparison of a TDMA- and CDMA-based UWB local positioning system. In: *2013 International Conference on Indoor Positioning and Indoor Navigation (IPIN)*, 1–9. Montbeliard-Belfort: IEEE.

20 Schatzberg, U., Banin, L., and Amizur, Y. (2014). Enhanced WiFi ToF indoor positioning system with MEMS-based INS and pedometric information. In: *2014 IEEE/ION Position, Location and Navigation Symposium – PLANS 2014*, 185–192. Monterey, CA: IEEE.

21 Selmi, I., Samama, N., and Vervisch-Picois, A. (2013). A new approach for decimeter accurate GNSS indoor positioning using carrier phase measurements. In: *2013 International Conference on Indoor Positioning and Indoor Navigation (IPIN)*, 1–6. Montbeliard-Belfort: IEEE.

22 Tiemann, J., Schweikowski, F., and Wietfeld, C. (2015). Design of an UWB indoor-positioning system for UAV navigation in GNSS-denied environments. In: *2015 International Conference on Indoor Positioning and Indoor Navigation (IPIN)*, 1–7. Banff, AB: IEEE.

23 Vaghefi, R.M. and Buehrer, R.M. (2014). Improving positioning in LTE through collaboration. In: *2014 11th Workshop on Positioning, Navigation and Communication (WPNC)*, 1–6. Dresden: IEEE.

24 Xu, W., Wang, J., Shen, H. et al. (2016). Indoor positioning for multiphotodiode device using visible-light communications. *IEEE Photonics Journal* 8 (1): 1–11.

25 Yan, K., Zhou, H., Xiao, H., and Zhang, X. (2015). Current status of indoor positioning system based on visible light. In: *2015 15th International Conference on Control, Automation and Systems (ICCAS)*, 565–569. Busan: IEEE.

26 Zampella, F., Jiménez Ruiz, A.R., and Seco Granja, F. (2015). Indoor positioning using efficient map matching, RSS measurements, and an improved motion model. *IEEE Transactions on Vehicular Technology* 64 (4): 1304–1317.

27 Bartone, C. and Van Graas, F. (2000). Ranging airport pseudolite for local area augmentation. *IEEE Transactions on Aerospace and Electronic Systems* 36 (1): 278–286.

28 Caratori, J., François, M., and Samama, N. (2002). Universal positioning theory based on global positioning system – upgrade. *InLoc2002*, Bonn, Germany.

29 Duffett-Smith, P and Rowe, R. (2006). Comparative A-GPS and 3G-MATRIX testing in a dense urban environment. *ION GNSS 2006*, Forth Worth, TX.

30 ECC report 145. (2010). Regulatory framework for GNSS repeaters, St. Petersburg.

31 ECC report 168. (2011). Regulatory framework for indoor GNSS pseudo-lites, Miesbach.

32 Fontana, R.J. (2004). Recent system applications of short-pulse ultra-wideband (UWB) technology. *IEEE Transactions on Microwave Theory and Techniques* 52 (9): 2087–2104.

33 Fluerasu, A., Jardak, N., Vervisch-Picois, A., et al. (2009), GNSS repeater based approach for indoor positioning: current status. *ENC-GNSS2009*, Naples, Italy.

34 Fluerasu, A. and Samama, N. (2009). GNSS transmitter based indoor positioning systems – deployment rules in real buildings. *13th IAIN World Congress*, Stockholm, Sweden.

35 Glennon, E.P., Bryant, R.C., Dempster, A.G., et al. (2007). Post correlation CWI and cross correlation mitigation using delayed PIC. *ION GNSS*, Forth Worth, USA.

36 Im, S.-H, Jee, G.-I, and Cho, Y.B. (2006). An indoor positioning system using time-delayed GPS repeater. *ION GNSS 2006*, Forth Worth, TX.

37 Jardak, N. and Samama, N. (2010). Short multipath insensitive code loop discriminator. *IEEE Transactions on Aerospace and Electronic Systems* 46: 278–295.

38 Jee, G.I., Choi, J.H., and Bu, S.C. (2004). Indoor positioning using TDOA measurements from switched GPS repeater. *ION GNSS 2004*, Long Beach, CA.

39 Kanli, M.O. (2004). Limitations of pseudolite systems using off-the-shelf GPS receivers. *The International Symposium on GNSS/GPS*, Sydney, Australia.

40 Kaplan, E.D. and Hegarty, C. (2017). *Understanding GPS/GNSS: Principles and Applications*, 3e. Norwood, MA: Artech House Publishers.

41 Kee, C., Yun, D., Jun, H., et al. (2001), Centimeter-accuracy indoor navigation using GPS-like pseudolites. *GPS World*.

42 Kee, C., Jun, H., and Yun, D. (2003). Indoor navigation system using asynchronous pseudolites. *Journal of Navigation* 56: 443–455.

43 Klein, D. and Parkinson, B.W. (1986). The use of pseudolites for improving GPS performance. *Navigation, Journal of the Institute of Navigation* 31 (4): 303–315.

44 Kupper, A. (2005). *Location Based Services—Fundamentals and Operation*. Chichester: Wiley.

45 Madhani, P.H., Axelrad, P., Krumvieda, K., and Thomas, J. (2003). Application of successive interference cancellation to the GPS pseudolite near-far problem. *IEEE Transaction on Aerospace and Electronic System* 39 (2): 481–487.

46 Martone, M. and Metzler, J. (2005). Prime time positioning: using broadcast TV signals to fill GPS acquisition gaps. *GPS World* 52–59.

47 Parkinson, B.W. and Spilker, J.J. Jr. (1996). *Global Positioning System: Theory and Applications*. Washington, DC: American Institute of Aeronautics and Astronautics, Inc.

48 Rizos, C., Barnes, J., Wang, J., et al. (2003). LocataNet: intelligent time-synchronised pseudolite transceivers for cm-level stand-alone positioning, *11th IAIN World Congress*, Berlin, Germany.

49 Samama, N. and Vervisch-Picois, A. (2005). 3D indoor velocity vector determination using GNSS based repeaters. *ION GNSS 2005*, Long Beach, USA.

50 Samama, N. (2008). *Global Positioning – Technologies and Performance*. Hoboken, NJ: Wiley.

51 Takada, Y., Kishimoto, M., Kawamura, N. et al. (2003). An information service system using Bluetooth in an exhibition hall. *Annales des Telecommunications* 58 (3–4): 507–530.

52 Vervisch-Picois, A. and Samama, N. (2006). Analysis of 3D repeater based indoor positioning system – specific case of indoor DOP, *ENC-GNSS 2006*, Manchester, UK.

53 Vervisch-Picois, A. and Samama, N. (2009). Interference mitigation in a repeater and pseudolite indoor positioning system. *IEEE Journal of Specific Topics on Signal Processing* 3 (5): 810–820.

54 Vervisch-Picois, A., Selmi, I., Gottesman, Y., et al. (2010). Current status of the repealite based approach – a sub-meter indoor positioning system. *IEEE-NAVITEC 2010*, Noordwijk, The Netherlands.

55 Wang, Y., Jia, X., and Rizos, C. (2004). Two new algorithms for indoor wireless positioning system (WPS). *ION GNSS 17th International Technical Meeting of the Satellite Division*, Long Beach, CA.

56 Yang, C. and Morton, J. (2009). Adaptive replica code synthesis for interference suppression in GNSS receivers. *ION ITM*, Anaheim, USA.

57 Samama, N., Vervisch-Picois, A., and Taillandier-Loize, T. (2016). A GNSS-like indoor positioning system implementing an inverted radar approach simulation results with a 6/7-antenna single transmitter. In: *2016*

International Conference on Indoor Positioning and Indoor Navigation (IPIN), 1–8. Alcala de Henares: IEEE.

58 Samama, N., Vervisch-Picois, A., and Taillandier-Loize, T. (2016). A GNSS-based inverted radar for carrier phase absolute indoor positioning purposes first experimental results with GPS signals. In: *2016 International Conference on Indoor Positioning and Indoor Navigation (IPIN)*, 1–8. Alcala de Henares: IEEE.

10

Wide Area Indoor Positioning: Block, City, and County Approaches

Abstract

As usual debatable, the difference between a building and a block is here dictated by the fact that the deterioration of the positioning increases with the distance covered (accelerometer) or with the range (local radio). Very high quality equipment can reach very high positioning performances, but average ones fit quite well with this chapter.

This chapter also deals with the so-called "long-range" positioning systems. The problem we have to face is linked to the indoor environments (remember this is a book about indoor positioning). The selected technologies are radio based and are all in the telecommunication domain. Nevertheless, they have not been designed for positioning purposes and the usual dichotomy still exists: for positioning purposes, one needs to detect the first signal that arrives (assuming line-of-sight propagation), whereas for telecommunication purposes, multipath are an opportunity. Thus, the performance of these technologies is very poor outdoors, and even worse indoors. Note that in addition to the poor accuracy, the reliability is also at a very low level. One of the only advantages of these techniques is the reduced number of transmitters required.

Keywords *Long range radio; city range; county range; TV*

The classification described in Chapter 4 led to the following table concerning these technologies (Table 10.1).

10.1   Introduction

The advent of radio systems, at the beginning of the twentieth century, led to a fantastic era of wireless data transmission. The most famous such systems are mobile telecommunications or the global navigation satellite system (GNSS), but there are numerous other radio systems.

Indoor Positioning: Technologies and Performance, First Edition. Nel Samama.
© 2019 The Institute of Electrical and Electronics Engineers, Inc. Published 2019 by John Wiley & Sons, Inc.

Table 10.1 Main "block," "city," and "county" technologies.

Technology	Positioning type	Accuracy	Reliability	Range	Sensitivity to environment	Calibration needed	Positioning mode	Technique	Signal processing	Position calculation
Amateur radio	Absolute	>100 m	Low	County	High	None	Continuous	Distance(s)	Propagation modeling	∩ Circles
Radio 433/868/... MHz	Absolute	>100 m	Low	Block	High	None	Continuous	Physical	Propagation modeling	∩ Circles
GSM/3/4/5G	Absolute	>100 m	Low	City	High	None	Continuous	Distance(s)	Propagation modeling	∩ Circles
LoRa	Absolute	>100 m	Low	City	High	None	Continuous	Distance(s)	Propagation modeling	∩ Circles
Sigfox	Absolute	>100 m	Low	City	High	None	Continuous	Distance(s)	Propagation modeling	∩ Circles
Radio AM/FM	Absolute	>100 m	Low	County	High	None	Continuous	Physical	Propagation modeling	∩ Circles
TV	Absolute	>100 m	Low	County	High	None	Continuous	Physical	Propagation modeling	∩ Circles

The main advantages of radio systems compared to optical ones are as follows:

- They operate in "all weathers," even in cloud, rain, or fog.[1]
- They can, through the use of appropriate antennas, be used with different radiating patterns, ranging from very directive beams (e.g. radar) to almost omnidirectional schemes (e.g. 4G).
- By diffraction,[2] they can have a range that is much further than the horizon.
- They can, by diffraction, be received behind obstacles.

The physical phenomena are identical in principle and very similar in results for optical waves and radio waves, but in the case of radio systems, the wavelengths are better adapted to broadcasting. Coming back to positioning systems, many approaches are then possible, using either the power levels received, the time of flight, or the direction of arrival of the signals.

10.2 Amateur Radio

Let us use this section to describe another quite original approach proposed a few years ago by amateur radio users. The system described below is an extension of the NCDXF (Northern California DX Foundation, Inc) and IARU (International Amateur Radio Union) beacon system and its aim is to provide rough positioning. This system is intended to provide amateur radio with a means of estimating the radio propagation conditions in various frequency bands.

Here, the idea is to implement a transmitting cycle where the power level of the transmission is progressively increased. Let us imagine four possible power levels and the following cycle P_1, P_2, P_3, and P_4 with, as an example, P_1 equal to 1 mW, P_2 to 10 mW, P_3 to 100 mW, and P_4 to 1 W. The rough distance (accuracy is not being sought) can then be determined, evaluating the power level received. This is based on the fact that the power level decreases as the inverse of the square of the distance when propagating in free space. Of course, many phenomena can disturb this relation, but this can be acceptable as a first approximation (and remains quite efficient when the propagation is effectively carried out in free space with no buildings, hills, or obstacles).

Let us now imagine that you receive P_2 but not P_1; it means that you are at a distance somewhere between the range corresponding to P_1 and that between P_1 and P_2, i.e. with an uncertainty equal to the square root of 10 (typically 3). This is indeed used to evaluate the quality of the radio link and not for

---

1 Although heavy snow or very high humidity can cause some trouble at some frequency bands.
2 Diffraction refers to change in the directions and intensities of waves when passing by an obstacle or through an aperture.

Table 10.2 Summary of the main parameters for amateur radio.

Infrastructure complexity	Infrastructure maturity	Infrastructure cost	Terminal complexity	Terminal maturity	Terminal cost	Smartphone	Calibration complexity
None	None	Zero	Low	Integration	Low	Moderate	None

Positioning type	Accuracy	Reliability	Range	Sensitivity to environment	Positioning mode	In/out transition	Calibration needed
Absolute	>100 m	Low	County	High	Continuous	Moderate	None

positioning purposes in the case of the NCDXF and IARU systems, but this extension is very easy to implement.[3]

Let us now come back to our parameter table. The various parameters considered for this technology are given in Chapter 4 (Table 10.2).

It is a totally dedicated approach and requires a specific implementation. This would be possible with many radio systems, but is not really in the current logic. We mention it here because the approach is simple and intelligent, but performance is not what is expected today and indoor operation is not guaranteed.

10.3 ISM Radio Bands (433/868/… MHz)

As previously said, many approaches are possible, already described for WiFi or Bluetooth technologies. The main difference here is related to the fact that radio transmitters are often mobile (car or garage remote control, systems temporarily deployed for an event, etc.). When they are fixed, they are also often "proprietary" and thus dedicated to a given function that is not necessarily available to the general public. This does not matter if you want to make a positioning system, but it is then necessary to know where to position them on a map. This information is generally not available. The case of mobile transmitters is even more complex because it is of little use knowing that you are receiving a signal that cannot be located.

In addition, the regulation of ISM (Industrial, Scientific, and Medical) bands authorizes transmissions, under certain conditions, without making any request to the telecommunications regulatory authorities. Indeed, in many countries, these bands are free (maximum power levels, and sometimes duty cycles, must still be respected). This makes it very difficult to identify transmissions, even for fixed transmitters. The latter, in addition to not being

---

3 Note that this technique is used for indoor WLAN positioning in many cases considering the received power level throughout a building (in this case, taking into account a complex propagation situation). See Chapter 9 for details.

Table 10.3 Summary of the main parameters for radio 433/868 MHz.

Infrastructure complexity	Infrastructure maturity	Infrastructure cost	Terminal complexity	Terminal maturity	Terminal cost	Smart-phone	Calibration complexity
None	None	Zero	Low	Integration	Low	Easy	None

Positioning type	Accuracy	Reliability	Range	Sensitivity to environment	Positioning mode	In/out transition	Calibration needed
Absolute	>100 m	Low	County	High	Continuous	Moderate	None

listed, generally have the ability to change frequency, precisely in order to avoid potential interference with other systems in the same band (as they are free, they are usually crowded).

Let us now come back to our parameter table. The various parameters considered for this technology are given in Chapter 4 (Table 10.3).

Infrastructure complexity, maturity, and cost: These various parameters are perfectly controlled, both economically and technically. Many systems exist and are deployed.

Terminal complexity, maturity, and cost: The same remarks apply here for the terminal.

Calibration complexity and need: There is no need for "calibration" in the direction used until now in the book. It would rather be an initialization of the electronics. As the equipment is intended to be installed in many environments, it usually has multiple channels and must therefore "synchronize" with the transmitter and avoid interference with other systems using the same frequency band. Thus, some basic manipulations are often necessary.

Positioning type: it could be absolute, depending on the identification and localization of the transmitters.

Accuracy: Not so good, mainly depending on the range of the transmitters.

Smartphone: Not yet available and probably not in the near future because of the complexity of the approach. Many other technologies are much better indeed.

10.4 Mobile Networks

10.4.1 First Networks (GSM)

As explained in Section 3.2.4, in order to forward communications, the GSM network needed to have access to a database that keeps track of the mobile

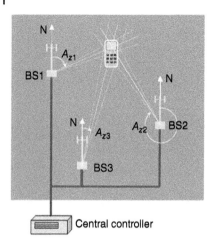

Figure 10.1 Angle of arrival principle.

locations. Indeed, the location is simply the identification of the base station providing the greatest power level to the receiver.

The next step, before the time-based solutions, is to come back to a very old method: measuring angles (see Figure 10.1). Indeed, in order to increase the capacity of a base station, operators have chosen to develop specific antennas that have the ability to determine the absolute direction of arrival of a signal, also called angle of arrival (AOA), relative to the antenna plane. This enables the channel used by a user in direction say D_1 to be used within the same base station by another user whose direction relative to the antenna of the base is D_2, sufficiently different from D_1. Thus, once again, this feature has been designed for telecommunications purposes.

Of course, as in the olden days, measuring the angle from two or three bases can be used to calculate a location. This is also very comparable to what was done by the sailors when measuring angles from landmarks in order to plot their location. In the case of wide area telecommunication networks, the idea is to carry out such measurements from three base stations. Assuming the accuracy of the angle measurement is around 1°, and the range around 1 km, one obtains an accuracy of position of about 100 m. The accuracy is not very good, and this approach is absolutely not applicable if the direct signal (line of sight) is absent, i.e. urban or indoor positioning is not intended to be really obtained with AOA.

The main disadvantage of this technique is that the antennas required are really complex and can only be implemented at the base station end. Furthermore, there is the need for the definition of a reference frame within which all the angles are calculated: thus, if different base stations have to be used together, this will require a precise orientation common to all the three bases. Note also that even if the direction of arrival is calculated in three dimensions, i.e. with in fact two angles, the way it has been imagined to be used is only in a two-dimensional positioning manner, i.e. considering only one angle for the AOA value. No real implementation of such an AOA positioning system is known, certainly because of the many constraints required in order to make the positioning possible and the complexity of the antennas and their deployment.

Therefore, power level measurements are not likely to provide good enough accuracy and an AOA method is too complex and really not acceptable for indoor areas. Quite logically, solutions implementing time measurements have been thought of. Different possibilities are open to us, such as direct time measurements or difference of time measurements. The main problem with time-based methods in telecommunication networks is that requirements, in terms of time precision, are once again not similar for telecommunication purposes and for positioning purposes. Telecommunication exchanges are based on protocols of transmission that include a synchronization feature, usually by specific heading data sent before the real data transmission, in order to define an identical "starting time" for both the transmitter and the receiver. For positioning purposes, one needs, as discussed in previous chapters, very good synchronization because the resulting localization is directly linked to it. Nevertheless, some methods have been proposed, as shown in Figures 10.2 and 10.3.

The basic idea of time of arrival (TOA) is to make direct time measurements between the mobile terminal to be located and various base stations. For similar reasons to GNSS systems, there is the need for three different measurements in order to calculate a two-dimensional position. As for GNSS, there is the need to know the bias of synchronization for each base station to a reference time (such as the global positioning system (GPS) time), since 10 ns of bias will directly lead to 3 m of error. As the base stations are in a network, it has been found to be easier to implement the time measurements at the base station end: thus, the

Figure 10.2 Time of arrival approach.

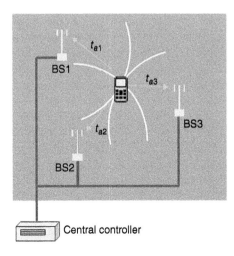

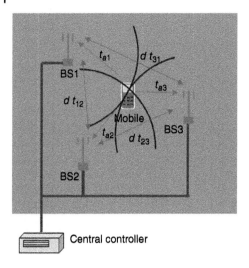

Figure 10.3 Time difference of arrival approach.

mobile sends data and the bases carry out measurements, use the synchronization bias, and finally calculate the mobile location. This location can then be sent back to the terminal, upon request.

Because of the poor time accuracy in telecommunication networks, the resulting accuracy is around 100 m, in the best cases: indeed, as the base stations are usually scattered all around the place, direct radio visibility is far from being usual. When multipath occurs, and multipath occurs very often in telecommunication networks, the accuracy drops dramatically to a few hundreds of meters. This is the typical indoor performance of such a technology.

A way of minimizing synchronization bias is to make differences of arrival times: this gives good results when the biases are of similar values for the two bases taken into account for the difference.

As with GNSS, considering differences of distances rather than the distances themselves leads to carrying out the intersection of hyperbolae rather than the intersection of circles for this two-dimensional problem. From a system of three base stations, one can obtain three equations when considering the time measurements, and only two when considering the differences. Theoretically, the two systems give the same solution, but practically, time difference of arrival (TDOA) allows less accurate time management. There was only one such system that included a new method based on multiple measurements and refinement of the positioning when many mobile terminals are in the vicinity of the one we want to locate. This system was called Matrix and was proposed by Cambridge Positioning System (see Figure 10.4).

This technique could also be used indoors, but because of the same limitations as for TOA and AOA, the results were not very good.[4] An advantage

---

4 However, it greatly depends on the application.

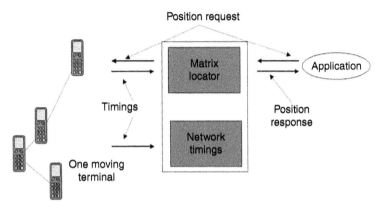

Figure 10.4 The matrix positioning approach.

of this technique compared to TOA is the possibility, at no expense, to have it implemented directly at the terminal end: this brings it closer to GNSS approaches.

To be complete on GSM-related positioning matters, other theoretical techniques should be mentioned. The first one is the combination of the Cell-Id with the so-called timing advance. As a matter of fact, within the GSM networks, the problem of the potential collision between two transmissions is a major concern. As the GSM is based on a time division multiple access (TDMA) scheme, it means that each transmission is allocated a time slot within a frame of eight time slots. The need for synchronization is then obvious in order to avoid simultaneous transmissions. The way synchronization is achieved through the wireless network is that each terminal transmits with a fixed delay of three time slots (compared to the first time slot received from the base station). Unfortunately, as mobile terminals can be found at quite different distances from the base station, transmissions may still overlap while the propagation delay from the base to the terminals is not taken into account. This collision problem is commonly dealt with by arranging guard times. However, as the maximum radius of a cell has been set to around 35 km, the corresponding guard time needed to avoid transmission overlaps (collisions) would be much too large to be handled with no further refinements, mainly because it would involve a drastic reduction in the network's capacities. Thus, the idea is to advance the transmission of any given mobile terminal (compared to the transmission time corresponding to three time slots after the receipt of the first time slot from the base station transmission) by the amount of its distance to the base station. This approach is called the timing advance (see Chapter 3) and is once again required by the network. In order to provide the terminal with the timing advance value, it is necessary that the base station permanently measures the so-called round trip time (more or less the propagation from base station to terminal and return). Then, the terminal transmission is "advanced" from the classical three time slot delay by this amount in order to reduce the required guard time and thus improve the global communication capabilities.

10.4.2 Modern Networks (3G, 4G, and 5G)

All the above techniques could have been implemented with GSM. The current mobile telecommunication system is based on universal mobile telecommunication system (UMTS),[5] 4G, or in the near future 5G networks. New names have been used such as OTDOA for "observed time difference of arrival," but no real differences exist with the above-mentioned techniques.[6] The new point in UMTS, compared to GSM, is that positioning was thought of at the beginning of the standardization.[7] Thus, the possibility to have positioning implemented is taken into account in the protocols and also the fact that this positioning could be achieved by means of different techniques, i.e. UMTS but also GNSS or even wireless local area network (WLAN): specific localization data are planned to be included in the protocols.

Concerning 5G, planned for 2020, some orientations have been given, as follows. A white paper states that 5G networks will have to be able to locate a device using triangulation, with an accuracy of 10 to less than 1 m in 80% of cases and less than 1 m in indoor areas, such as shops. Other "markets" have been identified such as autonomous cars or drones. The challenges are accuracy and latency time. The latter is expected to be as low as possible (close to a millisecond) to give the car all the reactivity it needs in relation to other vehicles, its environment, and traffic information provided by the smart city. As often, these orientations are really very high level and it is not enough to say that we will carry out triangulation for the performances to follow (as we have seen so far in the book). The question of geolocation in future networks therefore remains, in particular on the performance that will be possible indoors and on their reliability. However, as the basic idea is to allow very wide interconnectivity between the various wireless telecommunication networks, technical solutions mixing the many approaches described so far in the book could be considered depending on availability, for example, "symbolic WiFi" where the necessary data are accessible, LiFi or ultra wide band (UWB). In addition, of course, 5G-specific approaches when none of this is possible.

Let us now come back to our parameter table. The various parameters considered for this technology are given in Chapter 4 (Table 10.4).

---

5 Note that the "Global" term means total coverage of the Earth for GNSS and only local and not 100% for GSM, as Universal seems to be comparable to the GNSS meaning of Global.
6 The only difference concerns the way measurements are carried out. Please refer to further readings for details on the methods implemented in the UMTS networks.
7 "Positioning" here is considered as with GNSS, i.e. in terms of precise location of the terminal, and no longer in terms of "telecommunication like" positioning, i.e. approximation of the place the terminal is.

Table 10.4 Summary of the main parameters for GSM/3/4/5G.

Infrastructure complexity	Infrastructure maturity	Infrastructure cost	Terminal complexity	Terminal maturity	Terminal cost	Smart-phone	Calibration complexity
None	Existing	Zero	None	Existing	Low	Existing	None

Positioning type	Accuracy	Reliability	Range	Sensitivity to environment	Positioning mode	In/out transition	Calibration needed
Absolute	>100 m	Low	City	High	Continuous	Moderate	None

Infrastructure complexity, maturity, and cost: The necessary infrastructure is relatively complex, but it is not specific to positioning, and even more not specific to indoor positioning. This is the same questions as those of the late 1990s, when the various technologies competed for global positioning. Telecommunication networks are not made for this, but allow positioning. Thus, the infrastructure remains heavy in itself, but the positioning function does not imply any significant additional cost. As a result, the infrastructure can be considered to remain free, to exist and to be of reduced complexity.

Terminal complexity, maturity, and cost: The terminals are in principle completely standard. A "software" part must be included to take positioning into account, but this is part of the proposed terminal.

Calibration complexity and need: There is normally no calibration required at the terminal or user level. On the other hand, it is possible that calibrations are necessary at the level of buildings in certain cases. The basic idea is not to have one, but the accuracy sought, especially for the 5G (1 m) are likely to require more than a global model.

Positioning type: It can be absolute without any difficulty.

Accuracy: Quite difficult to say really. As a matter of fact, one should differentiate between what is announced, what is reported, and what is really achievable (the three are in decreasing order usually). In case there is no hybridization between technologies, it is rather poor (100 m). The fact to reduce the size of the cells is not always an advantage because the way connectivity is achieved often relies on the "best quality" of the radio link, meaning it is in majority not the closest stations that is considered. In case of hybridization (see Chapter 12 for a few details), it can be quite good, although 1 m seems to be rather difficult to reach in real and daily life conditions with a mass market terminal.

Reliability: Depending on the accuracy sought, it can be acceptable (i.e. if one accepts having a very poor accuracy. Otherwise, because of propagation aspects, it is rather low.

Range: This is the "good news" concerning mobile telecommunication networks. The large deployments mean there is a large coverage, hence a large range.

Smartphone: The terminal is "by definition" the smartphone.

Sensitivity to environment: Very high indeed because it combines a few difficulties. The use of radio waves, the environments where it is deployed, and the ranges sought. It makes the propagation modeling very complex for accurate positioning purposes.

Positioning mode: It is continuous.

In/out transition: Still the same comments. Such a transition is possible without difficulties as mobile signals are also available indoors, but positioning performances are highly dependent on the propagation conditions and they are much more difficult (for positioning purposes) indoors than outdoors. Thus, the "moderate" level considered for this technology.

10.5 LoRa and SigFox

From WPAN (wireless personal area networks) for very small ranges (a few meters) to telecommunication networks for wide ranges (a few tens of kilometers), the wireless systems encountered are, respectively, the WLAN for medium–small ranges (a few hundreds of meters) and the WMAN (wireless metropolitan area networks) for medium–large ranges (a few kilometers). The most widely known WMAN was called WiMax. Similar positioning systems can be envisaged such as WPAN (Bluetooth) and WLAN (WiFi), with an increased range. The LoRa and SigFox networks are of this type.

Unfortunately, the discrepancy between the telecommunication network and the positioning network is once more highlighted. The first one uses transmitters with greater range in order to reduce the number of such transmitters to be deployed, whereas the second one needs far more transmitters. This difficulty has been overcome for Bluetooth and WiFi because of the reduced size of the area being considered (a few hundred square meters). When dealing with a few square kilometers, the foreseen problem is that with only three transmitters, the size of the resulting positioning areas could still be quite large, thus requiring an increased number of transmitters (more than three). This does not seem to be a problem nowadays with LoRa and SigFox networks, but all the problems already described concerning time measurements, propagation modeling for positioning, and measurement error mitigation are still present. Several performance levels have been reported by the various actors implementing these networks (LoRa and SigFox), ranging from a few dozen meters to a few kilometers. This range of values recalls what happened 20 years ago in the field of mobile telephone networks. Good performances are possible in very special

Table 10.5 Summary of the main parameters for LoRa and SigFox.

Infrastructure complexity	Infrastructure maturity	Infrastructure cost	Terminal complexity	Terminal maturity	Terminal cost	Smart-phone	Calibration complexity
None	Existing	Zero	Low	Integration	Low	Easy	None

Positioning type	Accuracy	Reliability	Range	Sensitivity to environment	Positioning mode	In/out transition	Calibration needed
Absolute	>100 m	Low	City	High	Continuous	Easy	None

cases but are probably not significant from the technology. In addition, as usual, what about the reliability of both the positioning and its repeatability?

Let us now come back to our parameter table. The various parameters considered for this technology are given in Chapter 4 (Table 10.5).

Infrastructure complexity, maturity, and cost: It is being set up and deployed on a relatively large scale today. The technical maturity is there and markets exist for these networks. However, their use to offer a positioning service is not taken for granted and performance inside buildings is clearly not promoted.

Terminal complexity, maturity, and cost: Many terminals are available, not all of which necessarily offering localization. The question of the future of these networks arises at a time when 5G wants to extend its characteristics widely to the Internet of Things (or "Internet of Everything" as some are beginning to say).

Calibration complexity and need: The only need is, as usual, to know about the identifier and location of the transmitters and to implement the corresponding algorithms. No real difficulties.

Positioning type: Absolute.

Accuracy: Several different values have been reported, ranging from a few dozen of meters to kilometers. Indoors, values in between these two levels are bound to be the right ones: typically a few hundreds of meters.

Reliability: This is the major drawback of the technology. The dependency of the real performance is so high with propagation conditions that it cannot be at a good level. In addition, these networks are bound to be deployed in urban areas, where these propagation conditions are even worth.

Range: This is the strength of the technology. Ranges of several kilometers are common and this would significantly reduce the number of required transmitters. The problem of efficiency then becomes crucial.

Smartphone: Not yet available as the goal is to cope with "things" and sensors, but there are no foreseen difficulties in implementing LoRa or SigFox receivers in smartphones.

Sensitivity to environment: At a very high level, as explained above.

Positioning mode: Continuous.

In/out transition: The technology has not been designed for positioning purposes (as it is often the case with telecommunication systems) and thus no optimization has been carried out, neither outdoors nor indoors. However, apart from the positioning performance aspects, the system works in both environments.

10.6 AM/FM Radio

All the techniques that have been described in this section relative to the local area telecommunication systems could clearly be implemented by means of radio modules that are not intended to play any telecommunication role. Using WPAN, WLAN, or WMAN systems is done simply in order to reduce the cost of the positioning system. Thus, it should be easy to carry out the same measurements with amplitude-modulated/frequency-modulated (AM/FM) radio signals. The obvious advantage is the large availability of these signals worldwide. The only new requirement in order to allow positioning would be to list all the stations with the corresponding locations and radio characteristics (identifiers, frequencies, etc.). The range of these signals is quite long and the transmitters already exist. The main problem is clearly that nothing has been designed in order to cope with all the positioning-related matters (synchronizations, propagation modeling, etc.), leading to very poor performances. One can try to consider these signals as "opportunity signals," considering it should be possible to extract location information from them, but it appears not to be really useful indeed.

Let us now come back to our parameter table. The various parameters considered for this technology are given in Chapter 4 (Table 10.6).

Infrastructure complexity, maturity, and cost: Broadcast radio signals are probably one of the most usual signals worldwide. The coverage of people is of the

Table 10.6 Summary of the main parameters for AM/FM radio.

Infrastructure complexity	Infrastructure maturity	Infrastructure cost	Terminal complexity	Terminal maturity	Terminal cost	Smart-phone	Calibration complexity
None	Existing	Zero	Low	Integration	Low	Easy	None

Positioning type	Accuracy	Reliability	Range	Sensitivity to environment	Positioning mode	In/out transition	Calibration needed
Absolute	>100 m	Low	County	High	Continuous	Moderate	None

same kind of television signals and the maturity of the technology is at the best level. One can also consider the cost of the infrastructure as negligible as it is deployed for broadcast purposes. In the present case, as positioning is not a genuine feature (and even not a proposed one), no additional infrastructure is intended to be deployed. Thus, the cost can be seen as really zero.

Terminal complexity, maturity, and cost: AM/FM radio receivers are really off the shelf, very low price electronic systems. In order to propose positioning, one should slightly modify the way they run by adding new measurements that are indeed almost already available. It would just need a piece of software.

Calibration complexity and need: The main idea here is to implement nothing specific to positioning.

Positioning type: It should be absolute.

Accuracy and reliability: Quite poor, in association with an equally poor reliability. The basic idea is precisely the opposite of the one required for positioning indeed: one should be able to receive everywhere over the larger possible range with the simplest as possible piece of electronic, in all possible environmental conditions. As with the large majority of telecommunication systems, the most important characteristics required for positioning have not been implemented because of the corresponding increase in cost, complexity, and "radio" performance.

Range: The "county" level used here refers to the typical deployment of such system rather than to positioning-related capabilities.

Smartphone: Often available through the use of headsets, the cable of which plays the role of the antenna. Thus, AM/FM receivers are implemented.

Sensitivity to environment: At such a high level that only research papers have been issued concerning positioning with these kinds of signals.

Positioning mode: It could be continuous.

In/out transition: Possible with even worth performance indoors.

10.7 TV

As already suggested, every radio signal can be used for positioning purposes. If propagation time measurements are used, then the main constraint is the time synchronization and efforts must be carried out to try to overcome this problem. TV signals, for example, have been used in such a way. The system, called LuxTrace,[8] was divided into three parts:

- A mobile terminal that can be a mobile phone equipped with a TV tuner including a TV measurement module that receives TV signals and calculates pseudo ranges.

---

8 Proposed by Rosum Corporation in 2006.

Table 10.7 Summary of the main parameters for TV.

Infrastructure complexity	Infrastructure maturity	Infrastructure cost	Terminal complexity	Terminal maturity	terminal cost	Smart-phone	Calibration complexity
None	Existing	Zero	Medium	Integration	Medium	Future	None

Positioning type	Accuracy	Reliability	Range	Sensitivity to environment	Positioning mode	In/out transition	Calibration needed
Absolute	>100 m	Low	County	High	Continuous	Moderate	None

- A location server to calculate the position of the mobile terminal.
- A regional monitor unit that measures certain clock characteristics of TV signals and sends time correction data to the location server.

A communication channel is required between the TV measurement module and the location server and between the regional monitor unit and the location server. The TV signal range is typically 50–100 km.

Results conducted indoors reported a median position error of less than 50 m, whereas the 67th and 95th percentile values were 58 and 95 m, respectively. Outdoor results (with line-of-sight to TV transmitters) reported a median position error of less than 5 m, whereas the 67th and 95th percentile values were 4.9 and 13.6 m, respectively. As everybody can see, the deployment of such a system has not been tremendous and all these reported results are obtained in very specific environments that are not significant of a real use. It could have been a candidate, but it is not so easy to adapt a system to positioning as the required features are very specific and technically difficult to obtain.

Let us now come back to our parameter table. The various parameters considered for this technology are given in Chapter 4 (Table 10.7).

The same comments as for AM/FM radio apply. The only differences in the tables are related to the availability on smartphones and, correlatively, the terminal maturity. In fact, smartphones including TV tuners were available 10 years ago: the technical feasibility has been achieved. This is the "usage" that has completely changed: television is still seen on current smartphones, but the radio channel used for the transmission is Internet Data and no longer television broadcast. Thus, TV tuners are no longer implemented on smartphones, leading to the practical impossibility to implement TV signal-based positioning.

Bibliography

1 Tubbax, H., Wouters, J., Olbrechts, J. et al. (2009). A novel positioning technique for 2.4GHz ISM band. In: *2009 IEEE Radio and Wireless Symposium*, San Diego, CA, 667–670. IEEE.

2 Rauh, S., Lauterbach, T., Lieske, H. et al. (2017). Temporal evolution analysis of indoor-to-outdoor radio channels in the 868-MHz ISM/SRD frequency band. In: *2017 47th European Microwave Conference (EuMC)*, Nuremberg, 384–387. IEEE.

3 Montilla Bravo, A., Moreno, J.I., and Soto, I. (2004). Advanced positioning and location based services in 4G mobile-IP radio access networks. In: *2004 IEEE 15th International Symposium on Personal, Indoor and Mobile Radio Communications* (IEEE Cat. No. 04TH8754), Barcelona, vol. 2, 1085–1089. IEEE.

4 Kos, T., Grgic, M., and Sisul, G. (2006). Mobile user positioning in GSM/UMTS cellular networks. In: *Proceedings ELMAR 2006*, Zadar, 185–188. IEEE.

5 Liu, D., Sheng, B., Hou, F. et al. (2014). From wireless positioning to mobile positioning: an overview of recent advances. *IEEE Systems Journal* 8 (4): 1249–1259.

6 Omelyanchuk, E.V., Semenova, A.Y., Mikhailov, V.Y. et al. (2018). User equipment location technique for 5G networks. In: *2018 Systems of Signal Synchronization, Generating and Processing in Telecommunications (SYNCHROINFO)*, Minsk, 1–7. IEEE.

7 Witrisal, K., Hinteregger, S., Kulmer, J. et al. (2016). High-accuracy positioning for indoor applications: RFID, UWB, 5G, and beyond. In: *2016 IEEE International Conference on RFID (RFID)*, Orlando, FL, 1–7. IEEE.

8 Montilla Bravo, A., Moreno, J.I., and Soto, I. (2004). Advanced positioning and location based services in 4G mobile-IP radio access networks. In: *2004 IEEE 15th International Symposium on Personal, Indoor and Mobile Radio Communications* (IEEE Cat. No. 04TH8754), Barcelona, vol. 2, 1085–1089. IEEE.

9 Caffery, J. (2000). *Wireless Location in CDMA Cellular Radio Systems*. Kluwer Academic Publishers, IEEE.

10 Caffery, J.J. and Stüber, G.L. (1998). Overview of radiolocation in CDMA cellular systems. *IEEE Communications Magazine* 36 (4): 38–45.

11 Duffett-Smith, P. and Rowe, R. (2006). Comparative A-GPS and 3G-Matrix testing in a dense urban environment. *ION GNSS 2006*, Forth Worth, TX (September 2006).

12 Yang, F., Huang, J., Yao, S. et al. (2016). 3/4G multi-system of indoor coverage problems location analysis and application. In: *2016 16th International Symposium on Communications and Information Technologies (ISCIT)*, Qingdao, 376–380. IEEE.

13 Zhang, Y., Gao, R., and Bian, F. (2007). A conceptual architecture for advanced location based services in 4G networks. In: *2007 International Conference on Wireless Communications, Networking and Mobile Computing*, Shanghai, 6525–6528. IEEE.

14 Mayorga, C.L.F., Rosa, F.D., Wardana, S.A. et al. (2007). Cooperative positioning techniques for mobile localization in 4G cellular networks. In: *IEEE International Conference on Pervasive Services*, Istanbul, 39–44. IEEE.

15 Amineh, R.A. and Shirazi, A.A.B. (2014). Estimation of user location in 4G wireless networks using cooperative TDoA/RSS/TDoA method. In: *2014 Fourth International Conference on Communication Systems and Network Technologies*, Bhopal, 606–610. IEEE.

16 Fargas, B.C. and Petersen, M.N. (2017). GPS-free geolocation using LoRa in low-power WANs. In: *2017 Global Internet of Things Summit (GIoTS)*, Geneva, 1–6. IEEE.

17 Baharudin, A.M. and Yan, W. (2016). Long-range wireless sensor networks for geo-location tracking: design and evaluation. In: *2016 International Electronics Symposium (IES)*, Denpasar, 76–80. IEEE.

18 Randall, J., Amft, O. and Tröster, G. (2005). Towards LuxTrace: using solar cells to measure distance indoors. *Location and Context Awareness LoCA 2005*, Oberpfaffenhofen, Germany (May 2005).

19 Martone, M. and Metzler, J. (2005). Prime time positioning: using broadcast TV signals to fill GPS acquisition gaps. *GPS World* 16 (9): 52–59.

20 Moghtadaiee, V. and Dempster, A.G. (2014). Indoor location fingerprinting using FM radio signals. *IEEE Transactions on Broadcasting* 60 (2): 336–346.

21 Chen, L., Julien, O., Thevenon, P. et al. (2015). TOA estimation for positioning with DVB-T signals in outdoor static tests. *IEEE Transactions on Broadcasting* 61 (4): 625–638.

22 Rahman, M.M., Moghtadaiee, V., and Dempster, A.G. (2017). Design of fingerprinting technique for indoor localization using AM radio signals. In: *2017 International Conference on Indoor Positioning and Indoor Navigation (IPIN)*, Sapporo, 1–7. IEEE.

11

Worldwide Indoor Positioning Technologies: Achievable Performance

Abstract

This last category includes technologies that are either not intended to provide positioning or designed for completely different environments than indoors. Therefore, performances are generally highly disturbed indoors. The counterpart is clearly the total absence of the need for a complementary infrastructure. Among technologies that are not intended to provide positioning (at least alone), one can note the wide range of approaches: pressure sensors in order to provide an altitude or wired networks that are able, under certain conditions, to identify your location. An exception to that is the magnetometer based so-called magneto-inertial approach, potentially well adapted for indoor positioning, but indeed implementing a first kind of "hybridization." (*Chapter 12 will discuss, among others, a few hybridization approaches.*)

Keywords *World wide range approaches; pressure; GNSS; Magnetometer; magneto-inertial; Wired networks*

The classification described in Chapter 4 led to the following table concerning these technologies (Table 11.1).

11.1 Argos and COSPAS-SARSAT Systems

These two systems are based on the same technique (Doppler) for positioning but are designed for two different ways of use: Argos for scientific purposes and COSPAS-SARSAT for human life protection.

11.1.1 Argos System

The Doppler-based technique used for positioning is described in Section 3.3.2. Clearly, this system is not intended to provide indoor positioning, and the way the Doppler is used cannot be applied indoors (because of the fact that the

Indoor Positioning: Technologies and Performance, First Edition. Nel Samama.
© 2019 The Institute of Electrical and Electronics Engineers, Inc. Published 2019 by John Wiley & Sons, Inc.

Table 11.1 Main "worldwide" technologies.

Technology	Positioning type	Accuracy	Reliability	Range	Sensitivity to environment	Calibration needed	Positioning mode	Technique	Signal processing	Position calculation
COSPAS-SARSAT–Argos	Absolute	>100 m	Medium	World	High	None	Continuous	Frequency(ies)	A combination of	∩ Straight lines
GNSS	Absolute	100 m	Low	World	Very high	None	Continuous	Time(s)	A combination of	∩ Spheres
High-accuracy GNSS	Absolute	100 m	Low	World	Very high	Once	Continuous	Phase(s)	A combination of	∩ Spheres
Magnetometer	Orientation	A few degrees	Medium	World	Moderate	Several times	Continuous	Physical	Detection	Math functions $(\int, \int\int, \int\int\int, \ldots)$
Pressure	Relative	1 m	High	World	No impact	Several times	Continuous	Physical	Detection	Zone determination
Signaux radio opp	Absolute	>100 m	Low	World	High	None	Almost continuous	Physical	Propagation modeling	∩ Circles
Wired networks	Absolute	An address	Medium	World	No impact	None	Discrete	Fusion	Correlation	Zone determination

transmitters are in fact not moving). However, the interesting part is probably the fact that Doppler shift measurements are not often mentioned for indoor and this is a shame as it is a very powerful way to know whether a terminal is moving or not.

ARGOS is the result of the cooperation between France (Centre National d'Etudes Spatiales (the French Space Agency) [CNES]) and the United States (National Oceanic and Atmospheric Administration [NOAA]). Different beacons exist allowing various missions, from following the migrations of animals to the surveillance of the polar ice. The smallest beacons can weigh as little as 20 g. The transmitted signal has a frequency of 401.65 MHz, and each beacon is allocated a unique identification number. Figure 11.1 shows an overview of the ARGOS system.

The satellites have a polar orbit with a visibility of about 5000 km in diameter (the altitude of the orbit is between 830 and 870 km). The Earth is scanned a few times each day. Each time a satellite crosses over a global receiving station (there are currently two such stations), it downloads the data collected from ARGOS beacons. Other regional stations are also included in the distribution process in order to reduce the latency of the system. Finally, there are five processing centers located in Toulouse (France), Washington DC (United States), Lima (Peru), Tokyo (Japan), Jakarta (Indonesia), and Melbourne (Australia). Their goal is to

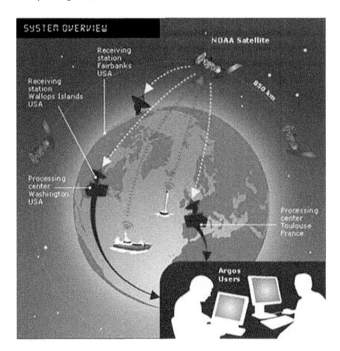

Figure 11.1 Overview of ARGOS system. Source: Copyright CLS.

process the raw data from the receiving station in order to make them available to users (a typical latency of less than 20 minutes). The accuracy of positioning is typically 300 m.

11.1.2 COSPAS-SARSAT System

COSPAS-SARSAT is also a Doppler measurement locating system (see Figure 11.2) aimed at providing assistance to mobile units in distress. Two frequencies are used leading to different positioning accuracies: 406 MHz (with an accuracy of 2 km) and 121.5 MHz (with an accuracy of 13 km). Each beacon is typically flown over 24 times a day. More than 10 000 people have been saved since 1982 in the maritime, aerial, and terrestrial domains. The COSPAS (COsmicheskaya Sistema Poiska Avarinykh Sudov, a satellite system for searching boats in distress) satellites have a quasi-polar orbit at an altitude of 1000 km, whereas SARSAT (Search and Rescue Aided Tracking Satellites) satellites orbit at 850 km (also with a quasi-polar orbit).

This program is a joint initiative between the United States (National Aeronautics and Space Administration [NASA]), France (CNES), Canada

Figure 11.2 Overview of COSPAS-ARSAT system. Source: Copyright: CNES/Ill./DUCROS David, 2002.

(Department of National Defense [DND]), and Russia (MORFLOT[1]) and started at the end of 1970s. The nominal constellation is composed of four satellites, and about 40 ground stations are in operation in 20 associated countries.

The principle of operation is based on the following parts:

- Reception and first processing of signals by the satellites
- Data transmission to ground stations
- Identification of people or equipment in distress
- Processing of data by mission control centers

Let us now come back to our parameter table. The various parameters considered for this technology are given in Chapter 4 (Table 11.2).

Infrastructure complexity, maturity, and cost: The case of these systems is a little particular here because they were really not designed for indoor positioning. The maturity of the system is at an excellent level. The associated costs are not intended to be considered by the general public.

Terminal complexity, maturity, and cost: The terminals are available in multiple forms that mainly depend on the use made of them. There are mobile terminals the size of portable global positioning system (GPS) receivers.

Calibration complexity and need: No calibration is required.

Accuracy: It is in the range of one to a few hundred meters. Some receivers are also equipped with modern global navigation satellite system (GNSS) systems and some are intended to be part of the SAR (Search and Rescue) Galileo program.

Reliability: It is at a high-level outdoors but probably not so good indoors; hence, the "medium" level is considered here.

Range: It is clearly a worldwide system.

Table 11.2 Summary of the main parameters for COSPAS-SARSAT and Argos.

Infrastructure complexity	Infrastructure maturity	Infrastructure cost	Terminal complexity	Terminal maturity	Terminal cost	Smart-phone	Calibration complexity
None	None	Zero	High	Existing	Medium	Almost impossible	None

Positioning type	Accuracy	Reliability	Range	Sensitivity to environment	Positioning mode	In/out transition	Calibration needed
Absolute	>100 m	Medium	World	High	Continuous	Easy	None

---

1 Ministry of Merchant Marine.

Smartphones: These devices are designed for very specific implementations and applications and are not really part of integration with a standard smartphone. In particular, the applications are generally carried out in relatively degraded weather conditions and thus require a much higher mechanical strength of the terminal than that offered by our modern smartphones.

Sensitivity to environment: There has been, to my knowledge, no specific work on indoor environments. However, the type of signals and the type of measurements performed suggest that such environments will not be very favorable.

Positioning mode: It is continuous.

11.2 GNSS

The advent of positioning over the past few years has clearly been due to the incredible success of GPS. Indoor positioning appears to be one of the most challenging problems for GNSS constellations as it represents their main current limitation. This was identified early on in many works carried out both on the receivers' detection capabilities and the so-called "local elements" or "local area augmentation systems" (LAAS). GNSS-based indoor positioning is not yet fully resolved, although interesting solutions have been proposed.

Of course, the first systems whose indoor capabilities were evaluated were the satellite navigation-based ones. The history of applications has shown that the success of satellite navigation was anything but planned by the original designers. Being the evolution of the TRANSIT[2] system, designed to fulfill military maritime applications (in order to allow aircraft or terrestrial vehicles a wider range of use), no one could have imagined, at the beginning, the indoor application. However, with the advent of modern telecommunication systems for personal users, a real need emerged. Unfortunately, for GPS[3], the power level of the signal received is too low for indoors as the signal margin allowed by the code correlation is about 10 dB, which is far too low to envisage the penetration of walls and other structures. One usually considers that, unless the building is wooden, attenuation generated by the structures when a radio signal penetrates, at 1.575 GHz, is between 15 and 30 dB. As it was not possible, at that time, to change the code's length because of the navigation message (see Ref. [1] for more details on the global structure and induced limitations of the signal used), the only possible direction for GPS was to develop more sensitive receivers. The various techniques are given below. For three or four years, from roughly 2000, a lot of effort was made on this approach, which finally appears not to be the definite answer to the indoor problem. Nevertheless, it enabled the GPS manufacturers to propose

---

2 TRANSIT was the first American satellite positioning system launched in the 1960s.
3 Here, the first system is GPS and not yet GNSS.

receivers that nowadays can work inside a car equipped with an athermic windscreen[4]: this was obviously not the first goal but still remains an interesting result.

The Galileo program had to be more efficient on indoor positioning. Therefore, the European Union decided to introduce the concept of "local elements" as a fundamental difference between Galileo and GPS. The timing of this decision was about the same as that when all manufacturers said that high sensitivity receivers would solve the problem: thus, a universal solution was foreseen at that time. Unfortunately, as often with the "great programs," once someone has the miracle solution, no other idea was investigated: then, when the high sensitivity approach, together with the almost only backup solution (namely ultra wide band [UWB]), appeared not to be the ultimate answer, no other acceptable solution was available.[5]

The solutions described below are intended to present the state of the art of the most "promising" approaches based on the use of satellite navigation signals. Of course, competition exists between solutions that require a local infrastructure and the others. Not having to add more relays or base stations (BSs) is better; however, although a great deal of technical and financial effort has been put into finding a solution, no results have been obtained yet. Thus, no-infrastructure solutions do not resolve the problem, although they do improve the indoor performance.

Let us now come back to our parameter table. The various parameters considered for this technology are given in Chapter 4 (Table 11.3).

Infrastructure complexity, maturity, and cost: Everything has of course been in place for years for GPS and GLONASS and is on the way to being so for BEIDOU and Galileo. Maturity is well established. As far as costs are

Table 11.3 Summary of the main parameters for GNSS.

Infrastructure complexity	Infrastructure maturity	Infrastructure cost	Terminal complexity	Terminal maturity	Terminal cost	Smart-phone	Calibrat complex
None	None	Zero	None	Existing	Zero	Existing	None

Positioning type	Accuracy	Reliability	Range	Sensitivity to environment	Positioning mode	In/out transition	Calibrat needed
Absolute	100 m	Low	World	Very high	Continuous	Easy	None

---

4 Such a windscreen introduces an attenuation of about 10 dB at 1.575 GHz, making the search for satellites fail in the first receiver technologies.

5 Only "marginal" solutions, developed by small teams were therefore alternatives, but they were not likely to be applied to a program such as Galileo.

concerned, things are a little different: it is the countries that are at the origin of the expenses relating to the various projects. Thus, it was the citizens of each country with a constellation who funded the project. Indeed, there is no subscription-type resourcing planned. The states' return on investment is achieved either through a dominant position of its industry (as in the case of the United States, for example) or through associated paid services (as in the case of Galileo and the planned value-added services).

Terminal complexity, maturity, and cost: The cost of the GNSS function is now reduced to a few dollars or even less. The main problem when dealing with the interior of buildings is the poor location performance and the availability of mapping (discussed in Chapter 13).

Calibration complexity and need: The system requires no calibration.

Positioning type: It is absolute, and this was the real first mass market product to propose such a common geographical reference frame worldwide.

Accuracy: Quite poor indoors as already stated in the previous chapters. Note that this is precisely the combination of the large availability of GNSS receivers and the poor indoor performances that led to this book. The indoor problem comes from the availability of GNSS outdoors, which raises the need for continuity of the positioning service.

Reliability: It is very good outdoors, once again, but very poor indoors. In many cases, it is not even available at all. Nevertheless, the very good aspect is the calculation, in real time, of an estimated accuracy. This latter is rather acceptable outdoors and not at all indoors, but it exists!

Range: The GPS was the first mass market system providing the users with a really global, i.e. worldwide positioning system.

Smartphone: All smartphones are nowadays equipped with a GNSS receiver.

Sensitivity to environment: This is the main problem with GNSS when dealing with indoor. Probably, more than other radio systems, the GNSS are sensitive to their environment. This is mainly due to both the very low level of transmitted power and to the fact measurements are based on times of flight between the satellites and the receiver.

Positioning mode: It is completely continuous by design (this was one strong specification of the GPS when evolving the first satellite-based positioning system, namely TRANSIT).

In/out transition: The satellites are not dependent on the fact that the receiver is outdoors or indoors. In that sense, the transition from outdoors to indoors is quite easy, although GNSS often do not work indoors.

11.3   High-Accuracy GNSS

In this book, high-accuracy GNSS is supposed to group high-accuracy techniques as well as the so-called "Assisted-GNSS," which is intimately linked to high accuracy and sometimes quite difficult to deinterleave from high accuracy. This section proposes two sections dedicated to the two approaches, respectively.

11.3.1 HS-GNSS

The search domain of a satellite signal is huge, both in frequency and in time. The frequency search is required in order to deal with the Doppler shift because of the motion of both the satellite and the receiver: as the correlation process is indeed a comparison of a local replica of the satellite code with the incoming satellite code, the replica must take this Doppler shift into account.[6] The time search is required in order to determine the propagation time shift between the transmission times from the satellite to the receiving time at the receiver's end. However, both searches are rather large: about ±10 kHz in frequency and as much as 1 ms (the duration of a complete code) in time. The steps of both searches are also rather small: a few Hertz for the frequency and a fraction of a chip[7] for the time: thus, the time required to lock on to a satellite is rather long.

Of course, this time is further increased when the receiver tries to find very low power signals, which are bound to occur indoors. Furthermore, with a very low signal, the search process is even more difficult because of the fact that the signal peak is not significantly higher than others. Thus, finding a way to cope with low signals could necessarily help in an indoor situation.

The various methods that have been developed toward this high sensitivity goal are, respectively, as follows:

- Complex electronic systems to allow direct frequency processing in order to find the frequency's peak at once,
- Multiple correlation in order to achieve parallelism,
- Long integration, either coherent or noncoherent in order to find very low pseudo random noise codes.

The first method can be achieved through the use of a Fourier transform. Unfortunately, this approach consumes a lot in terms of power supply and the complexity of the corresponding electronics. As the power consumption of a GNSS receiver is a major concern, other directions were investigated.

The second approach adopts a different philosophy: the idea is to try all the possibilities in the frequency and time domains at once, i.e. in parallel. The quasi-immediate electronic architecture would be to have as many processing channels as there are elementary possibilities. Let us consider a frequency

---

6 This allows the replica to have the right chip duration and then lead to a good quality correlation.

7 A "chip" is the duration of one bit of the code used for identifying (and carrying out the time of flight from the satellite) a satellite.

step of 10 Hz for a complete range of 10 kHz (i.e. ±5 kHz), which leads to one thousand possibilities. Let us also consider a GPS code of 1023 chips and an elementary time step of one chip, which leads to another one thousand possibilities. The complete search domain then consists of about one million possibilities. If it is possible to build an electronic device including one million parallel channels, then the treatment of all the possible combinations in time and frequency for one satellite can be achieved in one clock time duration. As a matter of fact, this should allow in one clock time all the correlation values to be output: the processing of the right location of the peak still has to be computed. Note that current receivers have typically between 14 and 20 channels, which are usually used by associating one channel to a given satellite. The search process is then conducted in a sequential mode. One of the first industrial realizations incorporated 32 000 parallel correlators. Further products have exhibited more than 200 000 correlators in parallel.

Another way to track very low signals is to use the characteristic of pseudo random noise features: the fact that it is not random at all and that if you know what you are searching for, it is possible to "integrate" the energy held in a code by repeating the correlation a few times in a row. Of course, this requires the correlation to be "followed" and "kept" as time goes on. This approach is called the "long integration." There are two kinds of long integration, depending on whether the integration is carried out in a time continuous manner or at some discrete times. Coherent long integration is limited, in GPS, by the global form of the signal. The navigation message is the main reason (the other one is the time required by the receiver electronics in order to carry out the correlations): because of the 50 Hz data rate, there is a 20 ms time interval during which the code remains identical (either the code or the inverse of the code, depending on the value of the data bit of the message) to itself. Thus, the receiver can achieve a "coherent" integration within a 20 ms period, at most.

Note that to carry out longer integration, the immediate possibilities would be to have either lower data rates for the navigation message, or even no navigation message at all. Both approaches have been considered in the Galileo program and in the GPS modernization program. Different navigation data rates have been proposed and also the so-called "pilot tones" that are signals without navigation data. The purpose of these signals is clearly to help low-level signal detection and acquisition.

A typical high-sensitivity GNSS (HS-GNSS) positioning system therefore does not require any further infrastructure, other than that of the GNSS. That it theoretically needs nothing in addition to the current constellation is the major advantage of this approach. Unfortunately, although greatly improving the receivers' performances in difficult environments, the determination of the correlation peak remains too difficult when the power level received is very low. Thus, false detections are possible and degrade the positioning. Nevertheless, positioning is still possible although imperfect. As the basic principle of the positioning is to carry out classical time measurements, the accuracy cannot be improved compared to an outdoor configuration with good reception conditions.

11.3.2 A-GNSS

A typical Assisted-GNSS (A-GNSS) positioning system is provided in Figure 11.3. It includes a lot of elements: an Assisted-GNSS server, a special handset that includes the specific "assisted" processing capabilities, and the specific telecommunication protocols for assisted data exchange. Here, the basic idea is to "assist" a GNSS receiver to allow it both to find a location in difficult environments (in the same sense as for HS-GNSS) and dramatically reduce the Time To First Fix, which is a major concern for applications such as LBS (location-based services) for personal users. Solutions implemented are thus quite similar to those developed for HS-GNSS on the one hand and for hybridization for GSM-like positioning on the other hand. In addition, using the transmission capabilities of the telecommunication network allows an immediate improvement in the future performances of pilot tones. The Assisted-GNSS server, located at the base station of the telecommunication network, acquires the GNSS constellation navigation message and transmits it to the Assisted-GNSS receiver. Thus, it can simply remove the navigation message from the received signal (coming from the satellites). In such a way, a coherent integration method can be applied and the 20 ms limitation no longer applies. Of course, the use of a HS-GNSS chip is possible and the fact that the A-GNSS server has to acquire the constellation makes it possible both to give an initial location for the mobile positioning, which is rather near the mobile

Figure 11.3 A typical Assisted-GNSS configuration (BS stands for base station).

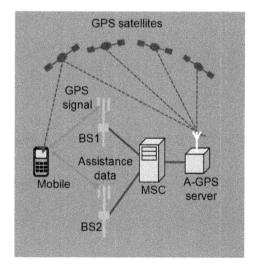

location and a good enough time to facilitate the reduction of the TTFF.[8] As already described in the GSM hybridization approach, another possibility implemented is to transmit information about which satellites to look for first.

The basic principle of the positioning is to carry out time measurements such as standard GNSS receivers. The availability of assisted data helps in reducing the TTFF down to a few seconds when leaving an obstructed area, but gives no further answer to indoor positioning, unlike HS-GNSS, for example. Furthermore, although not requiring any additional local infrastructure for buildings, this method is only made possible when the assisted server is deployed and the mobile terminal must be compatible with assisted data (i.e. it is not a current standard receiver). Some Assisted-GNSS providers propose to furnish assisted data on a worldwide basis: in this way, both hardware and complete software suites are available throughout the world for a rapid deployment. In this competitive world, the US companies are the most important ones, both in technological advances and in business development.

A-GNSS systems are facing the fact that the deployment is quite expensive and telecommunication operators want to be sure that potential users will show enough interest in order to cover the investments. The US situation is very different from that of European countries, for instance. Indeed, the The Federal Communication Commission (FCC) recommendation is a strong invitation to telecommunication operators to implement a location finding solution: A-GNSS is, from this point of view, an interesting approach that at least is currently possible (unlike other techniques, such as pseudolites or repeaters). In European countries, the regulation is not based on an obligation but on a "best effort" to be carried out by both telecommunication providers and terminal manufacturers. This is apparently not the most efficient way to stimulate industrial development. Furthermore, A-GNSS has shown that it provides a real added value compared to "just" GNSS, although this is still not the final answer to indoor positioning.

Let us now come back to our parameter table. The various parameters considered for this technology are given in Chapter 4 (Table 11.4).

Infrastructure complexity, maturity, and cost: This combination of high-sensitivity and Assisted-GNSS receivers has been used for many years on all smartphones. Thus, maturity, cost, and necessary infrastructure are already in place.

Terminal complexity, maturity, and cost: The same applies to terminals that already integrate these technologies.

Calibration complexity and need: As with conventional GNSS, no calibration is required. This is of course one of the strong points of these approaches: the self-calibration of the device makes it very easy for everyone to use.

---

8 TTFF stands for time to first fix and corresponds to the time required by the receiver in order to provide the user with its first positioning, typically after a turn on or a reset action.

Table 11.4 Summary of the main parameters for high-accuracy GNSS.

Infrastructure complexity	Infrastructure maturity	Infrastructure cost	Terminal complexity	Terminal maturity	Terminal cost	Smartphone	Calibration complexity
High	None	Zero	High	Existing	Very high	Future	None

Positioning type	Accuracy	Reliability	Range	Sensitivity to environment	Positioning mode	In/out transition	Calibration needed
Absolute	100 m	Low	City	Very high	Continuous	Difficult	None

Positioning type: Absolute, as with GNSS.

Accuracy: This is where things get a little more complicated. The interest of these technologies was originally to allow the provision of an indoor position: quite quickly, this objective was set aside because the expected performance was not achieved. However, real progress has been made in terms of reception when a few decibels of reception power were missing (through an athermal windscreen, for example) or when it was a question of obtaining a position very quickly (i.e. reducing the TTFF).

Reliability: In terms of reliability, in environments where a receiver is located in indoor environments, the reliability of these technologies is no better than that of conventional GNSS (we should probably say "older GNSS" as the deployment of high-sensitivity receivers is so important). It may even be worse in some cases: indeed, the increase in receiver sensitivity sometimes allows interference signals (such as reflected paths, for example) that were not previously received to be detected, leading to the deterioration of the positioning. Inside, the latter not being good anyway, it is not catastrophic.

Range: It is here that we see the limit of the proposed classification. Although the system's coverage is global, indoor coverage is almost zero.

Smartphone: Already available on almost all the smartphones.

Sensitivity to environment: At the highest level imaginable. It is the combination of the very low received power level and the type of measurement (time of flight) that makes the system particularly sensitive to its propagation environment.

Positioning mode: It is continuous, when available.

In/out transition: This transition was the intended goal, but could not be achieved.

11.4 Magnetometer

As accelerometers, such as gyroscopes and odometers, are primarily relative sensors, there could be the need for absolute ones in order to allow an absolute positioning, such as GNSS fixes. Nevertheless, accelerometers are sometimes designed in order to provide an inclinometer: in such a case, it is possible to define the horizontality of the mobile. This is a first approach of an absolute sensor as this is achieved without the need for any former attitude. Another important parameter is the absolute orientation of a terminal[9]: in applications where the discovery of the environmental world is required, this feature is a must. For instance, in a museum, the electronic guide should certainly take advantage of the fact that it knows what the visitor is looking at. This is also important when one wants to be oriented when taking a first step: one must probably know which direction to follow. With current GNSS receivers, one needs to start moving before this information is relevant.

Magnetometers are sensitive to the Earth's magnetic field and thus are available all around the world, without any calibration required. The main direction is the magnetic north, which is slightly different from the geographical north (this must be taken into account, at least by staying in the same referential, either magnetic or geographical). The difference is the declination, experimentally discovered by Christopher Columbus during his travels to "India."

In recent years, some work has focused on the development of systems based on the use of the magnetic field, following an approach similar to that of WiFi in its "fingerprint" version. It is then a question of carrying out a local mapping of the magnetic field, typically at the level of a building. This calibration is made necessary by infrastructure elements (walls, doors, metal elements, cables, etc.) that locally modify the Earth's field. It is in fact these modifications that are used and considered to be specific to each structure.

Another approach using magnetic field measurement is called "magneto-inertial."[10] The principle of the approach is based on a basic equation that links the evolution of the magnetic field of the sensors in their own reference frame when the latter is moving relative to the terrestrial reference frame (see Refs. [2–4]). This approach requires the measurement of various quantities such as the Earth's magnetic field (measured by magnetometers), the rotation speed of the system relative to the Earth's reference frame (via gyrometers), and finally a matrix (dB/dX) relating to the spatial variations of the magnetic field (which is estimated from several magnetometers spatially distributed locally in the system). It is important to note that the system is integrated and portable. In its current version, it is the same size as a satellite navigation receiver. The reported performance indicates a relative accuracy error of about 1% of the distance traveled. However, this approach requires the integration of several sensors (which will be discussed in Chapter 12) and only works well if the magnetic field gradient is not too low (e.g. in large halls).

---

9 Note that GNSS signals do not provide this information, unless in dynamic mode.
10 This technology has been developed in particular by the Sysnav company.

Table 11.5 Summary of the main parameters for magnetometers.

Infrastructure complexity	Infrastructure maturity	Infrastructure cost	Terminal complexity	Terminal maturity	Terminal cost	Smart-phone	Calibration complexity
None	None	Zero	Low	Existing	Low	Existing	Light

Positioning type	Accuracy	Reliability	Range	Sensitivity to environment	Positioning mode	In/out transition	Calibration needed
Orientation	A few degrees	Medium	World	Moderate	Continuous	Already exist	Several times

Let us now come back to our parameter table. The various parameters considered for this technology are given in Chapter 4. In this case, the table remains relative to the base magnetometer, the sensor allowing a user to be provided with an absolute orientation in the terrestrial reference frame (Table 11.5).

Infrastructure complexity, maturity, and cost: As with all inertial systems, there is no infrastructure. More precisely in the case of magnetometers, it is the land itself that is the infrastructure: it is thus present in an "innate" way.

Terminal complexity, maturity, and cost: All these elements are very mature. Costs are extremely low and integration into any terminal is very easy.

Calibration complexity and need: A calibration is necessary, but in some cases, it can be performed automatically (depending on the modernity and complexity of the magnetic sensor used).

Positioning type: Such a sensor does not allow positioning but provides terminal orientation information. However, it should be noted that such data are sometimes just as important as the position. The example of automotive GPS is a very good example: when you did not know the initial orientation of your vehicle (this is the case when you turn on your receiver at a position that is not that of the last recording, where you turned it off last time), the latter is unable to tell you which your current orientation is. In general, it makes a hypothesis, and if it is not the right one, suggests after a few meters[11] to "turn around."

Accuracy: The accuracy is given in degrees. It depends not only on the quality of the sensor but also on the stabilization systems that are incorporated into it. It is typically a few degrees in the case of today's smartphones.

Reliability: The main difficulty indoors is the presence of many elements that can disturb the measurement. It is not the sensor that is concerned but its

---

11 Indeed, even without the presence of a magnetometer, the GNSS receiver calculates the direction of travel in an absolute reference frame (in fact, the three-dimensional velocity vector). Thus, we can consider that the latter is a dynamic magnetometer, i.e. an electronic compass when the receiver is in motion).

environment: metal parts specific to buildings (concrete reinforcement, electrical cables, etc.) or office or home supplies (offices, beds, cabinets, shelves, etc.) are all disruptive to the Earth's magnetic field. Errors, in a corridor, for example, can reach several tens of degrees quite conventionally.

Range: Geographical coverage is the good news because it is really worldwide. Of course, the horizontal component of the Earth's magnetic field is zero at the poles, but indoor positioning in these areas is not really the main objective of the book.

Smartphone: Already available on almost all current smartphones.

Sensitivity to environment: The comments proposed in the "reliability" section also apply here.

Positioning mode: It can work continuously (and even at a rather high data rate: tens of Hertz are commonly reached).

In/out transition: The transition did not present any difficulties other than that related to the change of environment with the potential disruptions described above.

11.5   Pressure Sensor

One of the most important differences between outdoor navigation and indoor navigation is certainly the third dimension. As a matter of fact, outdoor navigation is primarily a planar system. Having knowledge of the altitude is interesting regarding general information but is really not required for most applications. Indoors, the problem is totally different as the fact of knowing the floor level is of primary concern (for emergency aspects, for instance, or in order to cope with floor maps that are basically not identical from one floor to another). In that sense, barometers can greatly help in allowing quite an accurate determination of the altitude. This can be achieved through the use of so-called micro-altimeters whose accuracy is typically 1 m on a local and time-limited extension scale. Once this time has elapsed, meteorological fluctuations are bound to occur and will cause large bias to the measurements. The idea is then to reset the altimeter when entering the building at a known location (and known absolute altitude relative to any referential) and to use the micro-altimeter to determine the floor level. An accuracy of 1 m is enough to achieve such a goal. With an increased accuracy of a fraction of a meter, one could also imagine determining whether the mobile lies on the floor or is really handheld, to evaluate if the user is standing up or lying down.

The variation in air pressure can be given by the relation $\Delta P = \rho g Z$, where ρ is the air density, g the acceleration of gravity, and Z the altitude. The calibration at the bottom floor of the building is intended to allow the calculation of the air density ($\rho = P/rT$) under the specific current conditions. Some GPS

Table 11.6 Summary of the main parameters for pressure sensors.

Infrastructure complexity	Infrastructure maturity	Infrastructure cost	Terminal complexity	Terminal maturity	Terminal cost	Smart-phone	Calibration complexity
None	None	Zero	Low	Existing	Low	Easy	Light

Positioning type	Accuracy	Reliability	Range	Sensitivity to environment	Positioning mode	In/out transition	Calibration needed
Relative	1 m	High	World	No impact	Continuous	Easy	Several times

receivers are currently equipped with micro-barometers and allow floor-level determination without ambiguity.

Let us now come back to our parameter table. The various parameters considered for this technology are given in Chapter 4 (Table 11.6).

Infrastructure complexity, maturity, and cost: As with inertial sensors, there is no infrastructure. Meteorology is our partner.

Terminal complexity, maturity, and cost: In the present case, namely the determination of a floor of a building, it is called a "micro-barometer." These are available at extremely low cost and can be easily integrated.

Calibration complexity and need: Calibration is the point to be seriously studied. It is a question of triggering a kind of "reset" when passing close to a reference point. It is then necessary to know both the fact of being very close to this point, but also the characteristics (altitude, for example) of the latter. It is thus the implementation of the calibration that is to be processed, more than a calibration protocol of the sensor itself. Note that it is possible (and perhaps desirable) to have several of these reference points in order to "make the calibration more reliable."

Positioning type: In the continuity of the above discussion, it should be noted that the measurement provided is typically differential, i.e. relative to the previous measurement. It is the pressure variation that is provided and useful, not the pressure value itself (hence the need for the above-mentioned calibration). It is in this sense that the positioning (which is not one in fact) is qualified as "relative."

Accuracy: The current accuracy of determining an altitude, once reset to zero (and therefore in differential) and is of the order of 1 m. This is sufficient to determine the floor of a building.

Reliability: Once calibrated, the sensor is reliable over a few hours under temperate weather conditions.

Range: It is supposed to be available and to work worldwide.

Smartphone: Already available on some smartphones. There should be no difficulties to implement it on every smartphone if needed.

Sensitivity to environment: Unless one considers pressurized spaces where "meteorological" conditions are modified with respect to the external atmosphere, the only constraint is the slow but potentially real change in atmospheric conditions. Thus, a calibration will have a typical validity of a few hours under typical temperate conditions (hence the potential interest to multiply the calibration points).

Positioning mode: It can work in a continuous manner.

In/out transition: The transition presents no problem unless in pressurized places. The differential principle of measurement implies the same types of conditions throughout the building.

11.6 Radio Signals of Opportunity

Everything that was said in Chapter 10 remains applicable to just about every radiated radio signal in our environment. The era of big data could provide all the information needed to identify and recover the data needed to make these signal sources for a positioning system. The problem seems to be of the "just have to" type. Without going back on what was said in Chapter 4, or revealing what will be said in Chapter 12, the reality is actually much simpler than that. Data is one thing, measurement reliability is another. Without this reliability, or at least the knowledge of the level of nonreliability, all treatments are useless unless the performance is present without effort. Unfortunately, this is not the case with indoor positioning.

Let us now come back to our parameter table. The various parameters considered for this technology are given in Chapter 4 (Table 11.7).

Table 11.7 Summary of the main parameters for radio signals of opportunity.

Infrastructure complexity	Infrastructure maturity	Infrastructure cost	Terminal Complexity	Terminal Maturity	Terminal cost	Smart-phone	Calibration complexity
None	Existing	Zero	Medium	Integration	Medium	Near future	None

Positioning type	Accuracy	Reliability	Range	Sensitivity to environment	Positioning mode	In/out transition	Calibration needed
Absolute	>100 m	Low	World	High	Almost continuous	Moderate	None

Everything that has been written in Chapter 10 remains valid with a difference that concerns the extent of data needed for the different issuers. In addition, it is very likely that similar transmissions will occur in different parts of the world, making the detection of a position in fact ambiguous (although to achieve a position, it would be necessary to have several simultaneous emissions and the presence of several similar emissions in different parts of the world could be questioned). Assuming this happens, one approach could then be to detect the language of the presenter (not that of a singer because then we would find ourselves almost all the time either in England or in the United States!) in order to determine at least the country, but this is a first approach to merging systems (subject of Chapter 13).

11.7 Wired Networks

You have certainly realized that when surfing on the Internet, some pop-up windows or advertisements sometimes appear on the screen offering you either goods or services that are suited to your location. It can be promotional offers in your city or region or proximity services. Note that we are well aware of this aspect concerning our usual preferences that are issued from our permanent "profiling" on the web: but we are now dealing with our location and no longer with the sites we visit. How is it possible the network "knows" about your connection location?

This is indeed due to the way the addresses of the Internet Protocol (IP) are attributed to the sites and users of the web: although it is not directly achieved on a geographical basis, blocks of addresses are attributed to regional entities, the Regional Internet Registry (RIR). Then, each RIR assigns blocks to the users. As some RIR databases are public, it is thus possible to have an idea of the geographical location of an IP address. Nevertheless, this is not really accurate as the assignation is global. In order to reach a better geolocalization, one can also mix additional data, such as, for example, your IP address with your postal address when you register for an account on any website.

Thus, when connecting to any given site, the protocol establishes the connection knowing, with a more or less good precision, about the respective addresses of the two nodes connecting to each other. In such a way, both have the ability to "know" the geographical location of the other.

As a matter of fact, only from the IP address, these locations are not known very accurately, but this is at least a geographical area where the user is. This positioning technique can be seen as a sort of "wired network Cell-Id" approach.

Table 11.8 Summary of the main parameters for wired networks.

Infrastructure complexity	Infrastructure maturity	Infrastructure cost	Terminal complexity	Terminal maturity	Terminal cost	Smart-phone	Calibration complexity
None	Existing	Zero	None	Existing	Zero	Not applicable	None

Positioning type	Accuracy	Reliability	Range	Sensitivity to environment	Positioning mode	In/out transition	Calibration needed
Absolute	An address	Medium	World	No impact	Discrete	Impossible	None

Let us now come back to our parameter table. The various parameters considered for this technology are given in Chapter 4 (Table 11.8).

Infrastructure complexity, maturity, and cost: The Internet network is certainly the best developed, whether by fiber or cable. The technological advances for its development are permanent and the performance levels achieved are impressive. Costs are also low (and depend mainly on the uses that everyone can make of the network, and thus on the services which they wish to access).

Terminal complexity, maturity, and cost: All communicating terminals have the ability to access this network. Here again, the costs are independent of the access and are more related to the services available. All this is at maturity.

Calibration complexity and need: It is carried out by the network and does not require any user intervention.

Positioning type: It provides the user with an absolute positioning, as long as the coordinates of the reference components are defined in absolute coordinates too.

Accuracy: Accuracy is achieved in two successive steps. The first one concerns the initialization of the network and is present in all cases, but is not very good (typically from a few hundred meters to a few kilometers). The second depends on the data that the user himself will make available on the network, provided with their full consent. This is likely to provide the precise address of the Internet access, but it must be understood that it is only obtained by the fact that the user provides this information.

Reliability: At a very good level as far as the network part is concerned. The reliability of the postal address is more limited because many situations can lead to errors, sometimes significant (incorrect address entry, no address entry, etc.).

Range: It works worldwide.

Smartphone: This terminal is not the one that is primarily targeted, although it is quite possible that it connects to a cable network via a local radio link and can then obtain its location.

Sensitivity to environment: If we talk about the "network socket," it is associated with a physical and nondisplaceable position (without electrical work). Thus, once the wired network is installed in the building, it is no longer impacted by the environment.

Positioning mode: It is discrete in the sense it gives only a single position, the same for all terminals connecting to the access plug.

In/out transition: This criterion is not really applicable in the case of wired networks.

Bibliography

1 Kaplan, E.D. and Hegarty, C. (2017). *Understanding GPS: Principles and Applications*, 3e. Artech House.

2 Vissière, D., Martin, A., and Petit, N. (2007). Using spatially distributed magnetometers to increase IMU based velocity estimation in perturbed areas. *Proceedings of the 46th IEEE Conference on Decision and Control*.

3 Dorveaux, E., Vissière, D., Martin, A.P., and Petit, N. (2009). Iterative calibration method for inertial and magnetic sensors. *Proceedings of the 48th IEEE Conference on Decision and Control*.

4 Dorveaux, E., Boudot, T., Hillion, M., and Petit, N. (2011). Combining inertial measurements and distributed magnetometry for motion estimation. *Proceedings of the American Control Conference*.

5 Ripka, P. (2001). *Magnetic Sensors and Magnetometers*. New York: Artech.

6 Avila-Rodriguez, J.A., Wallner, S. and Hein, G.W. (2006). How to optimize GNSS signals and codes for indoor positioning. *ION GNSS 2006*, Forth Worth, TX (September 2006).

7 Bartone, C. and Van Graas, F. (2003). Ranging airport pseudolite for local area augmentation. *IEEE Transactions on Aerospace and Electronic Systems* 36 (2): 278–286.

8 Carver, C. (2005). *Myths and Realities of Anywhere GPS – High Sensitivity versus Assisted Techniques*. GPS World.

9 Eissfeller, B. (2004). In-door positioning with GNSS – dream or reality in Europe. *International Symposium European Radio Navigation Systems and Services*, Munich, Germany.

10 Francois, M., Samama, N., and Vervisch-Picois, A. (2005). 3D indoor velocity vector determination using GNSS based repeaters. *ION GNSS 2005*, Long Beach, CA (September 2005).

11 Im, S.-H., Jee, G.-I., and Cho, Y. B. (2006). An indoor positioning system using time-delayed GPS repeater. *ION GNSS 2006*, Forth Worth, TX (September 2006).

12 Jee, G.I., Choi, J.H. and Bu, S.C. (2004). Indoor positioning using TDOA measurements from switched GPS repeater. *ION GNSS 2004*, Long Beach, USA (September 2004).

13 Kaplan, E.D. and Hegarty, C. (2017). *Understanding GPS: Principles and Applications*, 3e. Artech House.

14 Kee, C., Yun, D., Jun, H. et al. (2001). *Centimeter-accuracy Indoor Navigation Using GPS-like Pseudolites*. GPS World.

15 Kiran, S. (2003). A wideband airport pseudolite architecture for the local area augmentation system. Ph.D. dissertation. School of Electrical and Computer Engineering, Ohio University, Athens.

16 Parkinson, B.W. and Spilker, J.J. Jr. (1996). *Global Positioning System: Theory and Applications*. American Institute of Aeronautics and Astronautics.

17 Progri, I.F., Ortiz, W., Michalson, W.R., and Wang, J. (2006). The performance and simulation of an OFDMA pseudolite indoor geolocation system. *ION GNSS 2006*, Forth Worth, TX (September 2006).

18 Rizos, C., Barnes, J., Wang, J. et al. (2003). LocataNet: intelligent time-synchronised pseudolite transceivers for cm-level stand-alone positioning. *11th IAIN World Congress*, Berlin, Germany (October 2003).

19 Samama, N. and Vervisch-Picois, A. (2005). Current status of GNSS indoor positioning using GNSS repeaters. *ENC GNSS 2005*, Munich, Germany (July 2005).

20 Suh, Y.-C., Konish, Y. and Shibasaki, R. (2002). Assessing the improvement of positioning accuracy using a GPS and pseudolites signal in urban area," www.chikatsu-lab.g.dendai.ac.jp/s_forum/pdf/2002/10_suh.pdf.

21 Sun, G., Chen, J., Guo, W., and Ray Liu, K.J. (2005). Signal processing techniques in network-aided positioning – a survey of state-of-the-art positioning designs. *IEEE Signal Processing Magazine* 22 (4): 12–23.

22 Syrjärinne, J. and Wirola, L. (2006). Setting a new standard – assisted GNSS receivers that use wireless networks. *Inside GNSS* 1 (7): 26–31.

23 Van Diggelen, F. and Abraham, C. *Indoor GPS Technology*. Global Locate, Inc. www.gmat.unsw.edu.au/cr/gmat4910/globallocate.pdf.

24 Teunissen, P. and Montenbruck, O. (2017). *Springer Handbook of Global Navigation Satellite Systems*. Springer.

25 Misra, P. and Enge, P. (2006). *Global Positioning System: Signals, Measurements, and Performance*, 2e. Lincoln, MA: Ganga-Jamuna Press.

26 Cui, X., Li, Y., Wang, Q. et al. (2018). Three-axis magnetometer calibration based on optimal ellipsoidal fitting under constraint condition for pedestrian positioning system using foot-mounted inertial sensor/magnetometer. In: *2018 IEEE/ION Position, Location and Navigation Symposium (PLANS)*, 166–174. Monterey, CA: IEEE.

27 Willenberg, G.-D. and Weyand, K. (1997). Three-dimensional positioning setup for magnetometer sensors. *IEEE Transactions on Instrumentation and Measurement* 46 (2): 621–623.

28 Renaudin, V., Afzal, M.H., and Lachapelle, G. (2010). New method for magnetometers based orientation estimation. In: *IEEE/ION Position, Location
and Navigation Symposium*, 348–356. Indian Wells, CA: IEEE.

29 Hellmers, H., Norrdine, A., Blankenbach, J., and Eichhorn, A. (2013).
An IMU/magnetometer-based Indoor positioning system using Kalman
filtering. In: *International Conference on Indoor Positioning and Indoor
Navigation*, 1–9. Montbeliard-Belfort: IEEE.

30 Camps, F., Harasse, S., and Monin, A. (2009). Numerical calibration for
3-axis accelerometers and magnetometers. In: *2009 IEEE International
Conference on Electro/Information Technology*, 217–221. Windsor, ON:
IEEE.

31 Hellmers, H., Eichhorn, A., Norrdine, A., and Blankenbach, J. (2016).
IMU/magnetometer based 3D indoor positioning for wheeled platforms
in NLoS scenarios. In: *2016 International Conference on Indoor Positioning
and Indoor Navigation (IPIN)*, 1–8. Alcala de Henares: IEEE.

32 Wu, F., Liang, Y., Fu, Y., and Ji, X. (2016). A robust indoor positioning
system based on encoded magnetic field and low-cost IMU. In: *2016
IEEE/ION Position, Location and Navigation Symposium (PLANS)*, 204–212.
Savannah, GA: IEEE.

33 Song, J., Jeong, H., Hur, S., and Park, Y. (2014). Improved indoor position
estimation algorithm based on geo-magnetism intensity. In: *2014 International Conference on Indoor Positioning and Indoor Navigation (IPIN)*,
741–744. Busan: IEEE.

34 Brzozowski, B., Kaźmierczak, K., Rochala, Z. et al. (2016). A concept of
UAV indoor navigation system based on magnetic field measurements. In:
2016 IEEE Metrology for Aerospace (MetroAeroSpace), 636–640. Florence:
IEEE.

35 Blankenbach, J. and Norrdine, A. (2010). Position estimation using artificial generated magnetic fields. In: *2010 International Conference on Indoor
Positioning and Indoor Navigation*, 1–5. Zurich: IEEE.

36 Pasku, V., De Angelis, A., Dionigi, M. et al. (2016). A positioning system
based on low-frequency magnetic fields. *IEEE Transactions on Industrial
Electronics* 63 (4): 2457–2468.

37 Wang, Q., Luo, H., Zhao, F., and Shao, W. (2016). An indoor
self-localization algorithm using the calibration of the online magnetic
fingerprints and indoor landmarks. In: *2016 International Conference on
Indoor Positioning and Indoor Navigation (IPIN)*, 1–8. Alcala de Henares:
IEEE.

38 Kim, S.-E., Kim, Y., Yoon, J., and Kim, E.S. (2012). Indoor positioning
system using geomagnetic anomalies for smartphones. In: *2012 International Conference on Indoor Positioning and Indoor Navigation (IPIN)*, 1–5.
Sydney, NSW: IEEE.

39 Song, J., Hur, S., Park, Y., and Choi, J. (2016). An improved RSSI of geo-magnetic field-based indoor positioning method involving efficient database generation by building materials. In: *2016 International Conference on Indoor Positioning and Indoor Navigation (IPIN)*, 1–8. Alcala de Henares: IEEE.

40 Li, B., Gallagher, T., Dempster, A.G., and Rizos, C. (2012). How feasible is the use of magnetic field alone for indoor positioning? In: *2012 International Conference on Indoor Positioning and Indoor Navigation (IPIN)*, 1–9. Sydney, NSW: IEEE.

41 Kim, B. and Kong, S. (2016). A novel indoor positioning technique using magnetic fingerprint difference. *IEEE Transactions on Instrumentation and Measurement* 65 (9): 2035–2045.

42 Binghao, L., Harvey, B., and Gallagher, T. (2013). Using barometers to determine the height for indoor positioning. In: *International Conference on Indoor Positioning and Indoor Navigation*, 1–7. Montbeliard-Belfort: IEEE.

43 Jeon, J., Kong, Y., Nam, Y., and Yim, K. (2015). An indoor positioning system using Bluetooth RSSI with an accelerometer and a barometer on a smartphone. In: *2015 10th International Conference on Broadband and Wireless Computing, Communication and Applications (BWCCA)*, 528–531. Krakow: IEEE.

44 Gaglione, S., Angrisano, A., Castaldo, G. et al. (2015). GPS/Barometer augmented navigation system: Integration and integrity monitoring. In: *2015 IEEE Metrology for Aerospace (MetroAeroSpace)*, 166–171. Benevento: IEEE.

45 Bolanakis, D.E. (2016). MEMS barometers in a wireless sensor network for position location applications. In: *2016 IEEE Virtual Conference on Applications of Commercial Sensors (VCACS)*, 1–8. Raleigh, NC: IEEE.

46 Xu, Z., Wei, J., Zhu, J., and Yang, W. (2017). A robust floor localization method using inertial and barometer measurements. In: *2017 International Conference on Indoor Positioning and Indoor Navigation (IPIN)*, 1–8. Sapporo: IEEE.

47 Dammann, A., Sand, S., and Raulefs, R. (2012). Signals of opportunity in mobile radio positioning. In: *2012 Proceedings of the 20th European Signal Processing Conference (EUSIPCO)*, 549–553. Bucharest: IEEE.

48 Navratil, V., Karasek, R., and Vejrazka, F. (2016). Position estimate using radio signals from terrestrial sources. In: *2016 IEEE/ION Position, Location and Navigation Symposium (PLANS)*, 799–806. Savannah, GA: IEEE.

49 Yang, C., Nguyen, T., Venable, D. et al. (2009). Cooperative position location with signals of opportunity. In: *Proceedings of the IEEE 2009 National Aerospace & Electronics Conference (NAECON)*, 18–25. Dayton, OH: IEEE.

50 Webb, T.A., Groves, P.D., Cross, P.A. et al. (2010). A new differential positioning method using modulation correlation of signals of opportunity. In:

IEEE/ION Position, Location and Navigation Symposium, 972–981. Indian Wells, CA: IEEE.

51 Nanmaran, K. and Amutha, B. (2014). Situation assisted indoor localization using signals of opportunity. In: *2014 International Conference on Indoor Positioning and Indoor Navigation (IPIN)*, 693–698. Busan: IEEE.

12

Combining Techniques and Technologies

Abstract

This chapter provides a brief overview of current approaches to combining and merging or, more generally, a brief overview of the various signal processing methods that are often applied to positioning in order to propose better quality systems. Many books and articles are available on the subject, and we will try to describe the main points here. Note that many specialized books on these issues are available and that the purpose of this book is not to go into technical details on these methods but simply to provide some elements for discussion. In order to understand the expected improvements on the one hand, and the limitations on the other, it is important to keep in mind some lines of the tables presented in Chapter 4, as the basic idea is systematically based on a willingness to combine complementary technologies in the sense of a criterion (or criteria) that will be sought to be optimized.

Keywords *Filtering; Fusion; Hybridization; Estimation; Collaborative approaches*

The main parts of this chapter will successively discuss the filtering, merging, and processing techniques commonly used and then describe in more detail some potential collaborative approaches in which the system seeks to determine the relative positioning of the various "actors." We will conclude with a new discussion.

12.1   Introduction

Faced with the real difficulty, as shown in the previous chapters, of performing the indoor positioning function satisfactorily, more complex systems have been devised. In particular, the first idea is to couple two complementary systems. This was first of all the case with GNSS (Global Navigation Satellite

Indoor Positioning: Technologies and Performance, First Edition. Nel Samama.

Systems) with an inertial system: complementarities are particularly relevant, GNSS to adjust the inertial, and the inertial to ensure the continuity of positioning during the phases of GNSS unavailability. Such a system is in fact the ideal combination, and if it had been enough, this book would not have existed. The difficulty then comes from the implementation on the one hand and the performance obtained on the other. Current mass distribution inertial sensors do not offer sufficient performance to allow good continuity[1]. In addition, in the case of hand-held mobile devices, inertial sensors are in a very "noisy" and "biased" situation and performance deteriorates very quickly. This example is very good because it makes us understand the logic applied by the engineers: as the measurements are noisy, we will try to eliminate it and thus model the noise. This is how positioning becomes a matter of signal processing. Moreover, after a few "classic" attempts, we realize that the potential practical situations are innumerable and that a statistical approach should lead to better results. All this is quite commendable, but here is how the initial problem is gradually being forgotten in favor of "technology." It must be acknowledged, however, that along the way, substantial improvements are being made, mitigating the remarks that the initial difficulties are still there.

The term "fusion" is used by signal processing and data scientist communities, whereas the term "hybridization" is used by positioning communities. For our part, we will use both without any real distinction. The ways of combining several technologies are diverse: it can be a simple juxtaposition (the so-called "loose integration") or a more or less "deep" interleaving (called "tight integration") of the data from each. In any case, with this being the fundamental point, it is necessary to have a method that makes it possible to determine the framework for using, or on the contrary not using, the respective data. It is this point that is often overlooked in, leading to "the more data, the better" that often comes up against a much less sympathetic reality.

The nuances in the "depth" of mergers are based on the type of data that are merged. A wide range is indeed possible: to take up the case of inertial and GNSS, it is simple to consider the two systems independently, to conduct position estimates with each, and then to create a merger, of very high level here, in order to choose the one that is considered the best. A little more sophisticated would be the technique of applying an algorithm, formula, or method to extract a mixture of the two positions. However, it is possible to go much deeper and, ultimately, return to the basic (the so-called "raw") measurements of the two systems in order to produce a new system of interleaved equations to be solved. This is the case for pseudo-distance and Doppler measurements for GNSS and angular inertial acceleration (called angular rate) or acceleration for the example considered[2].

---

1 Note that we always come back to the definition of words. What is "good" continuity? The aim here is to obtain typical accuracies of around 1 m over periods of several hours: this is not achieved today on the mass market.

2 Doppler and acceleration can be very usefully coupled in such a way.

Another element leading to significant differences concerns the type of baseline measurements that are carried out. The latter can aim for a small margin of error (under good conditions) and then the calculation of the position is "easy," whereas if the latter is more noisy, the calculation will require more "finesse." Of course, we must not forget the last category where the measurements are of poor quality, inherently, and for which it is essential to find "tricks" for calculation or adequate estimation methods in order to have a chance of obtaining a more or less acceptable positioning.

As we have seen in all the chapters, it is clear that the environments in which these indoor positioning systems are intended to be deployed are generally the major difficulty and that this creates an incredible diversity of situations. It is in this context that we must understand the extraordinary diversity of treatment approaches proposed. We will briefly describe the main ones in the following paragraph.

12.2 Fusion and Hybridization

This paragraph is divided into four parts: the first one deals with possible strategies for combining technologies, the second with the basic question of the optimal choice of merged data to obtain the best position, the third part with the importance of classification and estimators, and the last one with filtering.

12.2.1 Strategies for Combining Technologies

When we decide to couple technologies, it is generally complementarities that are sought. The latter can take many forms depending on the criterion (or criteria) that we are trying to improve. The most frequently used criteria are coverage and accuracy. Thus, the example already mentioned of the coupling between GNSS and inertial would make it possible to cover both outdoors and indoors. The classification used[3] for the book makes it possible to organize this reflection: if we are looking for maximum coverage in order to propose a continuity of the positioning function in all environments, we will have to combine technologies from the top of Table 4.12 with technologies from the bottom of the table. If, on the contrary, we are only interested in the interior, then the combination will be between the technologies at the bottom of the table. It should be noted that the optimization criterion will certainly focus more on pure performance aspects, such as accuracy or reliability in this latter case.

Thus, if we stay inside buildings, a combination of bluetooth low energy (BLE) and micro-barometric sensors would allow us to reliably obtain the floor level in the building using the sensors and focus on two-dimensional positioning in the floor (with Bluetooth). We could also imagine the deployment of two complementary systems to improve accuracy. This would be the case of a symbolic

---

3 By "range."

positioning system that reliably provides the room in which you are located and then a second system like UWB (ultra wide band) or Lidar (for example) to finely define the position in the room.

It can be seen that the different approaches above actually only "juxtapose" technologies: the complete system only uses one after the other or one when the other is not available. In reality, things are often not this way because measurement uncertainties disrupt the theory. What happens if the part provided by the symbolic technology is not the right one? It is likely that the second technology, with a false assumption, will be severely disrupted. It is therefore necessary to go further in the fusion of data and in fact to mix data from the two (or more) technologies. This then leads to a fundamental question: how to choose these data and in particular how to assign them a weight in accordance with their "interest" in the reliability and quality of the final result.

12.2.2 Strategies for Choosing the Optimal Data

The aim here is to understand a classic technique for optimizing data processing. This technique is the least squares method. In order to describe it, we have chosen to consider the case of the GPS (global positioning system) receiver which, in general, acquires more satellites than necessary to calculate a position. Thus, the question arises about the choice of the optimal set of N data in a set of M measurements. The least squares method makes it possible to use all available data by optimizing the calculation.

In introduction, let us take the case of a position calculation based on a distance measurement technique and express a classical method of resolution in the case of intersections of spheres when the synchronization of the terminal clock with those of the transmitters is an unknown problem[4].

The basic method for calculating a position is given below. It is based on a Taylor series development[5] of the first order of the ranges obtained from at least four transmitters in order to cope with the three spatial coordinates, plus the clock time bias of the terminal. The system to be resolved is composed of the four range expressions (ρ_i) given below:

$$\rho_i = \sqrt{(x_i - x_r)^2 + (y_i - y_r)^2 + (z_i - z_r)^2} + ct_r \tag{12.1}$$

where (x_r, y_r, z_r) is the position of the receiver (searched for), (x_i, y_i, z_i) is the position of transmitter "i" (supposedly available to the terminal), and t_r is the terminal's clock bias with respect to the transmitters' reference time. Usually, an iterative process is implemented. At the first iteration of the process, only an estimated value of the propagation time can be used (based on an estimation of the terminal's position) and must be checked at the end of this first iteration. In case a mismatch is observed, a new iteration is then required, and so on.

---

4 This is typically the case of a UWB technology indoors or a GNSS one outdoors.
5 This is achieved for linearization purposes.

Thus, the principle of the resolution is to state an initial estimated position for the terminal. This initial position is noted $(\hat{x}_r, \hat{y}_r, \hat{z}_r)$. Furthermore, the solution vector of positioning is not only a spatial vector but also includes the time bias of the internal clock of the terminal, leading to a real first estimate of the form $(\hat{x}_r, \hat{y}_r, \hat{z}_r, c\hat{t}_r)$. The last time coordinate is multiplied by the speed of light in order to obtain a homogeneous four-coordinate vector, i.e. where all coordinates are given in meters.

This first position is then used to estimate the relative displacement from one iteration to the next: this latter quantity is going to be used as convergence criteria for the algorithm. The difference between the real position and the estimated one is then represented by the vector $(\Delta\hat{x}_r, \Delta\hat{y}_r, \Delta\hat{z}_r, c\Delta\hat{t}_r)$.

One can define the function f as follows:

$$\rho_i = \sqrt{(x_i - x_r)^2 + (y_i - y_r)^2 + (z_i - z_r)^2} + ct_r = f(x_r, y_r, z_r, t_r) \qquad (12.2)$$

It is also possible to define the function f for the estimated position as being:

$$\hat{\rho}_i = \sqrt{(x_i - \hat{x}_r)^2 + (y_i - \hat{y}_r)^2 + (z_i - \hat{z}_r)^2} + c\hat{t}_r = f(\hat{x}_r, \hat{y}_r, \hat{z}_r, \hat{t}_r) \qquad (12.3)$$

Considering that the actual position of the receiver is given by

$$\begin{cases} x_r = \hat{x}_r + \Delta x_r \\ y_r = \hat{y}_r + \Delta y_r \\ z_r = \hat{z}_r + \Delta z_r \\ t_r = \hat{t}_r + \Delta t_r \end{cases} \qquad (12.4)$$

then it follows

$$f(x_r, y_r, z_r, ct_r) = f(\hat{x}_r + \Delta x_r, \hat{y}_r + \Delta y_r, \hat{z}_r + \Delta z_r, c\hat{t}_r + c\Delta t_r) \qquad (12.5)$$

Considering that near the convergence $(\Delta\hat{x}_r, \Delta\hat{y}_r, \Delta\hat{z}_r, c\Delta\hat{t}_r)$ will be small in comparison with $(\hat{x}_r, \hat{y}_r, \hat{z}_r, c\hat{t}_r)$, a first-order Taylor series development is possible and leads to the next formula.

$$f(\hat{x}_r + \Delta x_r, \hat{y}_r + \Delta y_r, \hat{z}_r + \Delta z_r, \hat{t}_r + \Delta t_r)$$
$$= f(\hat{x}_r, \hat{y}_r, \hat{z}_r, \hat{t}_r) + \frac{\partial f(\hat{x}_r, \hat{y}_r, \hat{z}_r, c\hat{t}_r)}{\partial \hat{x}_r}\Delta x_r + \frac{\partial f(\hat{x}_r, \hat{y}_r, \hat{z}_r, c\hat{t}_r)}{\partial \hat{y}_r}\Delta y_r$$
$$+ \frac{\partial f(\hat{x}_r, \hat{y}_r, \hat{z}_r, c\hat{t}_r)}{\partial \hat{z}_r}\Delta z_r + \frac{\partial f(\hat{x}_r, \hat{y}_r, \hat{z}_r, c\hat{t}_r)}{\partial \hat{t}_r}\Delta t_r \qquad (12.6)$$

When considering the intermediate variable

$$\hat{r}_i = \sqrt{(x_i - \hat{x}_r)^2 + (y_i - \hat{y}_r)^2 + (z_i - \hat{z}_r)^2} \qquad (12.7)$$

the equality gives

$$f(\hat{x}_r + \Delta x_r, \hat{y}_r + \Delta y_r, \hat{z}_r + \Delta z_r, \hat{t}_r + \Delta t_r)$$
$$= \hat{r}_i + c\hat{t}_r - \frac{x_i - \hat{x}_r}{\hat{r}_i}\Delta x_r - \frac{y_i - \hat{y}_r}{\hat{r}_i}\Delta y_r - \frac{z_i - \hat{z}_r}{\hat{r}_i}\Delta z_r + c\Delta t_r \qquad (12.8)$$

As the partial derivatives are given by

$$
\begin{cases}
\dfrac{\partial f(\hat{x}_r, \hat{y}_r, \hat{z}_r, c\hat{t}_r)}{\partial \hat{x}_r} \Delta x_r = -\dfrac{x_i - \hat{x}_r}{\hat{r}_i} \Delta x_r \\[4mm]
\dfrac{\partial f(\hat{x}_r, \hat{y}_r, \hat{z}_r, c\hat{t}_r)}{\partial \hat{y}_r} \Delta y_r = -\dfrac{y_i - \hat{y}_r}{\hat{r}_i} \Delta y_r \\[4mm]
\dfrac{\partial f(\hat{x}_r, \hat{y}_r, \hat{z}_r, c\hat{t}_r)}{\partial \hat{z}_u} \Delta z_r = -\dfrac{z_i - \hat{z}_r}{\hat{r}_i} \Delta z_r \\[4mm]
\dfrac{\partial f(\hat{x}_r, \hat{y}_r, \hat{z}_r, c\hat{t}_r)}{\partial \hat{t}_r} \Delta t_r = c
\end{cases}
\tag{12.9}
$$

the final relationship between estimated and actual pseudo ranges is then given by

$$
\rho_i = \hat{\rho}_i - \frac{x_i - \hat{x}_r}{\hat{r}_i} \Delta x_r - \frac{y_i - \hat{y}_r}{\hat{r}_i} \Delta y_r - \frac{z_i - \hat{z}_r}{\hat{r}_i} \Delta z_r + c\Delta t_r
\tag{12.10}
$$

or also

$$
\hat{\rho}_i - \rho_i = \frac{x_i - \hat{x}_r}{\hat{r}_i} \Delta x_r + \frac{y_i - \hat{y}_r}{\hat{r}_i} \Delta y_r + \frac{z_i - \hat{z}_r}{\hat{r}_i} \Delta z_r - c\Delta t_r
\tag{12.11}
$$

By defining a new set of intermediate variables as follows:

$$
\begin{cases}
\Delta\rho = \hat{\rho}_i - \rho_i \\[2mm]
a_{xi} = \dfrac{x_i - \hat{x}_r}{\hat{r}_i} \\[4mm]
a_{yi} = \dfrac{y_i - \hat{y}_r}{\hat{r}_i} \\[4mm]
a_{zi} = \dfrac{z_i - \hat{z}_r}{\hat{r}_i}
\end{cases}
\tag{12.12}
$$

The equation, for any given transmitter that has to be solved is now:

$$
\Delta\rho = a_{xi}\Delta x_r + a_{yi}\Delta y_r + a_{zi}\Delta z_r - c\Delta t_r
\tag{12.13}
$$

When considering the four transmitters required for a three-dimensional positioning, one has to deal with a system composed of four equations and four unknowns (being now the vector $(\Delta\hat{x}_r, \Delta\hat{y}_r, \Delta\hat{z}_r, c\Delta\hat{t}_r)$). The system can be fully described by

$$
\begin{cases}
\Delta\rho_1 = a_{x1}\Delta x_r + a_{y1}\Delta y_r + a_{z1}\Delta z_r - c\Delta t_r \\[2mm]
\Delta\rho_2 = a_{x2}\Delta x_r + a_{y2}\Delta y_r + a_{z2}\Delta z_r - c\Delta t_r \\[2mm]
\Delta\rho_3 = a_{x3}\Delta x_r + a_{y3}\Delta y_r + a_{z3}\Delta z_r - c\Delta t_r \\[2mm]
\Delta\rho_4 = a_{x4}\Delta x_r + a_{y4}\Delta y_r + a_{z4}\Delta z_r - c\Delta t_r
\end{cases}
\tag{12.14}
$$

Such equations take advantage of the matrix representation. Therefore, introducing the following

$$
\Delta\rho = \begin{bmatrix} \Delta\rho_1 \\ \Delta\rho_2 \\ \Delta\rho_3 \\ \Delta\rho_4 \end{bmatrix} \quad
H = \begin{bmatrix} a_{x1} & a_{y1} & a_{z1} & 1 \\ a_{x2} & a_{y2} & a_{z2} & 1 \\ a_{x3} & a_{y3} & a_{z3} & 1 \\ a_{x4} & a_{y4} & a_{z4} & 1 \end{bmatrix} \quad
\Delta x = \begin{bmatrix} \Delta x_r \\ \Delta y_r \\ \Delta z_r \\ -c\Delta t_r \end{bmatrix} \tag{12.15}
$$

the system is finally

$$
\Delta\rho = H\Delta x \tag{12.16}
$$

The solution of which is finally given by

$$
\Delta x = H^{-1}\Delta\rho \tag{12.17}
$$

The reader should understand that one is interested in making the Δx vector equal to zero in order to find the position of the terminal which, in this case, is, for the last iteration, the position considered as initial estimate. It is thus an iterative method that has to be given a convergence criterion (as it is usually not possible to reach zero because of measurement uncertainties).

12.2.2.1 Least Squares Method

We are now back to our initial question (i.e. the selection of the best data), in the simple case where the number of transmitters is greater than four. This is an over determined system and the choice of the four best transmitters is made. The least squares method allows you to avoid having to make a choice by choosing all available transmitters. The idea is expressed mathematically, keeping the previous example, in the following form.

When more than four transmitters are available, the linearization method can be specified by

$$
\Delta\rho = \begin{bmatrix} \Delta\rho_1 \\ \Delta\rho_2 \\ \cdots \\ \Delta\rho_n \end{bmatrix} \quad
H = \begin{bmatrix} a_{x1} & a_{y1} & a_{z1} & 1 \\ a_{x2} & a_{y2} & a_{z2} & 1 \\ \cdots & \cdots & \cdots & \cdots \\ a_{xn} & a_{yn} & a_{zn} & 1 \end{bmatrix} \quad
\Delta x = \begin{bmatrix} \Delta x_r \\ \Delta y_r \\ \Delta z_r \\ -c\Delta t_r \end{bmatrix} \tag{12.18}
$$

where $\Delta\rho$ is an $N\times1$ vector, H an $N\times4$ matrix, and Δx an 4×1 vector. The relation $\Delta\rho = H\Delta x$ is still valid. Given the fact that measurements are noisy, it is possible to introduce a residual vector calculated as follows:

$$
r = H\Delta x - \Delta\rho \tag{12.19}
$$

The basic idea of the least square method is to minimize the sum of the square of the residuals. This sum has the form:

$$
r_1^2 + \cdots + r_n^2 = (H\Delta x - \Delta\rho)^2 = (H\Delta x - \Delta\rho)^{\mathrm{T}}(H\Delta x - \Delta\rho) \tag{12.20}
$$

Minimizing this quantity is achieved when its gradient is zero, leading to the expression:

$$\nabla(r_1^2 + \cdots + r_n^2) = 2\Delta x^{\mathrm{T}} \cdot H^{\mathrm{T}} \cdot H - 2\Delta \rho^{\mathrm{T}} \cdot H = 0 \qquad (12.21)$$

that is finally:

$$\Delta x = (H^{\mathrm{T}} \cdot H)^{-1} \cdot H^{\mathrm{T}} \cdot \Delta \rho \qquad (12.22)$$

This method allows a cost function to be minimized, here the sum of the squares of the residues, and not to make the calculation too complex. Indeed, another approach would have been to test all combinations of four transmitters in order to extract the best combination. This is not actually possible because it would require the ability to compare the results to the "true" position, which is not available in practice[6].

It is also possible to use this method with data from several technologies and thus to find an optimal combination. Different versions of this method exist depending on the characteristics of the model to be solved (see Eq. (12.19)). In this example, it is a linear model. Nonlinear methods are also available.

12.2.3 Classification and Estimators

Thus, in many cases, combining or merging several technologies shifts the problem from a material and computational approach to a classification and decision-making approach. Indeed, it is then necessary to identify the cases of use of such or such data, or of the fusion of such data. The choice of the optimal algorithm must be made with full knowledge of the facts. This often amounts to classifying the situation in which the terminal finds itself: in fast motion, in vertical motion, stationary but not immobile, with or without masking, etc.

This is where "data science" and "decision-making" techniques come in. They make it possible to provide a "best estimate" based on the availability of sometimes partial data by analyzing the most likely class in which the terminal is located. This often involves learning phases of the various possible environments, and this is where things get tricky. Not all cases can be considered and "learned," and it is therefore necessary to use decision algorithms that are able to make decisions also in "unlearned" cases, with the risk of failure that this entails. The more situations are close to cases learned, the more effective these methods become: hence, the importance of defining them properly and the difficulty to make them always efficient.

In general, all these approaches are based on the determination of estimators that are evaluated according to the various data available and that lead to a decision. The latest developments in the fields of Big Data (because the merged data

---

6 Indeed, if we knew the true position, there would be no point in trying to measure it or calculate it in order to get it, would there?

are often numerous and the possible combinations very broad), data mining, or neural networks are increasingly being used.

This is where this book should help "data scientists" by providing them with a basic understanding of the main physical factors influencing data quality so that they can take them into account in their models.

12.2.4 Filtering

Another way of looking at it, in parallel or in concert, is to set global rules that the terminal or measurements are most likely to follow. This would be, for example, the limitation of a pedestrian's possible acceleration, or even his speed, leading to limitations on the possibilities of varying his position between two moments. This can also be considered in relation to the current operating parameters of the system. For example, if the pedestrian's travel speed is 2 m/s, it is "unlikely" that he or she moved more than 2 m in the next second. This then makes it possible to measure the distance between two successive positions and thus potentially extract an estimate of the probable noise of the measurements. These new data can in turn be used to refine the models.

In further improved versions, such approaches are based on two-step mechanisms: the first is to define, from a present observed or estimated state, what the future state should be taking into account the observation parameters (noise, environment, etc.). This is the case of Kalman filters in particular, whether they are simple or extended, or more generally Bayesian filtering methods, including particle filters, for instance. The incredible power of these techniques is based on the simplicity and reality of the basic premise. The future state cannot be random depending on the present state. The difficulty then is to define the "right" operating parameters of these filters because we understand that depending on them, the operation will be efficient or useless. Thus, all the techniques overlap: a good quality classification of the situation in which the terminal is located will lead to a correct estimation of the parameters to be taken into account, which in turn will allow a good prediction of the future state, and so on.

Moreover, once the prediction of the future state has been made, it is possible to criticize it because once the future state has been reached, it is possible to evaluate the quality of the prediction. This element is particularly useful: it allows in particular the prediction to be compared with the measurements and thus to provide the user with a data set that provides an estimate of the quality of the prediction. This could well lead to a virtuous circle... This is true in many situations (GNSS would function much less fluidly without these filters), but not in all. The main criticisms are based on the sensitivity of the performance of these filters to their operating parameters, which in turn depend on the situations considered.

12.3   Collaborative Approaches

Another way to take a step back from the previous chapters is to imagine a collaborative system in which exchanges between the various nodes allow all the nodes to be positioned[7]. As always, such an achievement is highly dependent on performance targets. We are seeking here to lay the foundations for what could be a precise positioning system (typically in the order of a few decimeters if a potential value is to be given), which does not require a specific infrastructure and is inexpensive in terms of embedded technologies and low energy consumption. For the first case cited, this is due to the use of mature technologies, and in the second case, it comes from the fact that the proposed radio exchanges are carried out at very short distances and using signals that can be effectively coded (hence with reduced transmitted power).

Note that the previously mentioned environment-dependent limitations are still present with these solutions because propagation phenomena are their basis.

12.3.1   Approach Using Doppler Measurements to Estimate Velocities

We know from experience that one way to obtain precise radio measurements is to manage the phase of a signal (e.g. carrier phase): this will again be the case either through the use of signals already transmitted by existing terminals (and it will then be a matter of developing the receiver) or new signals to be transmitted at the terminals, such as compatible "GNSS" signals (for which a clean development will have to be provided). Such GNSS-like signals could be of very small amplitudes in order to manage the problem of potential interference.

We propose here to rely on Doppler measurements only, for example, from miniaturized transmitters installed on each terminal (i.e. "smartphone"). It is then possible to know the value of the projection of the difference in velocity vectors on the axis separating the two transmitters. Based on this single measurement between two terminals, we will try to define a new type of positioning: the latter is relative to the meaning in which the positioning is estimated "in relation to the other terminals." The term "relative location" is often used in a slightly different sense, as in the case of inertial systems, for example, where the computed position at a given time depends on the previous position. Therefore, the location is relative to a starting point. In this case, the concept is slightly different: the relative term applies well to a notion of relativity of some in relation to others. However, from this relative positioning, we will try to go back to an absolute positioning, sometimes made possible by the availability of absolute

---

7 Most of this paragraph is based on a publication proposed during the Scientific Days entitled "L'homme connecté" of the International Radio Science Union (URSI) France, held in 2014.

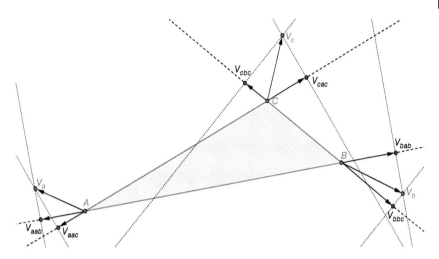

Figure 12.1 Problem definition.

data, such as distances or the coordinates of certain terminals. However, this objective of absolute positioning is not the quest.

The initial problem is thus schematized in Figure 12.1 in which A, B, and C are three transmitter and receiver terminals performing the Doppler measurements described above. We can also consider that A, B, and C are the communicating nodes of a network. A, B, and C have any velocity vectors V_a, V_b, and V_c, and we are interested in the various projections of these velocities on the AB, BC, and AC axes. The latter are rated V_{aab} and V_{aac} for V_a on AB and AC, respectively, V_{bab} and V_{bbc} for V_b on AB and BC, respectively, and finally V_{cac} and V_{cbc} for V_c on AC and BC, respectively.

Note that we remain within the framework of a two-dimensional structure for the simplicity (also relative as we will see) of the equation setting of the system. The basic idea being in fact to determine the shape of the triangle ABC, we now need to introduce some angles. Keeping similar notations to those in Figure 12.1, in Figure 12.2, we place the angles θV_{aab} and θV_{aac} that the velocity vector V_a makes with axes AB and AC, respectively, θV_{bab} and θV_{bbc} that the velocity vector V_b makes with axes AB and BC, respectively, and θV_{cac} and θV_{cbc} that the velocity vector V_c makes with axes AC and BC, respectively.

From this set, it is then possible to formalize the writing of the measured values that are the Doppler between the terminals. Within the framework of the three nodes A, B, and C considered here, this actually gives three measurements of relative velocities V_{ab}, V_{bc}, and V_{ca} between A and B, B and C, and C and A, respectively. Note that without trying to be extremely rigorous at first, the plane is still oriented: thus, V_{ab} represents the difference between the projections on the AB axis of V_b and V_a, in this order. The first system of equations of the

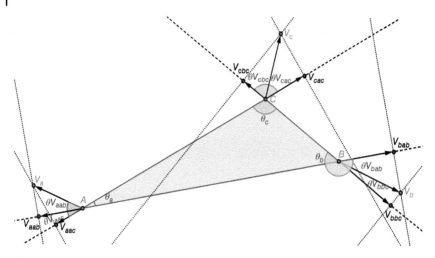

Figure 12.2 Definition of the angles.

approach is then given by the following formulas:

$$V_{ab} = V_b \cos(\theta V_{bab}) - V_a \cos(\theta V_{aab})$$
$$V_{bc} = V_c \cos(\theta V_{cbc}) - V_b \cos(\theta V_{bbc})$$
$$V_{ca} = V_a \cos(\theta V_{aac}) - V_c \cos(\theta V_{cac}) \tag{12.23}$$

A simplification of notations is possible if all projections are related to a reference axis. We will take the AB axis here. Let us then note that it is possible to write the following relationships by noting θ_a and θ_b the angles relative to the vertices A and B, respectively, of the triangle ABC:

$$\theta V_{abc} = \theta V_{aab} - \theta_a = \theta V_a - \theta_a$$
$$\theta V_{cac} = \theta V_{cab} - \theta_a = \theta V_c - \theta_a$$
$$\theta V_{cbc} = \theta V_{cab} - \theta_b = \theta V_c - \theta_b$$
$$\theta V_{bbc} = \theta V_{bab} + \theta_b = \theta V_b + \theta_b$$

We then write the system (12.23) in a simplified general form:

$$V_{ab} = V_b \cos(\theta V_b) - V_a \cos(\theta V_a)$$
$$V_{bc} = V_c \cos(\theta V_c - \theta_b) - V_b \cos(\theta V_b + \theta_b)$$
$$V_{ca} = V_a \cos(\theta V_a - \theta_a) - V_c \cos(\theta V_c - \theta_a) \tag{12.24}$$

This system of three equations has some eight unknowns at this stage: V_a, V_b, and V_c for the velocities, θ_a and θ_b for the shape of the triangle, and θV_a, θV_b, and θV_c for the projections of the velocity vectors on the AB axis. This is too much! A simple way to reduce the number of unknowns is to go back to the basic principle of the approach: analyze the relative positioning of the various terminals between them. Thus, it is possible to assume that we are only trying to know the relative displacement of the various nodes and therefore that absolute values are not our priority. Thus, we can take a node, node A in our case, as a reference. Then, V_a is zero and θV_a is also zero, but it is then necessary to consider that the speeds of terminals B and C are now relative to the speed of terminal A (if we want to keep the notion of relativity intact). The new system is written as follows (V_{bra} and V_{cra} are the speeds of B and C relative to A):

$$V_{ab} = V_{bra} \cos(\theta V_b)$$
$$V_{ac} = V_{cra} \cos(\theta V_c - \theta_a)$$
$$V_{bc} = V_{cra} \cos(\theta V_c - \theta_b) - V_{bra} \cos(\theta V_b + \theta_b) \tag{12.25}$$

By simplifying the notations by removing the latter "relativity" (i.e. V_b replaces V_{bra} and V_c replaces V_{cra}), the system that will serve as a basis is then described by the following simplified equations:

$$V_{ab} = V_b \cos(\theta V_b)$$
$$V_{ac} = V_c \cos(\theta V_c - \theta_a)$$
$$V_{bc} = V_c \cos(\theta V_c - \theta_b) - V_b \cos(\theta V_b + \theta_b) \tag{12.26}$$

This new system of three equations now has only six unknowns: V_b and V_c for velocities, θ_a and θ_b for the shape of the triangle, and θV_b and θV_c for velocity vector projections on the AB axis. The resolution of the latter is not yet achieved, however. We could continue in the search for new simplifying hypotheses in order to allow the system to be resolved: we reserve this ultimate approach for later, the aim here being to lay the foundations of a concept of relative localization. In this context, we will rather look at what happens when we add a fourth point, D.

It is then possible to isolate what is happening individually in the three triangles ABC, ABD, and ACD (knowing that the last triangle BCD is totally deduced from the first three). Let us write the nine (three times three) equations from the three triangles:

in ABC

$$V_{ab} = V_b \cos(\theta V_b) \tag{12.27}$$
$$V_{ac} = V_c \cos(\theta V_c - \theta_a ABC) \tag{12.28}$$
$$V_{bc} = V_c \cos(\theta V_c - \theta_b ABC) - V_b \cos(\theta V_b + \theta_b ABC) \tag{12.29}$$

in ABD

$$V_{ab} = V_b \cos(\theta V_b) \tag{12.30}$$
$$V_{ad} = V_d \cos(\theta V_d - \theta_a ABD) \tag{12.31}$$
$$V_{bd} = V_d \cos(\theta V_d - \theta_b ABD) - V_b \cos(\theta V_b + \theta_b ABD) \tag{12.32}$$

in ACD

$$V_{ac} = V_c \cos(\theta V_c - \theta_a ABC) \tag{12.33}$$
$$V_{ad} = V_d \cos(\theta V_d - \theta_a ABD) \tag{12.34}$$
$$V_{cd} = V_d \cos(\theta V_d - \theta_a ACD) - V_c \cos(\theta V_c + \theta_a ACD) \tag{12.35}$$

The notations require more precision, especially on the angles associated with the various vertices of the triangles considered. This is how the $\theta i ABC$ "writing" must be introduced, which is the angle associated with the vertex "i" in the triangle ABC, for example. Before moving on to the reduction of the number of equations, it should be noted that it is of course necessary in this case that B, C, and D are all three in the radio visibility (for measurements) of A. If D was not, then there would only be two triangles left available, challenging the ongoing development. Let us return to our equations to see that some of them are identical: this is the case for (12.30) and (12.27), (12.33) and (12.28), as well as (12.34) and (12.31). Our nine equations are therefore simplified into a system of six terminal equations, as follows:

$$V_{ab} = V_b \cos(\theta V_b)$$
$$V_{ac} = V_c \cos(\theta V_c - \theta_a ABC)$$
$$V_{bc} = V_c \cos(\theta V_c - \theta_b ABC) - V_b \cos(\theta V_b - \theta_b ABC)$$
$$V_{ad} = V_d \cos(\theta V_d - \theta_a ABD)$$
$$V_{bd} = V_d \cos(\theta V_d - \theta_b ABD) - V_b \cos(\theta V_b - \theta_b ABD)$$
$$V_{cd} = V_d \cos(\theta V_d - \theta_a ABD) - V_c \cos(\theta V_c - \theta_a ACD) \tag{12.36}$$

However, there are still about 10 unknowns. In summary, this first method of solving leads us to systems with 3 equations and 6 unknowns for three points, 6 equations and 10 unknowns for four points, and we would obtain 10 equations and 14 unknowns for five considered points. Such inflation is not good because the increase in the number of points leads to a similar increase in the number of unknowns. Thus, without additional data, the resolution is not possible.

12.3.2 Approach Using Doppler Measurements in Case Some Nodes Are Fixed

However, it should be noted that the approach described above could be interesting if we agree to add some "constraints," such as the fixed nature of some of the points. Let us take as an illustration the classic case of the search for the position of a terminal D from transmissions from three fixed transmitters A, B, and C. The situation remains very similar to what was discussed in Section 12.3.1, with a constraint on the immobility of the last three points mentioned. We could take the system of Eq. (12.36) and simplify it in order to solve it: the objective being to launch tracks; however, we suggest another approach, which consists in searching for Cartesian coordinates of D thanks to the redundancy of measurements of the projections of the velocity vector V_d from D to (AD), (BD), and (CD). The geometry of the approach is given, by way of example, in Figure 12.3.

The various nodes have Cartesian coordinates (x_i, y_i). Only the coordinates of D are not known. In addition, the Doppler measurements performed give the projections of V_d, the velocity vector of D, respectively, on the three axes already mentioned. The question then arises: is it possible to find both the coordinates of D and those of the end of V_d?

Without going into the details of the equations, relatively simple but with heavy writing, the principle of resolution could be as follows:

- Draw (or put it in equations) a perpendicular to (AD) passing through the point of (AD) that is at V_{ad} from the unknown point D.
- Draw (or put it in equations) a perpendicular to (BD) passing through the point of (BD) that is at V_{bd} of the unknown point D.

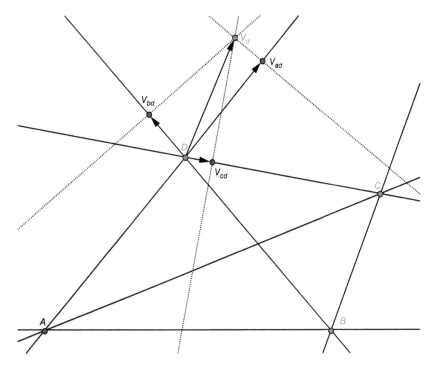

Figure 12.3 Cartesian geometry in the case of three fixed transmitters (A, B, and C) for 2D location.

- The intersection of these two lines is the end of V_d. The problem is that we do not know the coordinates of D and that this construction leaves an infinite number of possibilities for V_d.
- The measurement on (CD) takes place here by "fixing" point D. Indeed, the end of V_d is also at the intersection of the two previous lines with the perpendicular to (CD) passing through the point of (CD), which is at V_{cd} of D.

It is thus the fixity of the three points A, B, and C that makes it possible to find the coordinates of D and V_d, in amplitude and in direction. Although the equations of the system are easy to obtain, the same cannot be said for the resolution of the system.

The constraint of these methods based on Doppler measurements is that the latter must not be zero: and therefore terminal D must be in motion!

12.3.3 Approach Using Doppler Measurements to Estimate Angles

Now let us approach the problem by trying to calculate the angles. In the case of mobile nodes, it is useful to return to a more geometry- and measurement-oriented approach while maintaining the hypothesis of the relativity of things in relation to a reference terminal, here A. In this case, the figure becomes a little more complex, certainly, but we have the possibility to simply take into account the relationships between angles in the various triangles, which should allow us to find a system that can be solved. The new geometry is given in Figure 12.4 in which the angles are now referenced by a number to simplify the writing.

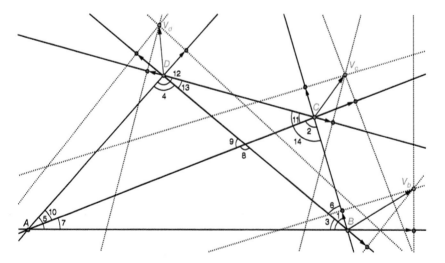

Figure 12.4 Geometry of the four-node system.

The system defining the basic equations for Doppler measurements is then as follows:

$$V_{bc} = V_{ab} \cos 1 - V_{ac} \cos 2 \tag{12.37}$$

$$V_{bd} = V_{ab} \cos 3 + V_{ad} \cos 4 \tag{12.38}$$

$$V_{bc} = V_{db} \cos 6 + V_{dc} \cos 14 \tag{12.39}$$

$$V_{cd} = V_{ac} \cos 11 + V_{ad} \cos 12 \tag{12.40}$$

to which we must add the relationships between the angles of the various triangles. Without going into detail, the latter are typically as follows:

$$1 + 2 + 7 = \pi \tag{12.41}$$

$$3 + 4 + 5 = \pi \tag{12.42}$$

$$6 + 13 + 14 = \pi \tag{12.43}$$

$$10 + 11 + 12 = \pi \tag{12.44}$$

$$4 + 9 + 10 = \pi \tag{12.45}$$

$$8 + 11 + 13 = \pi \tag{12.46}$$

$$2 + 6 + 9 = \pi \tag{12.47}$$

$$3 + 7 + 8 = \pi \tag{12.48}$$

$$8 + 9 = \pi \tag{12.49}$$

$$1 + 5 + 12 + 14 = 2\pi \tag{12.50}$$

We thus obtain the angles of a complete system with 14 equations and 14 unknowns. The fact of calculating the latter totally determines the geometry of the system, and therefore the positions of the various nodes. Thus, it would be possible, based only on the Doppler measurements of the four terminals relative to each other, to go back to their positions. In reality, things are a little more complex than they seem, particularly because the resolution of the complete system (Eqs (12.37)–(12.50)) is based on symmetrical functions in θ. Thus, we always have the possibility to start with a solution in "$+\theta$" or "$-\theta$." This is visible in Figure 12.5, which shows two of the possibilities.

The geometries $ABCD$ and $ABC'D'$ are equivalent in relation to the relative Doppler measurements if we consider the velocities V_b, V_c, and V_d associated, respectively, with terminals B, C, and D on the one hand and the velocities V_b', V_c', and Vd' associated, respectively, with terminals B, C', and D' on the other hand. However, this is not the only possible symmetry, as can be seen in Figure 12.6. This time, it is a question of taking into account a symmetry with respect to the AC axis and of noting that the geometries $ABCD$ and $AB'CD'$ are equivalent with respect to the relative Doppler measurements if we consider the speeds V_b, V_c, and V_d associated, respectively, with terminals B, C, and D on the one hand and the speeds V_b', V_c', and V_d' associated respectively with terminals B', C, and D' on the other hand.

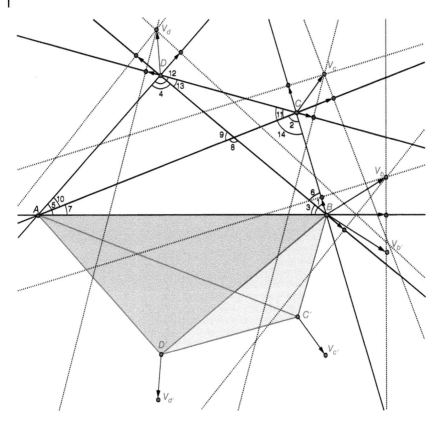

Figure 12.5 One symmetry of the problem.

Several system resolution techniques are available and it is possible to cite some classic approaches to resolving these ambiguities. The first is to have an absolute reference that allows the graph to be "oriented": this is particularly the case if the position of two nodes is known in an absolute way. The coupling with GNSS receivers of some nodes can of course be considered. A second, simplified possibility is to know a distance between two nodes, for example A and B (this is then UWB measurements, which could be implemented). In this case, it is possible to orient the graph on the AB axis by centering the relative reference mark on A and orienting it toward B (this does not completely solve the case of some of the symmetries described above). A last method, based on the same basis, consists in centering the system geometry on A and giving an arbitrary value to B on the AB axis. This distance is then counted in "base units" and allows the calculation of all positions also in base units. Thus, we obtain a resolution that is homothetic to reality. The value of the homothetic ratio (available possibly by using additional measurements: distances or positions of

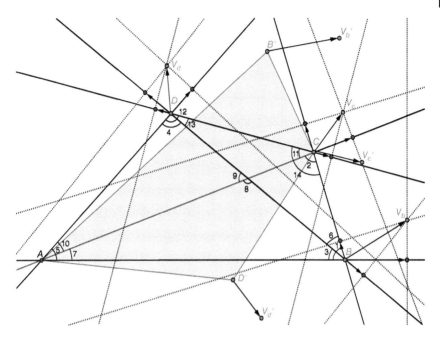

Figure 12.6 Symmetry of the problem with respect to the *AC* axis.

certain terminals) allows us to go back to the real absolute geometry. It should be noted that if we consider that *A* is the reference terminal, *A* will also be the homothetic center. Finally, let us note a last point on the geometry of the relative system *ABCD*: a circular symmetry clearly exists around each of the points considered, making here again the link with an absolute positioning. The "relativity" of the proposed approach then takes on a double dimension: the first concerns the Doppler measurements that characterize a relative displacement of the nodes, and the second concerns the calculation of the positions of the nodes that are relative to the definition adopted for the local geometric reference frame (here *A* as center and the *AB* axis as reference).

12.3.4 Approach Using Distance Measurements

A second approach, again based on relative measurements between two terminals, consists in using distance measurements. For example, we could consider flight time measurements between two nodes equipped with UWB technology (this to stay in the radio domain because we could also imagine using ultrasonic systems for accurate measurements). We do not discuss here the accuracy of the measurements or the inclusion of uncertainties in the calculations (steps that are of course essential).

We are here in a case that is easier to solve than the previous one from a geometric point of view. Let us consider a system of three nodes A, B, and C for which we have the following distance measurements:

- *dab* between A and B
- *dac* between A and C
- *dbc* between B and C

For a given measurement, it can be carried out by one or other of the nodes concerned, which may make it possible either to carry out averages or to keep the best measurement (although it is necessary to be able to estimate). If we place ourselves in two dimensions, in a reference frame (O, x, y) and define, according to a diagram identical to the one we presented in the Doppler approach, that A is in (0.0) and B in $(dab.0)$, then it comes quite simply that C is at the intersection of two circles: the first centered on A and of radius *dac* and the second centered on B and radius *dbc*. Several scenarios are likely to occur, but in general, as shown in Figure 12.7, we have two possible points of intersection. Without additional information, the choice is not possible. This once again characterizes the geometric symmetry specific to these relative measuring systems.

This is easily confirmed analytically. Suppose that C has as coordinates (xc, yc), then the latter check:

$$xc^2 + yc^2 = dac^2 \tag{12.51}$$

$$(xc - dab)^2 + yc^2 = dbc^2 \tag{12.52}$$

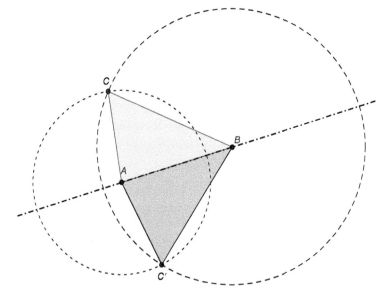

Figure 12.7 Ambiguity of position in a triangle for distance measurements.

The resolution of which is simple and gives

$$xc = \frac{1}{2dab}[dac^2 + dab^2 - dbc^2] \qquad (12.53)$$

$$yc = \pm\sqrt{dac^2 - xc^2} \qquad (12.54)$$

It appears here that yc can indeed take two values. In the case of a system with four nodes A, B, C, and D, the resolution remains relatively simple but always has this symmetry with respect to the axis used as a reference (AB in our case). Similarly, without attachment to a given position in an absolute reference frame, such as the GPS WGS84, for example, the resolution remains totally relative and rotations centered on each point do not change the general geometry. Our point of view is that this is not very important for many potential applications and uses: only this geometry, and certainly also the way it evolves, are useful.

12.3.5 Approach Analyzing the Deformation of the Network

Let us go one step further by completely changing the objective: it is no longer even a question of trying to find a geometry, but simply of analyzing how a network of nodes deforms by focusing on the "shape" that this deformation takes. The starting point remains Doppler-type measurements that allow us to say with a small margin of error (under "standard" conditions) if two nodes are coming close to each other or moving apart from each other. The Doppler amplitude gives information on the "speed" of this deformation.

Let us take the case of a simple network with three nodes: A, B, and C. The three associated measurements, respectively, mab between A and B, mbc between B and C, and mac between A and C, make it possible to accurately determine whether the corresponding distances are in approach, moving away or remain unchanged. This gives us three possibilities for each of the three links AB, BC, and AC. The three nodes being in the general case that interests us here totally independent, combinatory tells us that there are therefore 27 possible cases. Figure 12.8 lists these various cases.

The representation used using "="to characterize a link that remains unchanged, as well as "+" and "−," respectively, for an increasing link length and a decreasing link length, makes it easy to realize that if we do not establish a hierarchy between the nodes (and at this stage, there is no reason to establish one), then many cases are similar in the deformation that will be imposed on the triangle ABC. Remember that we are only looking at the shape of this deformation and not its amplitude. The dissimilar cases are thus represented in Figure 12.9 (maintaining the previous representation).

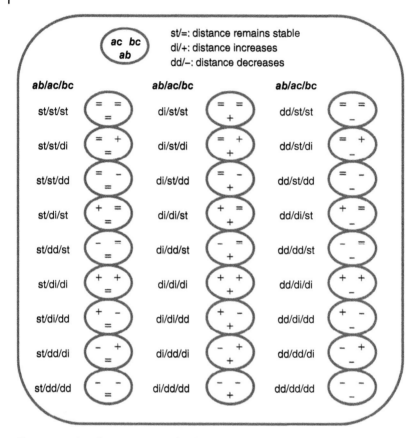

Figure 12.8 Set of possible cases of deformation of a three-node network.

It is also possible, but the complexity is increasing rapidly, to analyze the deformation of a four-node network. One can note that the remaining cases are quite reduced in number. Many technologies could be used in order to implement this kind of approaches, which are probably quite interesting in many real situations.

12.3.6 Comments

All that has been set out in Section 12.3 is only preliminary and much remains to be done to move toward the potential implementation of the proposed approaches. Among the subjects that would need to be addressed are the aspects related to the terminals themselves: which electronics, which processing, which codes, and how to manage the likely interference. In addition, which technologies could be used or combined to provide us with the best

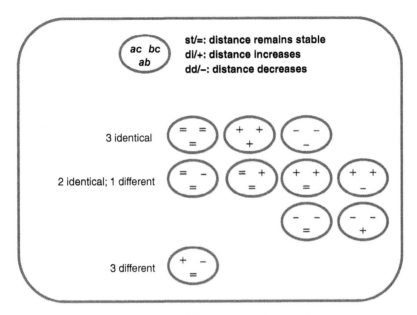

Figure 12.9 Set of different cases of deformation of a three-node network.

performance? Everything related to the signals themselves and the processing to be applied to them should also be addressed: proximity propagation between terminals, powers used, potential problem of radio nonvisibility of the various nodes, impact of clock drift, etc. The same applies to the performance that could be achieved in terms of accuracy (on the determination of geometry, for example), estimation of this accuracy, or impact of potential sources of error on these performances. In short, there is much more to be done than has been done, as is often the case!

Without going so far, it would also be necessary to address, always from a conceptual point of view as a first step, the (real) case of the third spatial dimension, which is not covered in this document (mainly for reasons of simplicity of presentation). And of course, if we are talking about the third spatial coordinate, the next step is naturally even more fundamental (and here again not covered here) and concerns the temporal coordinate. All this only makes sense if the dynamics of the network are taken into account. Extrapolating this dynamic to Doppler measurements (first proposed approach) or distance measurements (second approach) does not pose any technical difficulty and is a logical continuation of the descriptions detailed in the previous paragraphs. This is not the case with the approach on the study of network deformation: in the latter case, the methodology to be followed has yet to be invented.

12.4   General Discussion

This chapter could lead to two conclusions: the latest data processing methods will "revolutionize" the field and probably provide the long-awaited solution to the problem of indoor positioning (and more broadly to the problem of localization service continuity) and that much remains to be done. All this may be true, but it seems that we are constantly trying to make the solutions more and more complex. The basic idea is easy to follow: as the current methods do not give complete satisfaction, we will propose new ones by adding a new algorithmic layer, or by replacing them with new layers allowing a much larger amount of data to be processed. The underlying principle being of course that the more data is processed, the more precise the solution will be.

I personally do not fully agree with this principle. It is useful (and in general a source of progress) to multiply the work directions and the diversity of profiles of the researchers and industrial fields involved, but it is necessary to take a step back on the real performances that can be achieved. Positioning systems provide information about the present when we are trying today to predict the future, usually based on observation of the past. The problem when it comes to indoor positioning is that the future situation cannot, in some cases, be predicted because it depends fundamentally on many unplanned events (opening doors, presence of people, terminal posture, etc.). Thus, current statistical approaches are certainly useful in the context of global analyses of average situations, but probably not well adapted in the context of determining an individual position at a given time in a given unpredictable environment. As with the assisted and highly sensitive GNSS approaches in the GNSS field a few years ago, the progress generated by these new approaches is undeniable and undeniably useful, but they are not the definitive answer to the real problems because these are inherently singularities.

In fact, it is the promise that is not the right one: these approaches are presented as a general improvement of systems when in practice they are often only presented in "classic" cases. However, the expectation mainly concerns the singular situations mentioned above, in which these approaches are often not very effective.

According to the sentence "cum hoc sed non propter hoc,"[8] it is indeed very important not to confuse correlation and causality in the field of positioning because the physics of the propagation of phenomena comes into play. The identification of correlations is important and allows necessary classifications to be made, but if there is no link between these correlations and physical causes (identified or to be identified), it is dangerous, in any case as far as indoor positioning is concerned, to use it as a reliable argument.

---

8 "with this, however, not because of this."

This is the reason I think there is another direction for reflection: the simplification of approaches. The problem is that often the latter suggests that "simple" is equivalent to "easy" and thus to "not very effective." The search for the simplicity of a problem is on the contrary a much more delicate matter than the multiplication of layers of complexity that is often proposed. It is indeed essential to understand the causes of the problem's difficulties and this takes time (which we do not always have) and the result is often uncertain. This is why the current "Data Science" approaches are particularly useful because they allow results to be obtained relatively quickly and with satisfactory performance in the vast majority of cases in a "statistical" approach. The problem is that their assessment is generally carried out on rather classic cases and that difficult cases, say the 5% of sensitive cases, are not better managed than with standard approaches. This is, I believe, one of the explanations for the fact that we still do not have large-scale solutions for indoor positioning.

Bibliography

1 Herrera, J.C.A., Plöger, P.G., Hinkenjann, A. et al. (2014). Pedestrian indoor positioning using smartphone multi-sensing, radio beacons, user positions probability map and IndoorOSM floor plan representation. In: *2014 International Conference on Indoor Positioning and Indoor Navigation (IPIN)*, 636–645. Busan: IEEE.

2 Ifthekhar, M.S., Saha, N., and Jang, Y.M. (2014). Neural network based indoor positioning technique in optical camera communication system. In: *2014 International Conference on Indoor Positioning and Indoor Navigation (IPIN)*, 431–435. Busan: IEEE.

3 Lamy-Perbal, S., Guénard, N., Boukallel, M., and Landragin-Frassati, A. (2015). A HMM map-matching approach enhancing indoor positioning performances of an inertial measurement system. In: *2015 International Conference on Indoor Positioning and Indoor Navigation (IPIN)*, 1–4. Banff, AB: IEEE.

4 Ma, C., Wang, J., and Jianyun, C. (2016). Beidou compatible indoor positioning system architecture design and research on geometry of pseudolite. In: *2016 Fourth International Conference on Ubiquitous Positioning, Indoor Navigation and Location Based Services (UPINLBS)*, 176–181. Shanghai: IEEE.

5 Chen, Y., Chen, R., Pei, L. et al. (2010). Knowledge-based error detection and correction method of a multi-sensor multi-network positioning platform for pedestrian indoor navigation. In: *IEEE/ION Position, Location and Navigation Symposium*, 873–879. Indian Wells, CA: IEEE.

6 Pourabdollah, A., Meng, X., and Jackson, M. (2010). Towards low-cost collaborative mobile positioning. In: *2010 Ubiquitous Positioning Indoor Navigation and Location Based Service*, 1–5. Kirkkonummi: IEEE.

7 Zhen-Peng, A., Hu-Lin, S., and Jun, W. (2015). Classify and prospect of indoor positioning and indoor navigation. In: *2015 Fifth International Conference on Instrumentation and Measurement, Computer, Communication and Control (IMCCC)*, 1893–1897. Qinhuangdao: IEEE.

8 Kuusniemi, H., Bhuiyan, M.Z.H., Ström, M. et al. (2012). Utilizing pulsed pseudolites and high-sensitivity GNSS for ubiquitous outdoor/indoor satellite navigation. In: *2012 International Conference on Indoor Positioning and Indoor Navigation (IPIN)*, 1–7. Sydney, NSW: IEEE.

9 Tan, K.M. and Law, C.L. (2007). GPS and UWB Integration for indoor positioning. In: *2007 6th International Conference on Information, Communications and Signal Processing*, 1–5. Singapore: IEEE.

10 Itagaki, Y., Suzuki, A., and Iyota, T. (2012). Indoor positioning for moving objects using a hardware device with spread spectrum ultrasonic waves. In: *2012 International Conference on Indoor Positioning and Indoor Navigation (IPIN)*, 1–6. Sydney, NSW: IEEE.

11 Kaiser, S. and Lang, C. (2016). Detecting elevators and escalators in 3D pedestrian indoor navigation. In: *2016 International Conference on Indoor Positioning and Indoor Navigation (IPIN)*, 1–6. Alcala de Henares: IEEE.

12 Liao, J.-K., Chiang, K., Tsai, G.-J., and Chang, H.-W. (2016). A low complexity map-aided fuzzy decision tree for pedestrian indoor/outdoor navigation using smartphone. In: *2016 International Conference on Indoor Positioning and Indoor Navigation (IPIN)*, 1–8. Alcala de Henares: IEEE.

13 Gotlib, D., Gnat, M., and Marciniak, J. (2012). The research on cartographical indoor presentation and indoor route modeling for navigation applications. In: *2012 International Conference on Indoor Positioning and Indoor Navigation (IPIN)*, 1–7. Sydney, NSW: IEEE.

14 Park, Y. (2014). Smartphone based hybrid localization method to improve an accuracy on indoor navigation. In: *2014 International Conference on Indoor Positioning and Indoor Navigation (IPIN)*, 705–708. Busan: IEEE.

15 Nozawa, M., Hagiwara, Y., and Choi, Y. (2012). Indoor human navigation system on smartphones using view-based navigation. In: *2012 12th International Conference on Control, Automation and Systems*, 1916–1919. Jeju Island: IEEE.

16 Cankaya, I.A., Koyun, A., Yigit, T., and Yuksel, A.S. (2015). Mobile indoor navigation system in iOS platform using augmented reality. In: *2015 9th International Conference on Application of Information and Communication Technologies (AICT)*, 281–284. Rostov on Don: IEEE.

17 Garcia Puyol, M., Robertson, P., and Heirich, O. (2012). Complexity-reduced FootSLAM for indoor pedestrian navigation. In: *2012 International*

Conference on Indoor Positioning and Indoor Navigation (IPIN), 1–10. Sydney, NSW: IEEE.

18 Kajdocsi, L., Kovács, J., and Pozna, C.R. (2016). A great potential for using mesh networks in indoor navigation. In: *2016 IEEE 14th International Symposium on Intelligent Systems and Informatics (SISY)*, 187–192. Subotica: IEEE.

19 Galov, A. and Moschevikin, A. (2014). Simultaneous localization and mapping in indoor positioning systems based on round trip time-of-flight measurements and inertial navigation. In: *2014 International Conference on Indoor Positioning and Indoor Navigation (IPIN)*, 457–464. Busan: IEEE.

20 Sun, C., Kuo, H., and Lin, C.E. (2010). A sensor based indoor mobile localization and navigation using Unscented Kalman Filter. In: *IEEE/ION Position, Location and Navigation Symposium*, 327–331. Indian Wells, CA: IEEE.

21 Kulikov, R.S. (2018). Integrated UWB/IMU system for high rate indoor navigation with cm-level accuracy. In: *2018 Moscow Workshop on Electronic and Networking Technologies (MWENT)*, 1–4. Moscow: IEEE.

22 Ruotsalainen, L., Kuusniemi, H., and Chen, R. (2011). Heading change detection for indoor navigation with a Smartphone camera. In: *2011 International Conference on Indoor Positioning and Indoor Navigation*, 1–7. Guimaraes: IEEE.

23 Czogalla, O. and Naumann, S. (2016). Pedestrian indoor navigation for complex public facilities. In: *2016 International Conference on Indoor Positioning and Indoor Navigation (IPIN)*, 1–8. Alcala de Henares: IEEE.

24 Caruso, D., Sanfourche, M., Le Besnerais, G., and Vissière, D. (2016). Infrastructureless indoor navigation with an hybrid magneto-inertial and depth sensor system. In: *2016 International Conference on Indoor Positioning and Indoor Navigation (IPIN)*, 1–8. Alcala de Henares: IEEE.

25 Ma, S., Zhang, Y., Xu, Y. et al. (2016). Indoor robot navigation by coupling IMU, UWB, and encode. In: *2016 10th International Conference on Software, Knowledge, Information Management & Applications (SKIMA)*, 429–432. Chengdu: IEEE.

26 Yudanto, R.G. and Petré, F. (2015). Sensor fusion for indoor navigation and tracking of automated guided vehicles. In: , 1–8. 2015 International Conference on Indoor Positioning and Indoor Navigation (IPIN), Banff, AB: IEEE.

27 Exman, I. and Levi, E. (2014). Scalable cloud and smartphones for image based indoor navigation. In: *2014 IEEE 28th Convention of Electrical & Electronics Engineers in Israel (IEEEI)*, 1–4. Eilat: IEEE.

28 Glanzer, G. and Walder, U. (2010). Self-contained indoor pedestrian navigation by means of human motion analysis and magnetic field mapping. In: *2010 7th Workshop on Positioning, Navigation and Communication*, 303–307. Dresden: IEEE.

29 Tondwalkar, A. (2015). Infrastructure-less collaborative indoor positioning for time critical operations. In: *2015 IEEE Power, Communication and Information Technology Conference (PCITC)*, 834–838. Bhubaneswar: IEEE.

30 Taniuchi, D., Liu, X., Nakai, D., and Maekawa, T. (2015). Spring model based collaborative indoor position estimation with neighbor mobile devices. *IEEE Journal of Selected Topics in Signal Processing* 9 (2): 268–277.

31 Sridharan, M., Bigham, J., Phillips, C., and Bodanese, E. (2017). Collaborative location estimation for confined spaces using magnetic field and inverse beacon positioning. In: *2017 IEEE SENSORS*, 1–3. Glasgow: IEEE.

32 Giorgetti, G., Farley, R., Chikkappa, K. et al. (2011). Cortina: collaborative indoor positioning using low-power sensor networks. In: *2011 International Conference on Indoor Positioning and Indoor Navigation*, 1–10. Guimaraes: IEEE.

33 Thompson, B. and Buehrer, R.M. (2012). Characterizing and improving the collaborative position location problem. In: *2012 9th Workshop on Positioning, Navigation and Communication*, 42–46. Dresden: IEEE.

34 Zheng, S., Farley, R., Kaleas, T. et al. (2011). Cortina: collaborative context-aware indoor positioning employing RSS and RToF techniques. In: *2011 IEEE International Conference on Pervasive Computing and Communications Workshops (PERCOM Workshops)*, 340–343. Seattle, WA: IEEE.

35 Luo, Y., Chen, Y.P., and Hoeber, O. (2011). Wi-Fi-based indoor positioning using human-centric collaborative feedback. In: *2011 IEEE International Conference on Communications (ICC)*, 1–6. Kyoto: IEEE.

36 Thompson, B. and Buehrer, R.M. (2011). Cooperative indoor position location using reflected estimations. In: *17th European Wireless 2011 – Sustainable Wireless Technologies*, 1–6. Vienna, Austria: IEEE.

13

Maps

Abstract

In the vast majority of cases, in one way or another, the need to "locate" the calculated or estimated position on a map arises. This "positioning" can be done
through the user (in the case of a person; it is more complicated for an object)
if the latter has paper impressions of the place, as in the "old days." However,
we are all familiar with electronic maps, particularly democratized as part of our
road navigation systems or smartphone applications. The case of indoor maps is
a little particular for reasons that are ultimately quite similar to those relating to
the technologies that we have detailed in the book: the very great diversity of the
available media, and even the absence of these media. The mapping of a building
is not available in a central agency as road maps can be in some countries. Thus,
the work associated with data entry is enormous, even more than that required in
the 1980s and 1990s to produce the digital road maps used daily today. Another
common point between cartography and technology is the strong presence of the
major Internet players. Returning to interior mapping, the story has not yet been
told, as we will quickly see in this chapter.

Keywords   *Maps; Indoor maps; Recording tools*

Road mapping is a good example of the complexity underlying this field. In
many countries, national agencies are responsible for maintaining databases of
the road network, and much more: the detail of roads, of course, associated with
their size, the type of pavement, but also the presence of sidewalks, their characteristics, much more so concerning land, building areas, floodplains, ancillary
infrastructure such as fire hydrants, etc. It is also understood that not all these
data are necessarily stored or managed by a single organization, and sometimes,
they simply do not exist. Thus, when the major players of the time sought to produce a road map in order to propose what is nowadays called a "car GPS," they
were confronted with a task of considerable magnitude. Indeed, in addition to

Indoor Positioning: Technologies and Performance, First Edition. Nel Samama.
© 2019 The Institute of Electrical and Electronics Engineers, Inc. Published 2019 by John Wiley & Sons, Inc.

having to collect all this data, which is very diverse and certainly not uniform throughout the world, it was necessary to approach the concept of maps in a new way: in the context of a road navigation application, a road or path is not equivalent to a built-up area in which, by definition, the car cannot be found. Similarly, if we now come to the case of a route calculation, the basis of such a system, it must also be possible to describe the routes that can be used and those that cannot be used by cars[1]. Thus, we understand that it will sometimes be necessary, sometimes significantly, to "enrich" the existing bases in order to make this specific use of the car's navigation system. This chapter will thus modestly address these aspects and highlight the specific difficulties of the indoor environment.

13.1 Map: Not Just an Image

We have now understood that the map is not a simple image, but a set of objects (highways, small roads, paths, parking lots, areas that cannot be crossed, etc.) with properties (public/private, available for wheel chairs, etc.). In addition, these objects must also have attributes specific to the navigation application you wish to offer. For example, in order to provide a relevant itinerary, it is necessary to know the distribution of the prohibited directions in particular, but also the width of the lanes, their surface, the authorized speed, etc. This required significant investments and work for road mapping. However, the problem was not fully resolved because it was also necessary to verify that it was possible to access the roads. Of course, this is quite simple in the case of national roads, but much less obvious for local roads. Is it possible to propose an itinerary using such a path or not, the latter being in reality hindered by a chain or a barrier. This actually required battalions of people and systems to automatically capture environments, and thus in situ displacements. One could think of communities (Open Street Map contributors, Google Maps Reviewers, etc.) dedicated to map the world. Finally, these characteristics are constantly changing: new roads are being built, old ones destroyed, diverted, or traffic directions changed. All this must therefore be continuously updated.

Thus, we understand that this road mapping, although very rich databases of preexisting data were available, is much more than an image, but required enrichment specific to the navigation application. Finally, we also perceive the absolute need to "keep" this mapping up to date, and finally to have "local" knowledge of the real parameters[2]. All this makes road mapping an area in itself.

---

1 The case of the pedestrian (or an object) is very different because he can move everywhere, with more degrees of freedom than a car, which will lead to new difficulties in creating maps, as we will see in this chapter.

2 It should be noted that the "crowdsourcing" approaches implemented by the main actors in the field are particularly relevant by allowing all users to report data on their environments in order to provide a tool for real-time consolidation of mapping on a global scale.

The case of indoor mapping is even more complex because, although the dimensions are smaller, the diversity of formats is such that specific tools seem essential. In addition, data are generally not centralized and the necessary collection work is therefore absolutely gigantic and, to a large extent, remains to be done.

13.2 Indoor Poses Specific Problems

The indoor world is mainly that of individuals and objects and, as already mentioned, the latter are not subject to the same constraints as cars. In particular, traffic areas are generally two way, with no positioning constraints reserved for a given direction. Thus, it is difficult to deduce from the position any help on the movement. In addition, a pedestrian (for example) is likely to turn around almost instantly, or cross an open area in any way he or she wishes. On the other hand, as with cars, it is generally not possible for it to pass through a wall except when the wall has an opening, say a gate or a door. The question then remains as to whether, in the latter case, this door can be opened by the pedestrian or not (we will come back to this point). The fundamentals of indoor mapping are emerging: it will have to be very "fine."

Unlike cars, there are many "categories" of pedestrians. These categories can be linked to the physical characteristics of the pedestrian (age, physical ability, disability, etc.), but also to user profiles: general public, authorized personnel, outside worker, maintenance service, occasional repairer, etc. All these categories, presented here intertwined, lead to specific processing. This would of course be the case for a route calculation, for example, where stairs would be prohibited for a person in a wheelchair. Similarly, the various rooms of a company cannot be considered equivalent for an employee and a visitor. Such examples can be repeated at will. The attributes to be associated with each mapping entity are thus of different natures: some are related to geometry, others to displacement characteristics, and others to categorical considerations.

A traditional approach used in current applications is to define a "profile" for each user. This could be the case for indoor route calculation: "sporty" (taking the stairs), "fit" (taking them if there are fewer than three floors), "classic" (taking the elevator, except for one floor), "lazy" (never taking the stairs unless the elevators are unavailable), or "disabled" (physically unable to take the stairs and thus obliged to choose elevators or ramps).

The fact of approaching the case of the floors introduces a new difference with road navigation: the indoor world is really needs to take into account the elevation (we will see in the next paragraph the different ways of considering this problem) and this must be taken into account, both for positioning and for mapping.

Although remaining within the framework of a route calculation, let us note two elements:

- As in the case of roads, it is necessary to allocate to each entity of the map an attribute defining the travel speed that can be reached, on average, during its crossing. This has two objectives: the first is to reserve routes for traffic areas and, in particular, not to propose an itinerary that would go through a meeting room, for example[3]. The second is to provide an estimate of the travel time required to reach your destination. A fairly traditional approach is simply to assign different weights to the various entities and apply these weights to the length of the entity being crossed. Thus, if the weight of a meeting room is very high, the calculation algorithm will quickly abandon the path that crosses it.
- Unlike the road case, the calculation is based on a relatively simple graph network: the calculation is generally very fast and the possibilities of alternative routes are often reduced. The performance of the calculation algorithms is therefore very good.

A new constraint applies to the confidentiality of the mapping of a building. In a company, for example, it is not necessarily desirable for a visitor to have access to the location or geometry of storage rooms or computer rooms. Thus, it would probably be useful to design access rights that would define the level of detail of the mapping made available. These rights would be different depending on the function of the user: employee, maintenance, visitor, security service, delivery person, etc.

13.3   Map Representations

In terms of representations, two cases can be distinguished: the map itself and its representation for the user. In the first case, it is a question of proposing "professional" tools for data recording (see Section 13.4) adapted to the internal indoor and effective. The second case is different because it must take into account human behavior that is not naturally made to read a map. In particular, we do not have any natural internal sensors with absolute orientation, such as an integrated magnetic compass, for example. This is not a fundamental lack on the outside because we "perceive" our environment and we do not need to know our orientation in relation to an absolute reference in order to find our way in space. Orienting ourselves "relatively" to our environment (buildings, landmarks, etc.) is generally sufficient. The indoor world is different: let us take the case, classic in modern constructions, of a corridor with no windows to the

---

3 In a more complex vision, we could be tempted to couple the calculation of the route with the occupation of such rooms and allow the crossing if they are not occupied.

outside and let us suppose a building of several floors and organized in a cross. Our natural abilities do not allow us to have information about either the floor or the wing in which we are located without additional elements.

The representation of the indoor environment for the user thus becomes an important element in his appropriation of a positioning application, and even more in the case of a navigation and guidance application. The difficulty lies in the fact that the map is typically an assembly, a superposition, of geometric elements that are not necessarily visible to the user. In general, these are two-dimensional geometries (the floors) that can be represented as "stacked" in three dimensions. However, unlike the outside world, the user has only a very partial view of his inner environment because of the walls. Thus, the fact of having a complete representation of a floor, even the one in which one is located, is of limited interest because, in addition to being in an unknown place, one must then have an abstract representation of the floor. In some cases, a three-dimensional representation of the building is offered and one's location is represented by a dot: it is even worse for the user for the positioning applications. Probably, it could be interesting to use this 3D information for analytics.

Car navigation systems are an instructive example: although the real world is three dimensional, road mapping is typically two dimensional. A new distinction appears here within the representation of the environment for the user: the type of view. The first systems only offered a "top view" for which the point of view, i.e. the point of view from which the environment was viewed, was located above the user's physical point of presence, at altitude. A difficulty specific to the scale of representation then appeared: unlike paper maps for which it is sufficient to approach or move away from one's gaze in order to have a successively wide or more precise view, the electronic map on screen requires enlargements or reductions that make it possible to switch from a precise view to a wide view, but which then make one of the two aspects (precise or wide) disappear. However, the need (or habit of the elders) seems to be to have both at the same time. It is of course possible to multiply the views on the same screen, but this only complicates the graphic display. Another view then appeared in the form of an "oblique view," which presents an aspect of perspective that allows, to a certain extent, representations to be reconciled.

A perspective view is also a first approach to a three-dimensional representation. If we take the example of car navigation, some manufacturers propose to further enhance the restitution by including a 3D representation of the buildings. It should be noted that the interest for the user lies mainly in improving his ability to associate what he sees around him with the electronic representation offered to him. Figure 13.1 is an example.

The indoor environment is both simpler because the geographical coverage area is less important and more complex because user containment requires greater detail accuracy. The important thing is to provide the user with visual elements that should allow him to "understand" where he is on the map,

Figure 13.1 Oblique view and 3D buildings of an outdoor navigation system. Source: Courtesy of TomTom.

Figure 13.2 Indoor navigation system in augmented reality.

whatever it may be. That is the main difficulty. In this context, the contribution of augmented reality seems obvious (see Figure 13.2): by adding information about an image that is precisely what the user sees, the user intuitively locates himself in his surrounding space. We could think about using the camera on our smartphones as a portable navigation system. This would undoubtedly pose an "ergonomic" difficulty because it would require either a mechanical "assembly" to be defined or would deprive the user of one hand.

It should be noted that the addition of site-specific elements is also a problem when mapping the environment: it would be useful to include visual elements that are not generally available in conventional building descriptions. In addition, these elements, which are very useful when they are right, sometimes become disturbing when they are wrong. This leads to a need to check for potential changes on a regular basis, further complicating matters. These elements are more numerous: door color, elevator size, presence of billboards or screens, exact name of floors in elevators, wall textures, etc.

This shows that it is possible to distinguish the map, its representation, and the view provided to the user. This leads us to the various ways of taking into account the third dimension[4] of indoor environments. Three approaches are traditionally used:

- The 2D ½ consists in defining a position by its coordinates in two dimensions on a given floor. Thus, the determination of the floor makes it possible to return to a representation of the floor plan. The view can then either be flat or oblique.
- The (2 + 1)D allows us to keep the discrimination of the "altitude" of the terminal, but always for a given floor (we think here particularly of a storage place: several articles have the same 2D position but are stored one above the other). Thus, this representation is not equivalent to 3D.
- The 3D is a representation of space by three-dimensional coordinates.

13.4 Recording Tools

This raises the crucial question of entering building plans and the potential standardization of the format of these plans. The major players in the field of navigation, positioning, and related services have of course addressed this and there are standard formats for maps, but no standard input tool, nor standardized file format. More precisely, many computer-aided design software programs exist that allow you to enter plans, in two or three dimensions, and to convert the formats obtained without difficulty. However, they do not generally take into account the essential association in our case between the entry of an element and the navigation attributes specific to that element (see Figure 13.3). This is in particular the speed of movement possible in the element when moving, or the fact that this element is visible in a visible or private way, or the accessibility to the element (disability, safety, etc.). There are more examples of such attributes.

Traffic areas, such as meeting rooms or offices, should be identified as such. Let us consider, in Figure 13.3, the case of room "043": it is actually a large meeting room. Thus, if a route is requested from a guidance application between room 032 (on the left of the plan, in the middle) and room 042 (just below the 043), the latter must in no way propose going through the meeting room, otherwise the user will be placed in a delicate situation. To do this, several leads are possible: the first would be to consider part 043 as "not crossable." However, it does have two doors, one at each end. A second approach would be to assign each entity on the map an attribute of "travel speed." This would also provide the user with an indication of the travel time for the requested route. In this case, it would be easy to have a "speed" attribute that would be much higher in the case of travel zones than in the case of offices, even if the latter

---

4 It is interesting to note here that the discussion also applies to location systems, in similar terms.

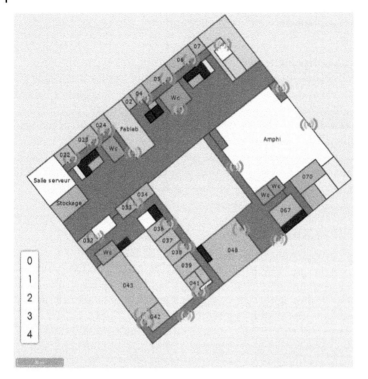

Figure 13.3 Example of indoor mapping with information on the use of a given room (displayed with different gray levels).

have two doors. A final case would be to have additional information on the use of the meeting room and to propose its crossing in the event that it is not occupied.

In addition, current software offers a very wide range of input options. This is of course an advantage for the designer who is given great freedom in his work, but it quickly becomes a disadvantage when it comes to standardizing formats. Such an example is common when entering interior plans: depending on whether the operator has entered a part as a polygon or a set of lines, the format conversion will not be equivalent. The basic possibilities being very numerous, the "converter" also becomes very complex to realize.

Thus, to date, there is no universal tool for entering an internal plan for navigation purposes. Each one adapts its approach by adding navigation-specific skills. The major players are of course present, such as Open Street Map or Google, but also smaller entities that offer specific tools, such as Map-Wize[5]. However, there are ongoing initiatives in this direction, without being fundamentally oriented toward navigation. This is the case with BIM. The BIM

---

5 https://www.mapwize.io/

is alternately the Building Information Modeling, the Building Information Model, or the Building Information Management. We can see that this is a major undertaking seeking to move building[6] design forward in the field of digital technology and data exchange between trades (energy, communication, construction, etc.). Navigation is not yet explicitly included, but it will come in time.

Among the functionalities required for such a tool, it is possible to mention some fundamental elements such as the possibility of creating the building by adding rooms, corridors, or any other element, then including it in its real environment (see Figure 13.4), sizing it at the right scale, rotating it, or resizing it.

Figure 13.5 then shows a way of approaching attribute entry: each element can be "edited" and thus characterized by a type, accessibility[7], travel speed, or emergency evacuation zone.

These attributes become essential when trying to calculate a route. Two types of profiles can then be used: the first is the user's profile and represents his individual choice of movement (slow, sporty, without stairs, etc.). The second is related to the structure which, in some cases, wishes to maintain the confidentiality of certain parts of its site (in terms of mapping). Thus, the user

Figure 13.4 Example of a tool for overlaying and dimensioning a building in its real environment (we use here the Open Street Map environment but all standards are possible).

---

6 In addition, not only buildings, as the whole field of construction is concerned.
7 Notion of the physical capacity necessary to enter the room, but also of the profile of users with access authorization or map confidentiality for sensitive sites, for example.

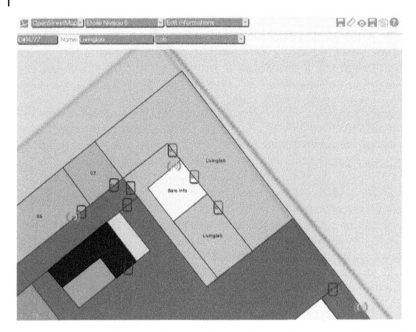

Figure 13.5 Interface for managing the attributes of a part of the cartography (here the Living Lab). Note that in this example, when zooming doors appear on the map.

may be a "visitor," an "employee," or an "administrator," with different access and vision of the site depending on the case. This raises a question about how maps are stored: does the user have temporary access to a "map server" or does he load them onto his terminal. Are the maps protected by any kind of encryption or not? Is the terminal provided to the user by the structure visited or is it his own terminal. All these questions refer to notions of data confidentiality that must be taken into account and which have a significant impact on the technical solutions deployed.

An "emergency" attribute can also be useful to specify whether a room (especially corridors) is an evacuation zone. In the event of an alarm, an "emergency" mode can then be activated, which will restrict the visibility of the map to useful travel areas only, but which will also take this parameter into account when calculating the route by reducing the number of possible destinations and using only dedicated traffic areas.

13.5   Some Examples of the Use of Indoor Mapping

In order to illustrate our discussion on mapping, we will briefly present some applications that certainly require the availability of the site plan.

13.5.1 Some Guiding Applications

If we are trying to guide a user, it will be necessary to have three pieces of information: his position, his destination, and the map. Its position is supposed to be obtained through the positioning system, the destination through an interface that will ask her/him where she/he wants to go, and finally the mapping would be included in the guidance application used[8].

A basic example of such a service is shown in Figure 13.6 as a screenshot of an application. The various steps are the opening of the application (here, it is automatic by reading an near field communication [NFC] tag), the determination of the user's position (here again automatic because it is linked to the reading of the tag), and then the request of the destination. To do this, the user activates the navigation screen (screen b) in Figure 13.6. The user must activate the "go to" tab. The third screen then provides all the possible destinations of the building and the user makes her/his choice by simply clicking on the destination. All kinds of classifications are possible at this stage: by floors, functionalities, activity areas, or any other appropriate choice for the place in question.

Once a destination has been selected, in this case room DB111, a route is proposed (see Figure 13.7) to the user according to the preferences that they will have previously filled in their profile. Note here that as the user is not

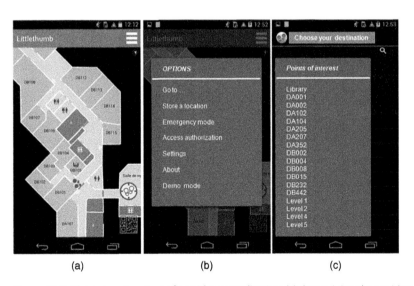

| (a) | (b) | (c) |

Figure 13.6 The successive steps of a guidance application: (a) determining the position, (b) choosing a destination, and (c) viewing a route (© LittleThumb).

---

8 It is not the purpose of this paragraph to go into the details of how these various tasks are carried out.

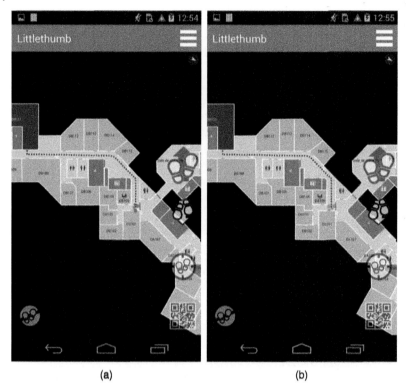

(a)   (b)

Figure 13.7 Graphical help for user orientation: (a) red feet wrong orientation, (b) green feet right orientation (© LittleThumb).

positioned continuously (see Chapter 5 on NFC positioning), the application cannot automatically know the direction viewed by the user. Thus, a graphic restitution trick is used: assuming that the user has not moved since the last "tag," the terminal compass is used to guide him in the right direction of departure. The interface feet, representing the user, will change from red to green when the user is in the correct orientation (screens a and b) in Figure 13.7).

The emergency mode described in the previous section can then be activated and the appearance of the map changes, revealing only the evacuation traffic areas (see Figure 13.8).

13.5.2 Some Services Associated with Mapping

Once the map is available, it is easy to imagine all kinds of applications: managing the booking of shared meeting rooms is a simple example. The coupling between the availability agenda of the rooms and the positions of the people to

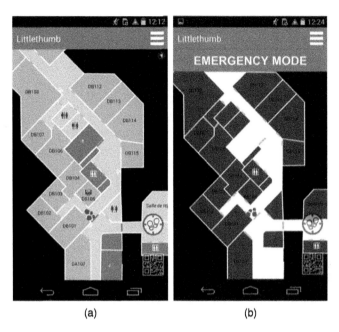

Figure 13.8 In case of an emergency mode (© LittleThumb).

be in the meeting allows the reservation to be made. It is of course necessary to know the start time and the expected duration of the meeting. Once booked, the room becomes unavailable. It is also possible to imagine an instant booking service in which a free room is booked and then immediately occupied, with an indication of its location being sent to the participants, who can then be guided to the room, again thanks to the mapping of the location.

In the continuity of this but in a different register, the management of events such as conferences or exhibitions is likely to benefit from the joint availability of the location of people and objects as well as the layout of buildings. Upon arrival, a participant can simply load the plan and agenda of the entire conference (which in this case extends over several days and sessions) on his/her smartphone. The conference application will allow her/him to manage her/his personal schedule by including appointments already scheduled and taking into account the travel times between two presentations that the participant wishes to attend. In addition, the application allows the user, at any time, to click on a presentation (or on an exhibitor's stand) in order to be directed to the corresponding room. Localization is an important element of the application, but so is mapping.

We will see in Chapter 14 other specific situations in which the availability of the location of people and objects indoors is likely to change the daily life.

13.6 Synthesis

Mapping is just as important as localization when it comes to providing a visualization or guidance service, but also to provide information to users with added value services. Indoor mapping must associate with each entity a set of attributes that are essential if it is to provide useful services. The current difficulties are multiple: lack of a standard for data entry, complexity of data entry, which often requires a physical visit to the site to remove certain ambiguities, the need to include visual elements to assist in user orientation, etc. All this is further amplified by the complexity and diversity of available map formats: the automation of the transition from current maps to "object-oriented" maps is therefore delicate. In addition, storage and representation of such maps is not so obvious and relies on choices of the actors. Indeed, maps are projections of the reality and projections involve hypothesis for the computation and then approximation on the values. For example, WGS84 (EPSG:4326) is used to provide GPS coordinates but not to display the position on web maps for which EPSG:3857 is much more appropriate. These representations are not valid for all kinds of applications (a web display is flat over a restricted coverage, for example, a street or part of a road). This results in a very significant amount of work today to digitize all the buildings.

One approach is to provide efficient tools to enable everyone to digitize their building because it is unlikely to imagine, as was the case for road mapping, that such work would be undertaken by a single entity. It is a kind of "crowd sourcing" that we should probably be able to create: it is in this sense that initiatives like IndoorOSM (indoor Open Street Map) are looking to go, but the path still seems to be long.

Bibliography

1 Pipelidis, G., Rad, O.R.M., Iwaszczuk, D. et al. (2017). A novel approach for dynamic vertical indoor mapping through crowd-sourced smartphone sensor data. In: *2017 International Conference on Indoor Positioning and Indoor Navigation (IPIN)*, 1–8. Sapporo: IEEE.

2 Pal, M., Thakral, A., Chawla, R., and Kumar, S. (2017). Indoor maps: simulation in a virtual environment. In: *2017 International Conference On Smart Technologies For Smart Nation (SmartTechCon)*, 967–972. Bangalore: IEEE.

3 Wen, C., Pan, S., Wang, C., and Li, J. (2016). An indoor backpack system for 2-D and 3-D mapping of building interiors. *IEEE Geoscience and Remote Sensing Letters* 13 (7): 992–996.

4 Xue, H., Ma, L., and Tan, X. (2016). A fast visual map building method using video stream for visual-based indoor localization. In: *2016*

International Wireless Communications and Mobile Computing Conference (IWCMC), 650–654. Paphos: IEEE.

5 Jeamwatthanachai, W., Wald, M., and Wills, G. (2016). Map data representation for indoor navigation a design framework towards a construction of indoor map. In: *2016 International Conference on Information Society (i-Society)*, 91–96. Dublin: IEEE.

6 Malla, H., Purushothaman, P., Rajan, S.V., and Balasubramanian, V. (2014). Object level mapping of an indoor environment using RFID. In: *2014 Ubiquitous Positioning Indoor Navigation and Location Based Service (UPINLBS)*, 203–212. Corpus Christ, TX: IEEE.

7 Lee, B., Lee, Y., and Chung, W. (2008). 3D map visualization for real time RSSI indoor location tracking system on PDA. In: *2008 Third International Conference on Convergence and Hybrid Information Technology*, 375–381. Busan: IEEE.

8 Wasinger, R., Gubi, K., Kay, J. et al. (2012). RoughMaps A generic platform to support symbolic map use in indoor environments. In: *2012 International Conference on Indoor Positioning and Indoor Navigation (IPIN)*, 1–10. Sydney, NSW: IEEE.

9 Yara, C., Noriduki, Y., Ioroi, S., and Tanaka, H. (2015). Design and implementation of map system for indoor navigation – an example of an application of a platform which collects and provides indoor positions. In: *2015 IEEE International Symposium on Inertial Sensors and Systems (ISISS) Proceedings*, 1–4. Hapuna Beach, HI: IEEE.

10 Bozkurt, S., Yazici, A., Günal, S., and Yayan, U. (2015). A survey on RF mapping for indoor positioning. In: *2015 23rd Signal Processing and Communications Applications Conference (SIU)*, 2066–2069. IEEE, Malatya.

11 Zhou, M., Wong, A.K., Tian, Z. et al. (2014). Personal mobility map construction for crowd-sourced Wi-Fi based indoor mapping. *IEEE Communications Letters* 18 (8): 1427–1430.

12 Schäfer, M., Knapp, C., and Chakraborty, S. (2011). Automatic generation of topological indoor maps for real-time map-based localization and tracking. In: *2011 International Conference on Indoor Positioning and Indoor Navigation*, 1–8. Guimarães: IEEE.

13 Kusari, A., Pan, Z., and Glennie, C. (2014). Real-time indoor mapping by fusion of structured light sensors. In: *2014 Ubiquitous Positioning Indoor Navigation and Location Based Service (UPINLBS)*, 213–219. Corpus Christ, TX: IEEE.

14 Chen, S., Li, M., and Ren, K. (2014). The power of indoor crowd: indoor 3D maps from the crowd. In: *2014 IEEE Conference on Computer Communications Workshops (INFOCOM WKSHPS)*, 217–218. Toronto, ON: IEEE.

15 Kaneko, E. and Umezu, N. (2017). Rapid construction of coarse indoor map for mobile robots. In: *2017 IEEE 6th Global Conference on Consumer Electronics (GCCE)*, 1–3. Nagoya: IEEE.

16 Deißler, T., Janson, M., Zetik, R., and Thielecke, J. (2012). Infrastructureless indoor mapping using a mobile antenna array. In: *2012 19th International Conference on Systems, Signals and Image Processing (IWSSIP)*, 36–39. Vienna: IEEE.

17 Babu, B.P.W., Cyganski, D., and Duckworth, J. (2014). Gyroscope assisted scalable visual simultaneous localization and mapping. In: *2014 Ubiquitous Positioning Indoor Navigation and Location Based Service (UPINLBS)*, 220–227. Corpus Christ, TX: IEEE.

18 Liu, K., Motta, G., Tunçer, B., and Abuhashish, I. (2017). A 2D and 3D indoor mapping approach for virtual navigation services. In: *2017 IEEE Symposium on Service-Oriented System Engineering (SOSE)*, 102–107. San Francisco, CA: IEEE.

19 Ma, L., Tang, L., Xu, Y., and Cui, Y. (2017). Indoor floor map crowdsourcing building method based on inertial measurement unit data. In: *2017 IEEE 85th Vehicular Technology Conference (VTC Spring)*, 1–5. Sydney, NSW: IEEE.

14

Synthesis and Possible Forthcoming "Evolution"

Abstract

After having presented the various technologies and the ways to implement them and then having quickly discussed the current approaches in order to merge them, this last chapter aims to make a synthesis of the "problem (In principle, at this stage of the book, you should no longer see the field of indoor positioning as a problem but as a very wide set of opportunities to move forward.)" of indoor positioning, but also on the interest of implementing solutions that, for many, already exist. We just have to be able to look at things from a very different angle from the one we all have without really knowing why: no doubt our history and our culture that will lead us in a direction whose outcome, if it exists, is still too far away. A change in the angle of vision should allow us (and this is my conviction) to move forward usefully and practically without further delay.

Keywords *Synthesis; Forthcoming evolution; Discussion on IA; Future life with positioning*

This chapter is divided into several parts: the first part provides a quick summary by proposing a review of some technology selections for specific use cases. The tables in Chapter 4 are repeated and "made to talk" in specific cases. Then, still in this first part, we compare three very different technological approaches in order to find the best one. It is then an opportunity to review the discussions we had in the book on the origin of the current difficulties and to propose a way out. In the second part, we seek to project ourselves into the future by considering that positioning is finally available in all environments and imagine the potential changes that this could bring to the daily life of a few people at university or in the organization of an outpatient department in a hospital.

Indoor Positioning: Technologies and Performance, First Edition. Nel Samama.
© 2019 The Institute of Electrical and Electronics Engineers, Inc. Published 2019 by John Wiley & Sons, Inc.

14.1 Indoor Positioning: Signals of Opportunity or Local Infrastructure?

The technologies studied range from accelerometers to the so-called symbolic wireless network systems (where positioning is given in the form of a room or set of rooms, rather than coordinates).[1] Without going into all the ones mentioned so far, it should be noted, however, that they include inertial systems (accelerometers, gyrometers, and magnetometers), image-based processing or analysis approaches (markers, displacement, SLAM – for simultaneous localization and mapping – or recognition), radio systems (from 3G, 4G, and 5G to opportunity radio signals via WiFi or Bluetooth, television, or FM radio), or optoelectronics such as Laser, Lidar, or LiFi (light fidelity). Other types of physical measurements are also present, such as those related to sound or ultrasound, atmospheric pressure, or infrared. RFID (radio-frequency identification) electronic tag systems are also included, as are the so-called "low-power and long-range" signals (LPWAN – low-power wide area networks) from SiFox or LoRa systems, for example. Wired networks are also taken into account because it is possible to determine a position by identifying, in some cases, the IP (Internet Protocol) address of a connection. We also considered proximity systems such as contactless cards or even bank cards. Finally, it seemed difficult to completely forget the global navigation satellite systems (GNSS), which are nevertheless at the root of the current questions on the continuity of the positioning service between outdoors and indoors.

In addition, for each of these technologies, a dozen parameters have been entered. In the first series, we find the maturity of the infrastructure to be deployed if necessary, the technological maturity of the associated terminal, the type of positioning (relative, absolute, or symbolic), the fact that the terminal can be a smartphone or a wearable device in the more or less short term, the sensitivity of the indoor positioning system to the environment, or the need for a calibration phase. In the second series, we find the positioning accuracy, the reliability of the latter, the positioning mode (from "continuous" to "requiring the user's action"), the type of processing performed (propagation modeling, image analysis or physical detection), the type of calculation performed (geometric intersection of spheres or planes, mathematical function such as integration, for example, or point proximity calculation), and finally the physical measurement performed.

14.1.1 A Few Constrained Selections

It is then on these parameters that it is possible to put some constraints in order to select the technologies that meet certain criteria. As a first

---

1 Most of this paragraph is based on a publication proposed during the Scientific Days entitled "Geolocation and Navigation in Space and Time" of the International Radio Science Union (URSI) France, held in 2018.

Table 14.1 Technologies for which infrastructure and availability on smartphones are almost assured.

Technologies	Infrastructure maturity	Terminal maturity	Positioning type	Smartphone	Sensitivity to environment	Calibration need
BLE	Existing	Software development	Absolute	Existing	High	A few times
Image markers	—	Software development	Absolute	Existing	Very high	Once
Image-relative displacement	—	Software development	Relative	Existing	High	A few times
Image SLAM	—	Software development	Relative	Existing	High	A few times
Pressure	—	Existing	Relative	Easy	No impact	A few times
RFID	Existing	Software development	Absolute	Easy	Low	—
Ultrasound	Existing	Integration	Absolute	Easy	Very high	—
WiFi	Existing	Software development	Absolute	Existing	High	A few times
Symbolic WLAN	Existing	Software development	Symbolic	Existing	Low	—

example, we are looking to extract technologies that respond to an existing infrastructure, a decametric accuracy, an easy integration on current smartphones even if it does not necessarily already exist, and a continuous positioning mode to look like GNSS. The result is nine technologies. Table 14.1 details the first six criteria and Table 14.2 details the next six.

It should be noted that we find the technologies that are most often used, such as Bluetooth low energy (BLE) or WiFi. It is also interesting to note that the criteria are general and that in a fusion approach, technologies from the same group will certainly have to be considered, for example, BLE and pressure sensor in order to determine a reliable 3D (three-dimensional) positioning. However, some of these technologies remain difficult to implement in an unstable environment, such as ultrasound, which will be useless in the midst of people on the move. Image processing techniques are very efficient but once again require restrictive operating conditions (good visibility, very sharp images, initial calibrations, and high data processing capabilities).

However, it would be possible to address other criteria. The second example considers an existing system on current smartphones, which has low environmental sensitivity and very high reliability. Tables 14.3 and 14.4 summarize the results obtained for the five technologies concerned.

le 14.2 Technologies offering almost continuous decametric precision.

chnologies	Accuracy	Reliability	Positioning mode	Signal processing	Position calculation	Physics used
E	A few meters	Medium	Almost continuous	Pattern matching	Math functions	EM waves
age arkers	<1 m	Medium	Almost continuous	A combination of signal processing approaches	Matrices calculus	Image sensor
age-relative placement	<1 m	Medium	Almost continuous	A combination of signal processing approaches	Math functions	Image sensor
age SLAM	<1 m	Medium	Almost continuous	A combination of signal processing approaches	Math functions	Image sensor
essure	1 m	High	Continuous	Detection	Zone determination	Physical sensor
ID	Decimeter	High	Almost continuous	Detection	Spot location	EM waves
tra sound	A few decimeters	Low	Continuous	Propagation modeling	∩ Spheres	Mechanic waves
iFi	A few meters	Medium	Continuous	Pattern matching	Math functions	EM waves
mbolic LAN	Dm	Very high	Continuous	Propagation modeling	Zone determination	EM waves

, electro-magnetic.

le 14.3 Technologies available on smartphones and with low sensitivity to their environment.

chnologies	Infrastructure maturity	Terminal maturity	Positioning type	Smartphone	Sensitivity to environment	Calibration need
r codes	Existing	Software development	Absolute	Existing	Low	—
edit cards	Existing	Existing	Absolute	Existing	—	—
FC	Existing	Software development	Absolute	Existing	—	—
R codes	Existing	Software development	Absolute	Existing	Low	—
mbolic LAN	Existing	Software development	Symbolic	Existing	Low	—

Table 14.4 Technologies offering very high reliability.

Technologies	Accuracy	Reliability	Positioning mode	Signal processing	Position calculation	Physics used
Bar codes	Decimeter	Very high	User action needed	Pattern recognition	Spot location	Image sensor
Credit cards	A few centimeters	Very high	Discrete	Detection	Spot location	Electronic
NFC	A few centimeters	Very high	User action needed	Detection	Spot location	EM waves
QR codes	Decimeter	Very high	Act user action needed	Pattern recognition	Spot location	Image sensor
Symbolic WLAN	Dm	Very high	Continuous	Propagation modeling	Zone determination	EM waves

QR, quick response; EM, electro-magnetic.

We obtain a very different list (only the symbolic WiFi is present in both selections). It thus appears that simple implementation criteria are likely to quickly eliminate many technologies. This is the first conclusion of the book.

14.1.2 Comparison of Three Approaches and Discussion

Let us now discuss the case of three very different technologies: inverted radar (Grin-Loc type described in Chapter 9), an near field communication (NFC)-based system and associated mapping (described in Chapter 5), and a cooperative approach between communicating nodes (described in Chapter 12). In all three examples, we consider that the system should apply to a pedestrian using his mobile device (nowadays, the smartphone).

14.1.2.1 Inverted GNSS Radar

A less traditional approach than those shown in the previous tables is based on the installation of GNSS-type signal transmitters indoors. These signals are transmitted by dual antennas, each transmitting the equivalent of a satellite signal (i.e. having a specific code). However, the two antennas are perfectly synchronized locally. The receiver is a current smartphone equipped with Android from version 7 (application programming interface, API 24) onward, allowing access to carrier phase measurements. The treatment simply consists of the difference in the phases of the two carriers. As the two antennas are separated by less than one wavelength (see Section 9.5.1), the measurement of this difference in distances at the two antennas is unambiguous. We show

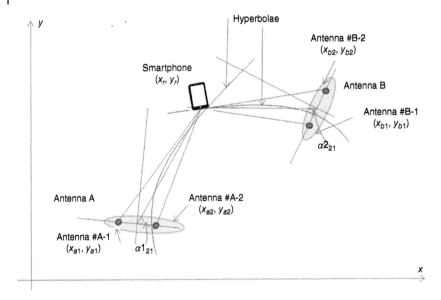

Figure 14.1 Principle of the inverted radar in two dimensions.

that the point location obtained for the same phase difference is characterized by a quadric and geometrically forms a hyperboloid of revolution. In two dimensions, this results in hyperbolas as shown in Figure 14.1.

Measurements made from a dual antenna alternatively determine the angle of view of the smartphone from the dual antenna. In order to position the smartphone, it is necessary to perform a second measurement (then a third three-dimensional measurement) that will provide a second angle from the second dual antenna. It should be noted that the measurements of the two dual antennas are totally independent and that they do not need to be synchronized: it is in fact the smartphone that establishes the temporal coherence of the system.

The advantages of such a system are real: perfect continuity with GNSS used outdoors, use of current smartphones, possible decimetric accuracy, and no synchronization required. However, it presents two difficulties in the current perception: the need for an infrastructure to be deployed and the use of radio signals that remain sensitive to the propagation environment.

14.1.2.2 NFC-Distributed System and Its Map

A very different approach is to include the user in the system and ask for a specific action. In this case, it is a question of "flashing" a quick response (QR) code or an NFC tag (see Chapter 5). These elements are identified, thanks to their unique identifier, geographically in an environmental mapping. The fact that it is necessary to approach relatively close to the tags makes the system

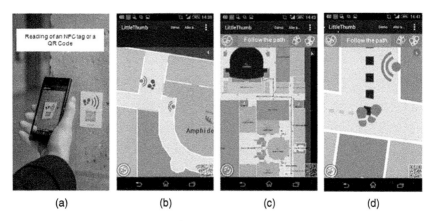

| (a) | (b) | (c) | (d) |

Figure 14.2 A user-initiated approach.

very accurate, but with a positioning that is in this case discontinuous in time and space. It seems that this constraint can be accepted in pedestrian sites of reduced complexity because the need in this case is mainly to have a digital map, to know where it is located and to obtain a particular route in return. All this is available.

Mapping is a major element of all systems. It is not an image, but a structure that allows the association of attributes to the various objects such as rooms, corridors, or technical rooms. These attributes range from a denomination to characteristics related to the possible speed of movement in the object. It will also be important to allow, in addition to the calculation of a route, the estimation of the associated travel time. In the example shown in Figure 14.2, the terminal's magnetometer, a smartphone here, is used to provide guidance to the user who may not know which way to go. The "small feet" characterizing the user then turn green. Then, the guidance is no longer carried out automatically: if the user feels the need, he will have to find a new tag to "realign" himself. Coupling is still possible with either WiFi or Bluetooth networks, or with a step counter (based on the use of accelerometers) to enrich the user experience.

This system has a low deployment cost, high reliability, and could be available immediately. However, it is not in the current mind that focuses on continuity with GNSS, particularly in their temporal and spatial aspects.

14.1.2.3 Cooperative Approach Between Communicating Terminals

In this third version, the system tries to calculate the relative positions of nodes that communicate with each other. Node A (bottom left in Figure 14.3) is the reference. All nodes estimate their Doppler relative to A when moving. In addition, they also consist of a dual receiving antenna that estimates the angle of arrival of signals received from other nodes. Not all nodes necessarily receive all the signals. The geographical resolution of the system is based

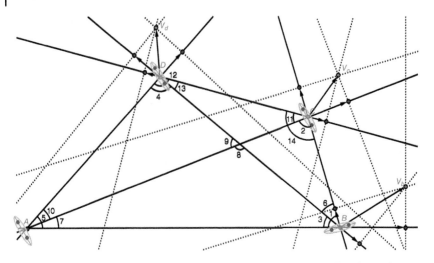

Figure 14.3 A cooperative model exchanging Doppler and relative angles of arrival.

here on the calculation of the various angles identified by numbers on the figure (reproduced in two dimensions for ease of reading). There are 14 of them and we need to obtain enough relationships to allow the calculation (see Chapter 12 for details).

However, this geometry has many symmetries, and this is where dual antennas play an important role. Even in static situations, it is possible to cross the signal measurements of the other nodes by pairs of two double antennas. This is an approach that resembles radar inverted in some aspects, but here with mobile entities whose coordinates are unknown.

Three technical difficulties arise. Arrival angle measurements are obtained in different reference frames, specific to each node. The attitude, i.e. the three-dimensional orientation, of each node is arbitrary. Finally, the calculation of hyperboloid intersections when we are in three dimensions is particularly complex in the general case. One solution is to couple these measurements with a real-time determination of the attitude of each node using other sensors. In the case of smartphones, for example, this calculation is made possible by the presence of accelerometers and magnetometers that allow an estimate of the receiver's attitude. However, positioning errors resulting from the inaccuracy of these measurements remain a subject of study.

In such an approach, it becomes difficult to imagine that measurements and calculations are managed by each node, which should also recover all the measurements made. This would generate particularly high telecom traffic. A supervisory entity would probably be more appropriate, but would in fact reduce the cooperative aspect, which can also be understood as "unsupervised." This remains open to discussion. One solution would then be to designate at the beginning of the process, the node that would have this role of supervision of a cooperative mini-network[2].

Another advantage of this approach is its ability to detect nondirect paths. It is common to name the paths between a transmitter and a receiver, regardless of the technology used, either LOS (line of sight) in the case of total absence of obstacles or NLOS (nonline of sight) in the opposite case. In order to maintain the nominal performance of a technique, it is generally necessary to be able to identify NLOS cases, which is a real technical difficulty that often requires the availability of high measurement redundancy. In this case, the angle of arrival measurements at any two nodes must have a spatial (or geometric) consistency that can be estimated. Thus, except in the case of nondirect paths that would result in the same values as they should have been (which is very unlikely but not impossible), NLOS cases should be identifiable.

We can see the complexity of this approach from the point of view of the organization of exchanges, measurements, calculations, and telecommunications. Some nodes would be supervisors of a mini-network but also potentially members of other mini-networks. This is still possible in theory, but the practical implementation is a challenge. In addition, the measurements will be affected by very variable errors depending on the environmental configurations of the various nodes. A mechanism should exist to estimate the importance of these errors and then eliminate the most damaging ones: the latter is an additional difficulty. Finally, such terminals are not yet available and neither are the associated measurement tools.

14.2 Discussion

One of the first commonly highlighted parameters is "accuracy." It should be noted that there is often a significant difference between the specifications given by development or research teams and the reality regarding this parameter. This is generally due to the choice of the selected test environments, which favor a "nice" one for the considered approach to the detriment of a "more difficult" environment (which could potentially be more representative). In addition, this parameter should be associated with a second parameter relating to the reliability of positioning. For example, when the latter depends very heavily on the environment in which the user is located, this reduces reliability, unless you have a measure of that environment. This may be an open path for current data science-oriented approaches.

If we return to the initial question in Section 14.1 as to whether it is better to have local infrastructure or opportunity signals, the answer is immediate:

---

2 Network engineers and researchers love this kind of problem where they will be able to propose sophisticated service architectures.

opportunity signals. However, not all the systems proposed so far with this orientation have found their place and the most deployed solutions today remain based on the deployment of specific terminals. The literature on approaches merging various technologies is very important, with inertial systems in particular. The latter have ideal characteristics: measurements available everywhere and at any time, which are not dependent on any infrastructure and are already present in smartphones. Unfortunately, they usually require calibration as the quality of the sensors is reduced[3], which is the case today on smartphones.

We have shown in a few examples that certain constraints (availability, maturity of technologies, reliability of positioning, etc.) are very strong. The guidelines that emerge from the work of the many communities involved clearly show that the practical difficulties of implementation present an obstacle to the industrial development of scientific and technical approaches. Of the three approaches presented in the previous paragraph, none seem to address the problem satisfactorily, either because of the need to deploy a specific infrastructure, or because of the need for user action and the fact that positioning is not continuous over time, or because of a network architecture that is not yet mature. These three systems are only a small part of the proposed solutions.

Let us imagine the reason for this lack of outbreak is elsewhere: in a very limited real interest or in the difficulty in extracting anticipated needs (which could potentially constitute orientations for researchers). The observation seems paradoxical: the expectation for an indoor positioning system is high, the solutions proposed are extremely numerous, but the majority of deployments are concluded by a gradual abandonment of the systems put in place and the associated applications. Our vision is that needs are not sufficiently well specified. For example, the required performance is certainly very different depending on the use that is planned. The guidance of visually impaired people in a hospital does not require the same characteristics as the detection and counting of people passing by a booth or a shop. However, synthesis does not exist and it is often the researchers themselves who determine the specifications of their system. The encounter between the unexpressed need and the solution developed elsewhere often does not come to fruition.

From time to time, this meeting takes place in special cases and then technical solutions are found. However, without a broader vision, the solutions remain specific. Writing a set, or sets, of specifications that would produce a sufficient and concrete market to trigger large-scale deployment is an important task. It should bring together many communities (users, industrialists, researchers,

---

3 Mass market MEMS sensors suffer from strong biases because of low-cost manufacturing approaches and temperature dependence.

etc.) and could be the subject of institutional work. For our part, we can launch some ideas for reflection.

First of all, it seems to us that the constraints should be operational rather than technical. The complexity of deployment, the sensitivity of performance to environments, and therefore the reliability of performance are key factors in the perception of the solution. The creation of a "common performance sheet" to compare technologies with each other would be very useful. The criteria for such a list would certainly be high level and user oriented. The criteria to be considered are probably also combinations of criteria, such as "accuracy and reliability" or "type of positioning and terminal." We also propose a classification that makes it possible to immediately perceive the complexity of a real deployment in the form of a "coverage" criterion, the same one that led to the organization of the book. The latter qualifies the typical scope of the proposed technology in the form of environmental data: room, set of rooms, building, set of buildings, site, etc. It would also be necessary to associate this criterion with a number of "components" to be deployed. Finally, a parameter on the energy efficiency of the approach, both for the possible infrastructure and for the terminal, would certainly be useful.

14.3 Possible Evolution of Everybody's Daily Life

The term "revolution" is often used nowadays: I personally used it in the last chapter of my previous book on the potential evolution of everyone's daily life if we consider the wide availability of location data. However, this term, by dint of being overused, gradually loses its strength. This paragraph will take up the daily framework of two categories, a student and a walk-in medical department in a public hospital. We will see how having people's positions, a map of places, and a route calculation tool could change the daily life a little (nothing revolutionary, therefore, but simplification and time and energy savings).

Let us imagine that the technical side is available and that positioning of people and objects indoors is as easy as outdoors where GNSS work perfectly: what could the changes be in our daily behavior? Take two situations, respectively, a day of a student and an outpatient medical service, let us try to imagine such modifications.

14.3.1 Student's Day

Many applications exist today to organize and manage everyone's travel. For example, it is possible to check that the weather is in accordance with the day's planned activities and thus choose the best possible means of travel, including all the constraints that can be addressed: public transport, car-pooling, self-service electric vehicle, need to transport equipment, etc. Similarly, "home groups" also make it possible to know the position of friends and to organize everyone's day even better.

14.3.1.1 Morning Session at the University

Of course, students always organize themselves the day before in order to be on time for their classes, but it turns out that today, exceptionally, they do not remember the exact schedule. As a result, when they arrive on campus, they can connect to the online schedule, available on the faculty's server. This allows them, after identification, to know not only which class is being taught but also where it is taking place. It should also be noted that this practice is useful because following a projector failure, the room that was mentioned in yesterday's schedule has had to be changed this morning. An indoor positioning system being available on campus, the student's position is determined (because she/he has accepted the principle), and by clicking on the course she/he must take, a map opens automatically and shows her/him a way to get there. Figure 14.4 illustrates the principle of this approach.

It should be noted that such a capacity to link agenda, position, and guidance is also very useful in many other cases that may occur on a campus: hosting external speakers, such as industrialists or visiting professors, conferences, or any other event. In fact, it would be useful to all people who are unfamiliar with the campus, especially because their passage is only ephemeral in the area under consideration.

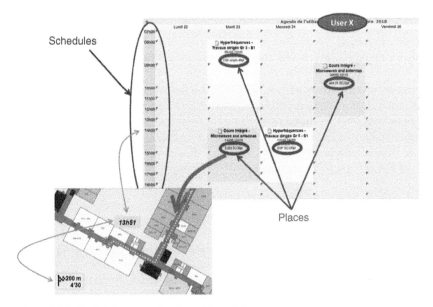

Figure 14.4 The link between the timetable and the course venue.

14.3.2 Improving an Outpatient's Visit to Hospital

14.3.2.1 Preparation of the "Journeys"

Ambulatory medicine is of definite interest in the care of patients: saving them time, optimizing their trip to the hospital, reducing the risk of nosocomial infection, reducing the cost of care, and above all improving the patient's "well-being." To do this, the hospital's outpatient service organizes, on an individualized basis, a "journey" for each patient that he or she must follow. This itinerary consists of a series of medical appointments, possibly an operation, which are generally supervised by the service, and a final visit (before returning home).

These paths constitute the heart of the functioning of the ambulatory service, for the satisfaction of the patient, the efficiency of the day, but also for the organization of the various hospital services involved. However, the observation is that appointments often shift over time and disrupt practitioners. In some cases, this is due to a poor consideration of the time needed for patients to travel, which sometimes goes hand in hand with the fact that they are often "lost" in the buildings.

14.3.2.2 Displacements of Patients and Automatic Rescheduling

Imagine the availability of a system, whatever it may be, to follow the patient on his or her journey, and possibly to send him or her, again in any way, some information on his or her next appointments or on the path to take to optimize his or her movements. We could imagine, as shown in Figure 14.5, equipping patients with mobile terminals that would allow such exchanges, including guidance guidelines.

In this case, these are watches, but it is quite possible to consider bracelets or even the patient's own smartphone.

In addition, the outpatient service also wants to be able to follow patients in order to optimize the use of doctors and technical facilities. Similarly, any delay by a patient is a risk of disorganization and potentially blocking the planned day. Thus, the department is also interested in a dashboard to monitor the smooth running of the day, i.e. the routes, of the various patients. Such an example is given in Figure 14.6 where it can be seen that one record is edited per patient in which the various appointments are listed, with the time and place. A color code allows ambulatory services to immediately see whether the patient is following his or her path normally (light gray in the figure), whether he or she is being consulted (gray in the figure), or whether a delay exists (dark gray in the figure), and therefore a risk of potential disorganization.

14.3.2.3 Reports – Analytics

In addition, the department also seeks to improve pathways: correct casting errors, reduce the time between appointments for the patient and thus improve

Figure 14.5 Some ideas for a mobile terminal and guidance instructions.

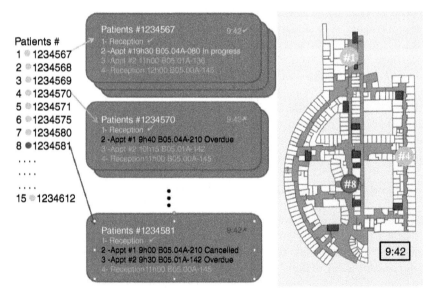

Figure 14.6 Real-time patient follow-up.

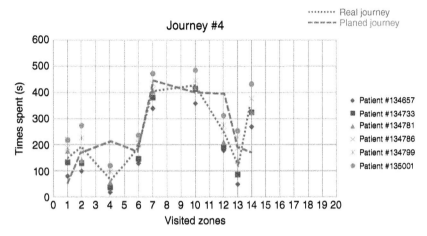

Figure 14.7 Real-time patient follow-up.

the performance of the hospital's technical facilities. To do this, it has, thanks to the patient follow-up tool and the recording of the progress of the days, the possibility of analyzing in delayed time the differences between what had been planned and what actually took place during the day. This allows for a loopback on the organization. Figure 14.7 gives an idea of some of the "analytics" that could be provided.

Finally, in an even better version, we could imagine a dynamic organization of appointments based on patient follow-up. When a delay (or a risk of delay) is detected, then the route is automatically recalculated. This involves a redistribution of slots for doctors and technical platforms, which is certainly complicated to implement. However, considering the increase in the number of ambulatory patients treated and the increasing size of hospitals, this approach seems to be of real potential interest.

14.3.3 Flow of People in Public Places

Let us continue a little bit in the current trend of generating data to produce analytics and now move on to a public place, such as an airport or a museum. With exceptions as usual, but these places are often visited by people who only pass through them occasionally. For example, a pedestrian guidance system would probably be very useful (there are some already installed around the world). However, this is not the direction we are going in because there is another need, which is also present in many other sectors such as crowd safety, for example: the management of people flows. Imagine the situation in which two planes simultaneously discharge their passengers into the airport, one ahead of schedule and the other behind. In such a situation, it is important to find a technique that allows the two populations not to cross, as this would inevitably lead to congestion.

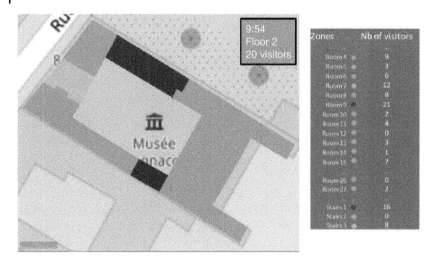

Figure 14.8 Visualization of congestion areas in a museum.

Having the position of a significant proportion of people then makes it possible, as is the case with car navigation systems on roads, to visualize congestion areas, as shown in Figure 14.8. This makes it possible to take the necessary accompanying measurements in real time. In addition, the analysis of these data over long periods of time should make it possible to establish congestion prediction models in order to no longer handle congestion, but to predict it and thus reduce it. This could be achieved, for example, by offering passengers a guidance system that takes into account these high-traffic areas, actually suggesting "alternative routes." It should be noted that it is essential to have a digitized map of the area at your disposal.

Figure 14.8 shows, in addition to the congestion zones, a dashboard that allows the situation to be monitored in real time (in a museum in this case). The latter then makes it possible to reorganize the visits in order to reduce the level of disappointment of visitors who cannot see "The Mona Lisa" because there are so many people in front of the famous painting.

To make the link with the technical part of the book, it should be noted that all this is possible already with the technologies described and can easily be implemented.

14.4 Internet of Things and Internet of Everything

Positioning is nowadays not only thought of in terms of people but also in terms of objects. Mobile equipment in a hospital could be located and its use optimized. Following parcels is already available with delivery companies:

positioning relies on restricted area RFID-based detection of the parcel. This is a valuable service provided to customers but remains quite limited to specific areas. Improvement is possible through permanent localization. This could certainly additionally allow dynamic pick up of parcels on a much larger scale than what exists today: if yours remains for some time in a warehouse close to you (or close to one of your displacements), you could be invited to collect it. The result is an economy of energy and time. Let us come back to any piece of equipment: knowing the place it will have to be available next time, one could reduce the energy needed for its displacement: there is no longer the need to take it back to the central office to put it in storage.

Pushing the fiction one step further, it is possible to imagine the vehicle displacement in cities in a completely different way from today. Imagine that pollution is at such a high level that cars are no longer allowed in cities[4]. Everybody is obliged to use these small electric cars that can be rented in a suburban car park: they are the only means to enter the city, apart from public transport systems. These cars are rented in a very different way from today: the rates depend on both the time one keeps the car and the distance traveled, and indeed depends mainly on the energy really used. Once your displacement finished, you can either keep the car at your disposal (and pay for it) or leave it wherever you want. Then, another person, knowing the exact locations of free electric cars could choose to rent it for a new displacement. Reduced fees are applied when you take the car back to a charging center. Free parking places are displayed on a screen in the car and information is available concerning all the services available in the vicinity of your route (charging centers as well). By programming your trip to a given destination in town, you could additionally make it available to any person who would like to share the cost of the car on the basis of a shared trip. Dynamic appointments could thus be organized in order to further decrease the total energy consumption of the global displacement of people in the city. This can of course be applied to movements of all professionals and parcels (or objects). This is already possible with outdoor deployed positioning technologies but remains limited because of the absence of real indoor solutions. At the end of this book, I hope many readers will have understood it is nevertheless possible: it will be up to them to implement such approaches for the sake of all of us (and not only for financial benefits).

14.5 Possible Future Approaches

The global positioning system (GPS) has paved the way for the positioning revolution, as portable clocks did about two centuries ago for the time revolution. It

---

4 It is just about to come.

is difficult to imagine what the future of positioning will be, but let us remember some "fundamentals":

- There are a large number of applications that could take advantage of positioning, with as many performance specifications.
- There are currently numerous technologies of indoor positioning, applying in almost all environments, but not having the same level of global performance[5].
- GNSS (GPS indeed) have been designed to allow positioning in environments where no infrastructure was available. The extension of their use in cities, for example, has been driven by the goal to reduce the positioning infrastructure only to existing satellites. This is not the way GPS was thought of initially.
- The advent of huge telecommunication capabilities makes the future easier to imagine.
- Ubiquity is sought by large communities (such as positioning, computer science, telecommunication, etc.).

This book has given the overall limitations of the different indoor positioning systems. It appears clearly that there is currently no single means of positioning that is acceptable for all applications; moreover, hybridization approaches are being thought of in order to fulfill all the specifications for many applications. This is even truer with applications concerning people, such as numerous location-based services. Therefore, the questions about the future positioning systems are quite natural: will the GNSS be extended to cover environments not yet available to these techniques? Will hybridization be developed to the point where it works transparently? Are networks of sensors the best solution? Is Artificial Intelligence or Big Data the next steps which will solve this problem adequately? Are Data Scientists our saviors?

For the next 10–15 years, it is probable that GNSS is going to take a growing part in our lives, but this is not so sure for the long-term future unless alternative solutions are found to overcome the real limitations of GNSS. The current methods, which consist of obtaining the absolute coordinate of a location with reference to a global frame with an accuracy indicator being given in meters, are maybe not the only solution. For example, the positioning methods dealing with the so-called "symbolic" algorithms that rely on local positioning given in terms of rooms or zones may be an interesting alternative. The accuracy is then an approximation of the number of individual rooms the user might be located in. In addition, the specification that is being sought is the reliability of the positioning: it is of less importance to have an accurate position but it is of uppermost importance to be able to rely on the position.

---

5 By performance, one has to understand positioning performance, but also the ease of deploying the system, the complexity of the terminal, etc.

There is another way one could imagine the future of positioning, either indoors or even outdoors. First of all, we can consider that for the near and mid-term future, GNSS is almost an ideal candidate for positioning where no infrastructure is available at all: at sea or in the desert, but also in the countryside where the coverage is very good. However, in cities or indoors, satellite-based positioning is not well suited, even considering all the efforts of the positioning communities to find a solution. However, in these environments, there are a lot of fixed objects distributed everywhere. That is either mobile objects, such as personal terminals, or fixed objects, such as doors, bus stops, or numbers on the front of houses. Usually, telecommunication features are associated with objects and people who need to exchange data, but one could easily imagine broadening the scope of telecommunication-equipped objects[6].

Once an object is able to communicate, it can also transmit its position, as long as it knows it. Depending on the range of the wireless transmission system used, all receiving electronic terminals located in the proximity of this object could, at least, know that they are not far away (at a maximum distance that is precisely the range of the telecommunication system of the object) from the object. This is indeed the basic concept of ubiquity in telecommunications: this allows the object to discover its environment. If the telecommunication capabilities grow sufficiently (this is one of the objectives of 5G), this could be the real next step for positioning. Imagine one mobile terminal that cannot achieve GNSS positioning (because it is indoors, for instance) but is equipped with a communication capability. It is linked to another device that can transmit information gathered by its environment[7]: gradually, the mobile terminal can obtain information about the geographical distribution of all the points in the network (this requires new approaches of data processing taking these aspects into account). As soon as one element in the network knows its own position, deductions can be carried out in order to determine the possible positions of other points. In such a way, it could be possible to obtain the positions of many points, knowing the exact locations of only a few. This is the philosophy of the described "collaborative = approaches" (see Chapter 12 and Section 14.1.2 for details): the problem is that the way it is imagined today would require so much energy that probably one should think of a simplification of the protocols for positioning purposes[8].

Furthermore, it is imaginable that most fixed objects could be defined by their location, either indoors or outdoors. Moreover, one can easily imagine that some fixed objects could obtain their location through the use of GNSS-based systems. In such a case, the ultimate positioning system would be able to exchange data as well as detect the proximity of two objects. Only fixed objects would be equipped with GNSS (because of the reduced cost and

---

6 In a similar way that RFID has been developed.
7 In this context, "environment" means the "telecommunication environment."
8 Note that some works are still going this way.

the automatic configurability of the infrastructure): the location of a device is obtained through the exchange with close objects that, most of the time, know their own location. When this is not the case, more sophisticated algorithms are required (maybe inducing an increased uncertainty of the positioning).

An example of such a system could be the positioning of a pedestrian through his mobile phone, which is linked to the mobile phone of the car driver going along the same street. As the car is equipped with a GNSS system, the mobile phone of the driver has an accurate location available. When passing close to the pedestrian, position data transfer is possible and gives information about the position to the pedestrian's mobile phone. The car is an obvious way to get the position, but one can think that bus stops, entrances of buildings, traffic lights, or also doors, windows, lights, etc. (for indoors) know their position. This idea brings us back to the initial aspects of GPS: positioning where no infrastructure is available and not in cities, buildings, etc.

14.6   Conclusion

The field of indoor positioning involves large communities and offers many technical solutions. The fundamental difference with the outside world where GNSS has taken over the monopoly is the very great diversity of situations. The need is regularly reported, but the necessary investments often come up against history: as the positioning was not available (and still is not really available), things were set up without it. What would be the acceptable investment for a need that is not actually fundamental at the moment[9]?

After an analysis of potential technologies based on the observation of high-level criteria, it seems to us that there is a lack of real specifications, as well as the use of low-level criteria that do not work well with building managers, who are essential partners in the problem. We therefore propose the concerted creation of a list of mixed criteria that allows a comparison under similar (and representative) conditions of the various technologies. Such an undertaking requires a coordination that we call for and in which we are ready to play our part. I hope and wish that this book will make a contribution to this.

Finally, and these will be my last sentences in this book, I think it is necessary to refocus on the real needs and stop looking for what is useless and leads nowhere. All the current unrest over ephemeral and useless keywords only uses up energy pointlessly, while there is so much to do for the good of all, in a useful way. Current "scientific" approaches are very often oriented toward adding complexity to problems that could not already be addressed when they

---

9 By "nonfundamental," we must understand "what we can do without," in particular because this is what we have always done, do without it!

remained "simple." The problem is undoubtedly that "simplifying" problems, in addition to not being fashionable, is not an easy thing and requires hindsight and humility toward our supposed intelligence, that is to say, time, which we think we have less and less. Let us slow down a little bit.

Bibliography

1 Abdel-Salam, M. (2005). Natural disasters inference from GPS observations: case of earthquakes and tsunamis. ION GNSS 2005, Long Beach, CA (September 2005).

2 Fuller, R. and Grimm, P. (2006). Tracking system for locating stolen currency. ION GNSS 2006, Forth Worth, TX (September 2006).

3 Heinrichs, G. (2003). Personal localisation and positioning in the light of 3G wireless communications and beyond. IAIN 2003, Berlin, Germany (October 2003).

4 Jensen, A.B.O., Zabic, M., Overø, H.M. et al. (2005). Availability of GNSS for road pricing in Copenhagen. ION GNSS 2005, Long Beach, CA (September 2005).

5 Kaplan, E.D. and Hegarty, C. (2006). *Understanding GPS: Principles and Applications*, 2e. Artech House.

6 Pateli, A., Fouskas, K., Kourouthanassis, P., and Tsamakos, A. (2002). On the potential use of mobile positioning technologies in indoor environments. 15th Bled Electronic Commerce Conference, Bled, Slovenia (June 2002).

7 Rohmer, G., Dünkler, R., Köhler, S. et al. (2005). A microwave based tracking system for soccer. ION GNSS 2005, Long Beach, CA (September 2005).

8 Schaefer, R.P., Lorkowski, S., and Brockfeld, E. (2003). Using real-time FCD collection for trip optimisation in urban areas. The European Navigation Conference, GNSS 2003, Austria, Graz.

9 Woo, D., Mariette, N., Salter, J. et al. (2006). Audio nomad. ION GNSS 2006, Forth Worth, TX (September 2006).

10 De Angelis, A., Moschitta, A., Carbone, P. et al. (2015). Design and characterization of a portable ultrasonic indoor 3-D positioning system. *IEEE Transactions on Instrumentation and Measurement* 64 (10): 2616–2625.

11 He, S. and Gary Chan, S.-H. (2016). Wi-Fi fingerprint-based indoor positioning: recent advances and comparisons. *IEEE Communications Surveys and Tutorials* 18 (1): 466–490.

12 Wang, G., Changzhan, G., Inoue, T., and Li, C. (2014). A hybrid FMCW-interferometry radar for indoor precise positioning and versatile life activity monitoring. *IEEE Transactions on Microwave Theory and Techniques* 62 (11): 2812–2822.

13 Kok, M., Hol, J.D., and Schön, T.B. (2015). Indoor positioning using ultra wide band and inertial measurements. *IEEE Transactions on Vehicular Technology* 64 (4): 1293–1303.

14 Yasir, M., Ho, S.-W., and Vellambi, B.N. (2014). Indoor positioning system using visible light and accelerometer. *Journal of Lightwave Technology* 32 (19): 3306–3316.

15 Chen, L.-H., Wu, E.H.-K., Jin, M.-H., and Chen, G.-H. (2014). Intelligent fusion of Wi-Fi and inertial sensor-based positioning systems for indoor pedestrian navigation. *IEEE Sensors Journal* 14 (11): 4034–4042.

16 Perttula, A., Leppäkoski, H., Kirkko-Jaakkola, M. et al. (2014). Distributed indoor positioning system with inertial measurements and map matching. *IEEE Transactions on Instrumentation and Measurement* 63 (11): 2682–2695.

17 Noh, Y., Yamaguchi, H., and Lee, U. (2018). Infrastructure-free collaborative indoor positioning scheme for time-critical team operations. *IEEE Transactions on Systems, Man, and Cybernetics: Systems* 48 (3): 418–432.

18 Zhuang, Y., Syed, Z., Li, Y., and El-Sheimy, N. (2016). Evaluation of two WiFi positioning systems based on autonomous crowdsourcing of handheld devices for indoor navigation. *IEEE Transactions on Mobile Computing* 15 (8): 1982–1995.

19 Chen, C., Han, Y., Chen, Y., and Ray Liu, K.J. (2016). Indoor global positioning system with centimeter accuracy using Wi-Fi. *IEEE Signal Processing Magazine* 33 (6): 128–134.

20 Gaber, A. and Omar, A. (2016). Utilization of multiple-antenna multicarrier systems and NLOS mitigation for accurate wireless indoor positioning. *IEEE Transactions on Wireless Communications* 15 (10): 6570–6584.

21 Liu, Q., Qiu, J., and Chen, Y. (2016). Research and development of indoor positioning. *China Communications* 13 (2): 67–79.

22 Dardari, D., Closas, P., and Djurie, P.M. (2015). Indoor tracking: theory, methods, and technologies. *IEEE Transactions on Vehicular Technology* 64 (4): 1263–1278.

23 Davidson, P. and Piché, R. (2017). A survey of selected indoor positioning methods for smartphones. *IEEE Communications Surveys and Tutorials* 19 (2): 1347–1370.

Index

Indoor Positioning: Technologies and Performance, First Edition. Nel Samama.
© 2019 The Institute of Electrical and Electronics Engineers, Inc. Published 2019 by John Wiley & Sons, Inc.